PHYSICAL FORCES
AND THE
MAMMALIAN CELL

PHYSICAL FORCES
AND THE
MAMMALIAN CELL

Edited by

John A. Frangos

Department of Chemical Engineering
Pennsylvania State University
University Park, Pennsylvania

ACADEMIC PRESS
HARCOURT BRACE JOVANOVICH, PUBLISHERS

San Diego   New York   Boston
London   Sydney   Tokyo   Toronto

Copyright © 1993 by ACADEMIC PRESS, INC.

All Rights Reserved.
No part of this publication may be reproduced or transmitted in any form or by any
means, electronic or mechanical, including photocopy, recording, or any information
storage and retrieval system, without permission in writing from the publisher.

Academic Press, Inc.
1250 Sixth Avenue, San Diego, California 92101-4311

United Kingdom Edition published by
Academic Press Limited
24–28 Oval Road, London NW1 7DX

Library of Congress Cataloging-in-Publication Data

Physical forces and the mammalian cell / edited by John A. Frangos.
    p.   cm.
    Includes bibliographical references.
    ISBN 0-12-265330-0
    1. Biophysics.   2. Cell physiology.   I. Frangos, John A.
QH505.P46   1993
599'.087–dc20   92-23514
                                                                           CIP

PRINTED IN THE UNITED STATES OF AMERICA
92 93 94 95 96 97   EB   9 8 7 6 5 4 3 2 1

CONTENTS

■ ■ ■ ■ ■

1

Techniques for Studying the Effects of Physical Forces on Mammalian Cells and Measuring Cell Mechanical Properties

Roger Tran-Son-Tay

2

Mechanochemical Transduction across Extracellular Matrix and through the Cytoskeleton

Donald Ingber, Seth Karp, George Plopper, Linda Hansen, and David Mooney

3

Mechanical Strain and the Mammalian Cell

Albert J. Banes

4

Hemodynamic Forces in Relation to Mechanosensitive
Ion Channels in Endothelial Cells

Peter F. Davies and Randal O. Dull

5

Effects of Flow on Anchorage-Dependent Mammalian
Cells—Secreted Products

François Berthiaume and John A. Frangos

6

Shear Stress Effects on the Morphology and Cytomatrix of Cultured Vascular Endothelial Cells

Peggy R. Girard, Gabriel Helmlinger, and Robert M. Nerem

7

Fluid Shear-Stress-Dependent Stimulation of Endothelial Autacoid Release: Mechanisms and Significance for the Control of Vascular Tone

Rudi Busse and Ulrich Pohl

8

Chronic Effects of Blood Flow on the Artery Wall

B. Lowell Langille

9

Fluid Stress Effects on Suspended Cells

Larry V. McIntire and Sridhar Rajagopalan

10

Physical Forces in Mammalian Cell Bioreactors
Eleftherios T. Papoutsakis and James D. Michaels

11

Gravity and the Mammalian Cell
Paul Todd

CONTRIBUTORS
■ ■ ■ ■ ■

Numbers in parentheses indicate the pages on which the authors' contributions begin.

Albert J. Banes (81), Department of Surgery, University of North Carolina at Chapel Hill, Chapel Hill, North Carolina 27599

François Berthiaume (139), Department of Chemical Engineering, Pennsylvania State University, University Park, Pennsylvania 16802

Rudi Busse (223), Department of Applied Physiology, University of Freiburg, D-7800 Freiburg, Germany

Peter F. Davies (125), Department of Pathology, Pritzker School of Medicine, University of Chicago, Chicago, Illinois 60637

Randal O. Dull (125), Department of Pathology, Pritzker School of Medicine, University of Chicago, Chicago, Illinois 60637

John A. Frangos (139), Department of Chemical Engineering, Pennsylvania State University, University Park, Pennsylvania 16802

Peggy R. Girard[1] (193), Biomechanics Laboratory and School of Mechanical Engineering, Georgia Institute of Technology, Atlanta, Georgia 30332

Linda Hansen (61), Department of Surgery, Children's Hospital, Harvard Medical School and Department of Pathology, Brigham and Women's Hospital, Boston, Massachusetts 02115

Gabriel Helmlinger (193), Biomechanics Laboratory and School of Mechanical Engineering, Georgia Institute of Technology, Atlanta, Georgia 30332

Donald Ingber (61), Department of Surgery, Children's Hospital, Harvard Medical School and Department of Pathology, Brigham and Women's Hospital, Boston, Massachusetts 02115

Seth Karp (61), Department of Surgical Research, Children's Hospital, Boston, Massachusetts 02115

B. Lowell Langille (249), Vascular Research Laboratory, Toronto Hospital and Department of Pathology, University of Toronto, Toronto, Ontario M5G 2C4, Canada

[1]*Present address:* School of Biology, Georgia Institute of Technology, Atlanta, Georgia 30332.

Larry V. McIntire (275), Cox Laboratory for Biomedical Engineering, Institute of Biosciences and Bioengineering, Rice University, Houston, Texas 77251

James D. Michaels (291), Department of Chemical Engineering, Northwestern University, Evanston, Illinois 60208

David Mooney (61), Department of Surgical Research, Children's Hospital and Department of Chemical Engineering, Massachusetts Institute of Technology, Cambridge, Massachusetts 02115

Robert M. Nerem (193), Biomechanics Laboratory and School of Mechanical Engineering, Georgia Institute of Technology, Atlanta, Georgia 30332

Eleftherios T. Papoutsakis (291), Department of Chemical Engineering, Northwestern University, Evanston, Illinois 60208

George Plopper (61), Program in Cell and Developmental Biology, Harvard Medical School, Boston, Massachusetts 02115

Ulrich Pohl (223), Department of Physiology, University of Lübeck, Lübeck, Germany

Sridhar Rajagopalan (275), Cox Laboratory of Biomedical Engineering, Institute of Biosciences and Bioengineering, Rice University, Houston, Texas 77251

Paul Todd[2] (347), Chemical Science and Technology Laboratory, National Institute of Standards and Technology, Boulder, Colorado 80303

Roger Tran-Son-Tay (1), Department of Mechanical Engineering and Materials Science, Duke University, Durham, North Carolina 27706

[2]*Present address:* Department of Chemical Engineering, University of Colorado at Boulder, Boulder, Colorado 80309.

PREFACE
■ ■ ■ ■ ■

It is increasingly evident that mechanical forces play a crucial role in the physiology of living systems. Over the past fifteen years there has been exponential growth in the research in understanding how mechanical forces act on various tissues. It is now known that mechanical forces regulate vascular diameter and caliber, bone remodeling, hearing, gastrointestinal motility, and skin growth. It appears that many of the tissue responses and mechanisms have some degree of commonality; however, a comprehensive and unifying review of these aspects is lacking.

The objective of this text is not only to present reviews of the current state of knowledge, but also to provide guidelines and techniques for further research. We focus here on how mechanical forces, such as hydrodynamic shear and elongational strain, play a role in the normal physiology of the cardiovascular system, blood cells, bone, and other mammalian tissues. The mechanisms of how such forces act on cells are discussed. In addition, the implications of physical forces to bioreactor design and tissue engineering are examined. The intended audience for this text is cardiovascular physiologists, hematologists, and bioengineers.

The chapters have been arranged in the order of methodology, theoretical mechanisms, basic studies on cultured cells, vascular physiology, and biotechnology applications. Chapter 1 discusses and evaluates the various experimental techniques and devices that can be used to study the role of mechanical forces on cells and tissues. The second chapter deals with a proposed mechanism of how hydrodynamic shear affects cells, with particular emphasis on endothelial cells. Chapter 3 reviews what is known about how elongational strain affects cells of various tissues. Chapters 4, 5, and 6 address how hydrodynamic shear affects cells, with particular emphasis on endothelial cells. Chapters 7 and 8 address the acute and chronic effects of blood flow on vascular physiology. Flow effects on blood cells and other suspended cells are discussed in Chapter 9. Chapter 10 deals with the importance of hydrodynamic forces in the design and operation of bioreactors. Chapter 11 deals with the role of gravity in the context of hydrodynamic forces.

The importance of understanding how mechanical forces regulate mammalian cell and tissue function is demonstrated by the exponential growth of research in the field. This timely text, contributed by a diverse group of pathologists, biochemists, and bioengineers, comprehensively examines mechanotransduction and serves as a convenient resource for the design of new research.

CHAPTER 1

■ ■ ■ ■ ■

Techniques for Studying the Effects of Physical Forces on Mammalian Cells and Measuring Cell Mechanical Properties*

■ ■ ■ ■ ■

Roger Tran-Son-Tay

I. INTRODUCTION

Enormous progress has been made in the last decade toward understanding the effects of physical forces on the structure and function of mammalian cells. However, the mechanisms by which shear stress and strain modulate cell morphology and metabolism are still not well understood. An understanding of the effects of shear stress and strain can be achieved through studies in which cell populations are exposed to controlled and well-defined flow or deformation environments. Very often, the results reported in the literature are contradictory because of the different methodologies used. For example, the type of cell, the substrate, the medium, and the nature of the flow environment or deformation must all be "controlled" in order to rationalize the behavior of a given cell.

The kind of experimental technique used depends largely on the type of cells and properties to be studied, and, in general, techniques are classified accordingly. However, consistent with an approach that focuses

*This work was supported by the National Institutes of Health through grant 2R01-HL-23728 and the National Science Foundation through grant BCS-9106452.

1

on the mechanical effects, experimental techniques will be classified here in terms of the nature of the physical force used to deform the cells. For example, techniques that load the cell by the action of hydrodynamic force are distinct from techniques that specifically cause membrane stretch.

The objectives of this chapter are to review the techniques used to study the effects of fluid flow and strain on mammalian cells and to discuss methods that allow mechanical properties to be determined. A range of instruments provide well-defined flow fields, and cells can be strained by flow or surface stretch deformations. Some of the most common instruments used for studying suspended and/or attached cells are described, and some of the problems that arise when using these instruments are outlined. It is my intention to address some questions that have been inadequately treated, and to illustrate the specific kinds of problems that are associated with each technique so that readers can make knowledgeable choices for their own particular applications.

II. FLOW DEFORMATION

The focus of this section is on instruments that produce well-defined flows. Turbulent flow effects and their relevance to cell cultures will be discussed in the next section. Many of the fluid shear devices were originally developed to study the rheological properties of suspensions and associated deformations of the individual suspended particles. In order to discuss the deformation produced by flow for a given cell, we must recognize that two general classes of culture cells are usually identified: (1) *anchorage-dependent* cells, which require attachment to a solid substrate for normal growth; and (2) *suspension cells*, which are able to grow freely suspended in the culture medium. The possible influence of hydrodynamic forces on a given cell is clearly dependent on whether it is attached to a substrate or freely suspended.

The present section introduces the fundamental concepts and assumptions needed for analysis of fluid motion. In the forthcoming sections the suspending fluid is assumed to be

1. A continuum. The fluid is treated as an infinitely distortable and divisible substance. The continuum assumption is valid in treating the behavior of fluids under most conditions. A fluid particle is just a linguistic convenience for specifying an arbitrarily small fluid element.

2. Incompressible. Flows in which variations in density are negligible are termed *incompressible* (i.e., the density remains constant throughout

the volume of the fluid and throughout its motion). In other words, there is no noticeable compression or expansion of the fluid. It may be easy to compress fluids with static devices, but it is very difficult to get compression through flow. The fluid behind pushes on the fluid in front, and the latter flows rather than compresses. In fluid dynamics, the question of when the density may be treated as constant involves more than just the nature of the fluid (i.e., liquid or gas). Actually, it depends mainly on a certain flow parameter (the Mach number). We then speak of incompressible and compressible flows, rather than incompressible and compressible fluids. Most liquid flows are essentially incompressible. However, water hammer and cavitation are examples of compressible effects in liquid flows. Whenever a valve is rapidly closed in a pipe, a positive-pressure wave is created upstream of the valve and travels at the speed of sound. This pressure may be great enough to cause pipe failure. This process is called *water hammer*. Cavitation occurs when vapor is formed at those points in the flow field where the local pressure falls below the saturation pressure; there, local boiling takes place without the addition of heat. For most gases, compressibility effects are negligible up to a Mach number Ma of about 0.3 (the maximum density variation is less than 5% for values of $Ma < 0.3$). The Mach number is defined as the ratio of the flow speed to the local speed of sound. The speed of sound in air at sea level under standard conditions is about 340 m/s, so that air behaves nearly as an incompressible fluid for speeds of up to about 113 m/s. For comparison, the speed of sound in water is about 1500 m/s.

3. Newtonian. The fluid exhibits a linear proportionality between the applied shear stress, and the resulting rate of deformation obeys Newton's law of viscosity. Water, the fluid of main concern here, is virtually a perfect Newtonian fluid.

The basic equations that enable us to predict fluid behavior are the equations of motion and continuity. The equations of motion for a real fluid may be developed from consideration of the forces acting on a small element of fluid, including the shear stress generated by fluid motion and viscosity. The derivations of these equations, called the Navier–Stokes equations, are beyond the scope of this chapter, and are listed for future reference. The equations of continuity are developed from the general principle of conservation of mass, which states that the mass within a system remains constant. The term real fluid means that we are dealing with situations in which irreversibilities (e.g., friction) are important. Viscosity is the fluid property that causes shear stresses in a moving fluid; it is also one way by which irreversibilities or losses are developed. Without viscosity in a fluid, there is no fluid resistance.

For incompressible fluid with constant viscosity μ, the equations of motion or Navier–Stokes equations are

$$\rho\frac{D\vec{V}}{Dt} = \rho\left(\frac{\partial\vec{V}}{\partial t} + \vec{V}\cdot\nabla\vec{V}\right) = -\nabla p + \mu\nabla^2\vec{V} \tag{1}$$

where $\vec{V}$ is the local fluid velocity, p is the dynamic pressure, ρ is the fluid density, μ is the fluid dynamic viscosity (commonly referred as *viscosity*), t is time, D/Dt is the material or total time derivative, and ∇ and ∇^2 are the nabla and Laplace operators, respectively (Happel and Brenner, 1986).

Simple flow experiments performed by Newton led him to two conclusions concerning fluid friction that are fundamental to all that is known of the mechanics of real fluids. The first comes from the observation that a fluid does not slide along a solid boundary surface, but rather adheres to it in all cases; This implies that the tangential flow component at a solid boundary vanishes (no-slip condition). In addition, of course, one must satisfy the kinematical condition that the normal velocity of the fluid is the same as that of the boundary. This latter condition holds regardless of whether the surface is fluid or solid, or whether the fluid is viscous. The second conclusion deduced by Newton relates to the force exerted by a fluid and a solid boundary surface on each other. He found, with his one-dimensional flow experiment, that the shear stress τ is linearly proportional to the shear rate $\dot{\gamma}$. For any flow field $\vec{V} = \vec{V}(x, y, z, t)$, and incompressible Newtonian fluid is defined as

$$\tau = \mu\dot{\gamma} \tag{2}$$

where τ is the stress tensor and $\dot{\gamma}$ is the rate of deformation tensor. The proportionality factor μ is the dynamic viscosity of the fluid and is independent of the flow geometry.

Viscous flow regimes are classified as laminar or turbulent on the basis of the internal flow structure. In the laminar regime, the fluid moves in layers, or laminas, one layer gliding smoothly over an adjacent layer with only a molecular interchange of momentum. Any tendencies toward instability and turbulence are damped out by viscous shear forces that resist relative motion of adjacent fluid layers. Flow structure in the turbulent regime is characterized by random three-dimensional motions of fluid particles superimposed on the mean motion. The nature of the flow, that is, whether laminar or turbulent, is determined by the value of a dimensionless parameter, the Reynolds number Re

$$\mathrm{Re} = \frac{\rho VL}{\mu} = \frac{VL}{\nu} \tag{3}$$

where V is the fluid velocity, L is a characteristic length descriptive of the flow field, and ν is the fluid kinematic viscosity. It is not the dynamic viscosity and the density that matter so much in the determination of the nature of the flow but their ratio: the higher the kinematic viscosity, the lower the Reynolds number.

Instruments are usually designed to provide simplified and well-defined flow fields to facilitate the analysis and interpretation of the data. For this reason, flows generated by the fluid shear devices described in this section are assumed to satisfy the following conditions (in addition to the continuum, incompressible, and Newtonian conditions mentioned above):

(4) laminar,
(5) steady, and
(6) fully developed.

These conditions and those stated earlier will be referred to as assumptions (1–6) in the text. Laminar flow is defined above. Steady flow occurs when conditions (e.g., velocity, density, pressure, and temperature) at any point in the fluid do not change with time. A flow is fully developed when the shape of the velocity profile no longer changes with increasing distance from the flow entrance. The flow may approach full development asymptotically, but it is said to be fully developed for practical reasons when it has achieved 99% of the final axial velocity. The distance downstream from the entrance to the location at which fully developed flow begins is called the *entrance length*.

There is no universally accepted classification of flow dimension, and consequently some clarification is necessary. A flow is commonly classified as one-, two-, or three-dimensional depending on the number of space coordinates required to specify the *velocity field* only. However, some investigators choose to classify a flow as one-, two-, or three-dimensional on the basis of the number of space coordinates required to specify *all* fluid properties. For this reason, the exact same flow field is sometimes defined as one-dimensional by one group and two-dimensional by another. For example, the laminar flow in a parallel-plate chamber is defined as one-dimensional by Truskey and Pirone (1990), whereas Levesque and Nerem (1989) defined it as two-dimensional; the velocity field is a function of the vertical distance y only, and the pressure distribution is a function of the axial distance x (Fig. 1). Whatever the classification, the motion of a fluid is completely determined when the velocity vector is known as a function of time and position. Knowledge of the velocity field permits calculation of the pressure distribution, shear stresses, and flow rate. The location of an individual particle is given by position coordinates through integration of the velocity, while forces and pressures are related to

Detailed view

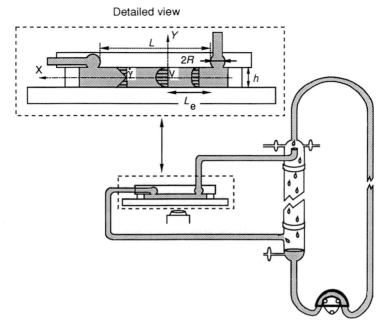

FIGURE 1   Parallel-plate flow channel. The flow channel chamber is connected to a continuous flow loop. The flow loop consists of an elevated reservoir that provides the required pressure drop across the chamber and a roller pump to return the outflow from the collecting reservoir back to the feeding reservoir. This is the flow channel device due to Frangos et al. (1988). The velocity profile is parabolic, and the shear stress is linear with y. Parameters are as follows: velocity field, $V_x = V_x(y)$, $V_y = 0$, $V_z = 0$; shear rate, $\dot{\gamma} = (\partial V_x/\partial y)$; lines of shear, straight lines parallel to the channel axis; shearing surfaces, plane surfaces parallel to the channel axis.

accelerations in the form of velocity derivatives. The parameter of importance in the determination of the flow field is the Reynolds number.

Flow is expected to affect cells in two ways: (1) the fluid moving over or around the cell will exert mechanical stresses on the cell membrane that will be transmitted throughout the cell, and (2) the motion of fluid will alter the concentration of chemical species in the immediate surrounding of the cell, leading to the mass transport of nutrients and waste products to and from the cell. A fundamental issue of importance is how a given applied stress, such as fluid shear, might generate a signal that induces cellular responses. Current thinking is that the shear force initiates a transmembrane event, such as altered-ion permeability or membrane polarization (Lansman et al., 1987; Lansman, 1988; Olesen et al., 1988; Grimm et al., 1988; Nakache and Gaub, 1988). Stress-activated channels have been identified in many cellular membranes and are the focus of much attention in present research.

It is important to recognize that for unsteady motions, one has, in general, both a translational speed U_0 and a frequency factor ω to contend with. Very often, and wrongly, investigators neglect the term $\partial V/\partial t$ in the equations of motion on the grounds that the Reynolds number is very small. However, a creeping flow can be unsteady. The so-called creeping flow (or Stokes) equations are the results of a complete omission of the inertial term $\vec{V} \cdot \nabla \vec{V}$ in the Navier–Stokes equations [Eq. (1)]. It can be shown that the nondimensional forms of the Navier–Stokes equations are (Happel and Brenner, 1986):

$$\mathrm{Re}^r \frac{\partial \bar{V}}{\partial \bar{t}} + \mathrm{Re}\, \bar{V} \cdot \bar{\nabla}\bar{V} = -\bar{\nabla}\bar{p} + \bar{\nabla}^2 \bar{V} \tag{4}$$

where the two most important parameters are the translational Reynolds number, or simply the Reynolds number $\mathrm{Re} = LU_0/\nu$, and the rotational Reynolds number $\mathrm{Re}^r = L^2\omega/\nu$. In linearizing Eq. (4) by letting Re go to zero, two situations can occur, depending on whether Re is an independent variable. For example, in the case of a propellerlike particle settling freely in a gravitational field, the particle simultaneously translates with velocity U_0 and spins with angular velocity ω. These two velocities are both dependent on the same physical variables and therefore are not independent variables. As a consequence, Re^r vanishes with Re and Eq. (4) reduces to the quasisteady, or simply, creeping-flow equations, which in dimensional form is

$$0 = -\nabla p + \mu \nabla^2 \vec{V} \tag{5}$$

On the other hand, in the case of a ball forced to oscillate in a magnetic field, ω and U_0 are independent (Tran-Son-Tay et al., 1990). Therefore, Re^r need not be small, even though Re is. In dimensional form, Eq. (4) then becomes

$$\rho \frac{\partial \vec{V}}{\partial t} = -\nabla p + \mu \nabla^2 \vec{V} \tag{6}$$

and is known as the *unsteady creeping-flow equation*. Of course, if $L^2\omega/\nu$ is also small, as in the case of low frequencies of oscillation, then Eq. (6) reduces to Eq. (5). To illustrate these points, let us consider the general case of an oscillating body with arbitrary shape, such as spherical or infinite plane surface. The body, which has a dimension L, oscillates with an angular frequency ω and amplitude a. The simplification of the general equation of motion [Eq. (1) or (4)] to Eq. (5) or (6) depends on the Stokes number $\alpha = L/\delta$, where $\delta = \sqrt{2\nu/\omega}$ is the Stokes shear wave layer thickness (Landau and Lifshitz, 1987; Tran-Son-Tay et al., 1990). Two important limiting cases are possible

1. If $\delta \gg L$, that is, $\alpha \ll 1$, and if the Reynolds number $Re = \omega aL/\nu$, is small, then the equations of motion reduce to the creeping-flow equations [Eq. (5)]. This is the case of low frequencies of oscillation.
2. If $\delta \ll L$, that is, $\alpha \gg 1$, and if $a \ll L$ then the nonlinear term in the Navier–Stokes equations can be neglected. These equations then reduce to the unsteady creeping-flow equations [Eq. (6)]. It is important to note that for this particular case the Reynolds number need not be small.

It should also be noted that the Womersley number W, commonly used in biological fields, is the Stokes number multiplied by a constant $\sqrt{2}$; $W = \alpha\sqrt{2} = L\sqrt{\omega/\nu}$. For flows between infinite plates and in cylindrical tubes, the characteristic length L is equal to the channel height h and the tube radius R, respectively. If W and Re are less than unity, viscosity dominates and the velocity profile is essentially parabolic (for flows between plates and in tubes) and the flow is said to be quasisteady. If W is above unity, inertial forces distort the velocity profile and the equations for steady flow cannot be applied.

As a remark, in the measurements of oscillatory shear properties, neglect of inertia can lead to serious errors at high frequencies of oscillation, particularly for the elastic modulus η'' (Tran-Son-Tay et al., 1990).

A. Parallel-Plate Flow Channel

1. Governing Equations

The parallel-plate flow channel is certainly the most widely used instrument for studying the effects of shear stress on anchorage-dependent mammalian cells. The flow between infinite parallel plates is often referred to as plane Poisuille flow. The velocity field reduces to one component in the direction of flow v_x

$$v_x = \frac{h^2 \Delta P}{8\mu L}\left[1 - \left(\frac{2y}{h}\right)^2\right] \tag{7}$$

where h is the distance between the plates, μ is the fluid viscosity, ΔP is the pressure drop between the inlet and the outlet located at a distance L, and y is the vertical distance from the origin taken at the centerline of the channel. A schematic of the flow channel is given in Figure 1.

The relationship between the pressure drop and volumetric flow rate Q is

$$\Delta P = \frac{12\mu QL}{wh^3} \qquad (8)$$

where w is the channel width.

From the velocity field [Eq. (7)] the shear rate $\dot{\gamma}$ across the channel gap is derived readily:

$$\dot{\gamma} = -\frac{dv_x}{dy} = \frac{12Q}{wh^3}y \qquad (9)$$

For a Newtonian fluid, the relationship between shear rate and shear stress τ is linear, and the shear stress across the flow channel gap is

$$\tau = \mu\dot{\gamma} = \frac{12\mu Q}{wh^3}y \qquad (10)$$

From Eq. (10), the shear stress is zero along the channel centerline and the maximum shear stress τ_s is at the surface of the plate

$$\tau_s = \frac{\Delta Ph}{2L} = \frac{6Q\mu}{wh^2} \qquad (11)$$

2. Flow Characteristics

The major design requirements for parallel-plate flow channels are as follows:

1. The major axial velocity profile must be fully developed one-dimensional flow.
2. The transverse velocity profile must be uniform across the width of the channel.

These are the characteristics of the flow between infinite parallel plates. The assumption of infinite plates requires that the gap of the channel h be much less than both its width w and length L; $h/w \ll 1$ and $h/L \ll 1$. Since the channel length is larger than the width, the first requirement can be easily satisfied by choosing an appropriate height/width ratio (h/w). In practice, a ratio value of at least 1/50 is used. The second requirement can be met by proper header design. The most practical solution requires the discharge of media from a large cylindrical chamber of radius R in which the pressure is uniform and the velocity is negligible (Fig. 1). This implies that the pressure drop along the plate must be much larger than (1) the pressure drop in the cylindrical chamber and (2) the kinetic energy in the inlet stream. These two conditions can be approximated, respectively, by the following equations (Colton, 1969):

$$\frac{w^2 h^3}{LR^4} \ll 4.7 \tag{12}$$

$$\frac{Qwh^3}{\nu LR^4} \ll 236.9 \tag{13}$$

In order to compare these two conditions, it is necessary to express Eq. (13) in terms of the Reynolds number. The Reynolds number in a flow channel is defined as

$$\text{Re} = \frac{\rho \bar{V} h}{\mu} = \frac{\rho Q}{\mu w} = \frac{\rho \Delta P h^3}{12 L \mu^2} = \frac{\rho \tau_s h^2}{6 \mu^2} \tag{14}$$

where $\bar{V}$ and ρ are the average velocity and density of the fluid, respectively. For the flow to be laminar, the Reynolds number Re must be less than 1400. However, by carefully preventing perturbations at the inlet, laminar flows can be obtained at higher Reynolds numbers. Substitution of Eq. (14) into Eq. (13) yields

$$\frac{w^2 h^3}{LR^4} \ll (236.9/\text{Re}) \tag{15}$$

Comparison of Eqs. (12) and (15) indicates that the second condition is more restrictive than the first for a flow with a Reynolds number greater than about 50. For Re < 50, the opposite is true. The minimum diameter size of the cylindrical chamber is given by the more restrictive criterion. For example, let us consider a typical flow channel with the dimensions $h = 0.025$ cm, $w = 2$ cm, and $L = 6$ cm. If a Reynolds number with a maximum value of 60 is desired, then R must be much greater than 0.04 cm [Eq. (15)]. The term "much greater" is very difficult to quantify, but it is commonly taken as 50–100 times greater. Taking the smallest value for the ratio, the radius of the cylindrical chamber must be larger than 2 cm.

The size of the inlet header is seldom specified in the literature. At low Reynolds numbers, the header diameter is certainly large enough; however, as this number increases, one has to be aware of the minimum header diameter required for having a uniform transverse velocity profile.

Other designs exist for ensuring one-dimensional flow at high Reynolds numbers. Two designs of special interest are the channel with a divergent entrance and a convergent exit due to Koslow et al. (1986) and the channel that combines a bell-shaped entrance with baffles due to Viggers et al. (1986). In addition to the particular design of their flow channel, Viggers et al. use a three-dimensional-flow finite-element analysis to calculate the wall shear stresses over the plate. They find that the velocity profile across the flow direction is essentially uniform across the central 80% of the coverslip area. They also find that shear stresses vary in the flow direction. The values of wall shear stresses given by their numerical analysis are higher than those predicted by the theory for fully developed flow between infinitely wide parallel plates [Eq. (11)]. However, it appears from their results that a better agreement could have been achieved with a channel having a larger width/depth ratio w/h. Viggers et al. use channels with $w/h = 15.8$ and 31.6 and refer to them as the lo-shear and hi-shear models, respectively. The larger ratio, that is, the smaller channel depth, can provide shear stresses over 200 dyn/cm^2. Shear stresses are given at three positions along the coverslip in the direction of flow: proximal edge, midline, and distal edge. At the highest flow rate used (20 cm^3/s) the lo-shear chamber shows a 16.5% variation in the wall shear stresses between the proximal and distal edges, whereas the variation in the hi-shear chamber is only 11.6%. Again at the highest flow rate, which is the most unfavorable case, the variation between the value of the mean shear stress (proximal, midline, and distal) and the value of wall shear stress given by Eq. (11) is 23.2% for the lo-shear chamber and drops to 13.4% for the hi-shear chamber. It is comforting to note that the smallest variation in the values of wall shear stresses is given by the channel with the largest width/height ratio ($w/h = 31.6$). As mentioned earlier, in practice, a ratio value of at least 50 is used to ensure one-dimensional flow.

For the velocity profile to be fully developed and parabolic over nearly the entire length of the channel, the entrance length for the flow L_e must be small compared to the channel length. The length required for the centerline velocity to reach 99% of its undisturbed value is (Bodoia and Osterle, 1961; Schlichting, 1979)

$$L_e = 0.044 h \, \mathrm{Re} \tag{16}$$

This length is known as the entrance length (some sources give slightly different numerical constants). However, the velocity field used to calculate the entrance length represents a first-order solution, valid only for very large Reynolds numbers. Thus, Eq. (16) deviates at low Reynolds numbers from the corresponding numerical solution of the full Navier–Stokes equations (Van Dyke, 1970; Schlichting, 1979). In addition, Eq. (16) is derived by assuming a uniform flow at the channel inlet, which may not be true experimentally.

To give an order-of-magnitude value for the maximum shear stress that can be generated at the wall, let us assume that the fluid is water and has a viscosity $\mu = 1$ cP and density $\rho = 1$ g/cm^3, and that the channel gap has a value $h = 220$ μm. From Eq. (14), the maximum shear stress is found to be equal to 1735 dyn/cm^2 (for the flow to be laminar, Re must be less than 1400) with a corresponding entrance length for the velocity of 1.1 cm. It is important to realize that the entrance length for shear stress development is much less than the entrance length for the velocity field development.

Also note that for a given flow geometry and flow rate, the shear stress at the surface is constant. In addition to its simplicity in design and operation, this constancy of surface shear stress is a primary feature that makes the parallel-plate flow channel well adapted for studying the effects of shear stresses on the structure and function of anchorage-dependent cells. However, if suspended cells (especially non-Newtonian suspensions) are studied with this instrument, it is difficult to interpret the data because both shear stress and shear rate vary over the cross section of the chamber (i.e., in the vertical direction). In the parallel-plate instrument, the nonuniform flow regions are near the ends of the flow chamber where the fluid flows into the plates from the upstream reservoir and out of the plates into the downstream reservoir. Consequently, these flow development effects must be taken into account, if the shear stress (usually calculated at the plate wall) makes use of the pressure difference between locations in the two reservoirs.

3. Experimental Problems

a. End effects   A simple method that allows one to compensate for end effects is to obtain flow data between plates with an identical gap width but different lengths, while keeping the upstream and downstream geometries constant. A simple subtraction then gives the wall shear stress free of end effects,

$$\tau_s = \frac{h(\Delta P_1 - \Delta P_s)}{2(L_1 - L_s)} \tag{17}$$

provided that the shorter plates have a section of appreciable length where the flow is fully developed. The subscripts l and s refer to the long and short plates, respectively.

More specifically, end effects can be ignored by direct measurement of the pressure drop between two given points in the fully developed laminar region. This requires that the pressure taps be outside the entrance-effect zone and that the distance between the pressure taps is known. This setup is designed for studying a limited range of shear stresses over a given length of test section. For different test lengths and especially higher flow rates, the pressure taps may be within the entrance or exit lengths of the chamber and new flow conditions must be specified.

The wall shear stress is often calculated from Eq. (7) by direct measurement of the rate of flow (Frangos et al., 1988; Truskey and Pirone, 1990). Two simple experiments are usually performed to verify that the flow is steady, laminar, and fully developed. The first is to check that Eq. (4) is valid, that is, for a given flow rate, the calculated and predicted pressure drops in the flow chamber must agree. The second is to use dye-visualization techniques to observe streamlines.

b. Other effects   Other physical effects include those caused by sedimentation, wall proximity, flow history, and unsteady flow rate. The flow chamber is usually operated in a horizontal position, so sedimentation effects could be important for the study of suspended cells and in situations where aggregation occurs. If the channel depth or gap is very small, wall effects cannot be neglected, either. Another problem with suspended cells is that the flow history is poorly defined if the cells must flow through tubing and inlet headers. The flow in a flow channel is readily generated by means of a motor-driven syringe or a peristaltic pump, but the flow velocity may not be constant. The condition of constant flow rate can be achieved by means of an inexpensive device involving two reservoirs. The main purpose of the upstream liquid reservoir is to act as a flow regulator, while the role of the downstream is to avoid fluctuation of the flow. This dual-reservoir arrangement is also more suitable for suspended cells, since cells could be exposed to excessive shear stresses in the circulation pump and auxiliary tubing over and above those in the test flow channel.

Finally, it is not clear what effects biochemical substances secreted by the upstream cells have on those downstream (Frangos et al., 1985, 1988; McIntire et al., 1987; Diamond et al., 1989).

4. Discussion

Designs for parallel-plate flow channels are, in general, very similar, but the flow in the chamber can be produced and controlled by different means. The most common method uses a hydrostatic pressure drop system (see for example, Eskin et al., 1984; Frangos et al., 1985, 1988). In this system, the parallel-plate flow chamber is connected at both ends to a reservoir forming a continuous flow loop system. Flow is produced by the hydrostatic pressure head due to the vertical distance between the upper and lower reservoirs. The pressure head is maintained constant by continuous pumping of the fluid medium from the lower to the upper reservoirs with a roller pump. Because of the pump, this system may not be suitable for studies on suspended cells. A schematic of the flow loop used by Frangos et al. (1988) is shown in Figure 1. In this system, the suspending medium was pumped back to the upstream reservoir at rates in excess of that flowing through the chamber. The excess drains down the glass overflow manifold, whose main purpose is to facilitate gas exchange with the medium, into the downstream reservoir.

Flow channels are used in a wide variety of investigations, particularly for studies on cultured mammalian cells. Reports on the effects of shear on cell morphology and alignment are not always in agreement even for similar cell lines and the same flow pattern. For example, the steady, uniform, laminar flow of a tissue culture medium at a low shear stress (10^{-2} dyn/cm^2, this is an extremely low value) over a monolayer of bovine kidney cells induces changes in morphology after less than 1 h (Krueger et al., 1971). However, when bovine carotid artery endothelial cells are subjected to a shear stress of 26 dyn/cm^2, detectable morphologic alterations occur only after 6 h of continuous exposure (Wechezak et al., 1985). In addition, Levesque and Nerem (1985) have reported that orientation begins 4 h after the start of shear stress, when bovine endothelial cells are exposed to shear stresses ranging from 10 to 85 dyn/cm^2. This behavior is somewhat different from that reported by Viggers et al. (1986), in which alignment of the cells with their longitudinal axes parallel to the direction of flow took 12 h and 36 h when the same type of cells were exposed to shear stresses of 60 dyn/cm^2 and 128 dyn/cm^2, respectively. Viggers et al. also reported a seemingly counterintuitive result that, after 6 h at 128 dyn/cm^2, bovine carotid artery endothelial cells pass through an intermediate phase in which they are aligned perpendicular to the flow direction. The authors speculate that this phase represents a form of accommodation in achieving orientation at such a high shear stress. Similar studies and results in other devices have been reported, but most of the work on cultured cells have been performed in a parallel-plate flow chamber.

Confluence is also found to affect the rate of alignment. The time of alignment for subconfluent bovine aortic endothelial cell cultures is 6.5 h, as opposed to 21–22 h for confluent cultures when exposed to a constant shear stress of 34 dyn/cm^2 (Eskin et al., 1984). To complicate the issue, the behavior of cells from two different species in response to shear stresses up to 50 dyn/cm^2 is quite different (Ives et al., 1983). Bovine aortic endothelial cells elongate and align within 24 h at all shear stresses, while human umbilical vein endothelial cells sometimes orient after a long period of exposure to shear stress (137 h). Similar observations are also described by Wechezak et al. (1985).

Even though reported studies on the specific effects of shear on cell morphology and alignment are not always in agreement, they clearly indicate two general responses to shear stress: (1) endothelial cells orient themselves predominantly with the direction of flow, and (2) cells become more elongated when exposed to higher shear.

In addition to shear stress level and cell type, the nature of the substrate (Ives et al., 1983; Levesque and Nerem, 1989) and the type of flow [pulsatile or steady (Levesque and Nerem, 1989)] are important in cell kinetics, cell alignment, and cell morphology. Shear flows can produce more than alterations in morphology. In fact, they have been found to have a wide range of effects and applications. Pulsatile and steady flow onset are reported to produce a sudden increase of prostacyclin production (Frangos et al., 1985, 1988). Flow is found to induce cell polarity and to lead to changes in the proteoglycan metabolism in endothelial cells (Grimm et al., 1988). It has even been reported that wall shear stress generated by fluid flow may regulate gene expression in endothelial cells (Diamond et al., 1989). Finally, the influences of shear stress on endothelial permeability have recently been studied by Jo et al., 1991. Their results indicate that endothelial permeability to albumin is extremely sensitive to wall shear stress. Shear flows have also been used as a means to study surface adhesion (Hochmuth and Mohandas, 1972; Truskey and Pirone, 1990).

In general, the main advantages of the flow channel are that (1) the device is simple in design and operation, (2) the wall shear stress is constant, (3) microscopic observations can easily be made, (4) the device can be integrated into a spectroscopic system or a digital videomicroscopy system to examine cell–surface contact (Desai and Hubbell, 1989; Grapa et al., 1990), (5) the flow system offers a large cell/volume ratio that makes it suitable for analysis of the effects of the shear stress on the metabolism of anchorage-dependent cells, and (6) permits continuous sampling of the cell-incubating medium or cell suspension. However, studies of the effects of shear on suspended cells are difficult to interpret in the parallel-plate device.

B. Cylindrical Tube

1. Governing Equations

Similar to the parallel plate flow channel, laminar flow through a straight cylindrical tube is one-dimensional (Fig. 2). The velocity profile in the z

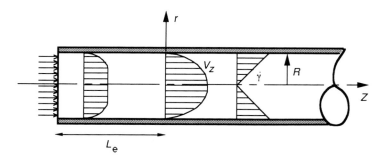

FIGURE 2   Cylindrical tube. Fully developed, laminar, viscous flow in tubes produces a
parabolic velocity profile. The shear stress varies linearly with the radial distance r. Parame-
ters are as follows: velocity field, $V_z = V_z(r)$, $V_r = 0$, $V_\theta = 0$; shear rate, $\dot{\gamma} = -(\partial V_z/\partial r)$; lines
of shear, straight lines parallel to the tube axis; shearing surfaces, concentric cylinders.

direction v_z is

$$v_z = \frac{\Delta P R^2}{4\mu L}\left[1 - \left(\frac{r}{R}\right)^2\right] \tag{18}$$

where ΔP is the pressure drop between two points located at a distance L
apart, R is the tube radius, r is the radial distance from the tube axis, and
μ is the fluid viscosity.

For one-dimensional laminar flows in a pipe, the pressure gradient
necessary to produce a given flow rate is proportional to the viscosity and
inversely proportional to the fourth power of the tube radius,

$$\frac{\Delta P}{L} = \frac{8\mu Q}{\pi R^4} \tag{19}$$

Thus, the average fluid velocity can be expressed in terms of the volumetric
flow rate Q or pressure drop ΔP

$$\bar{V} = \frac{Q}{\pi R^2} = \frac{\Delta P R^2}{8\mu L} \tag{20}$$

For a Newtonian fluid, the stress distribution in the tube is linear
with r

$$\tau = \mu\dot{\gamma} = -\mu\frac{dv_z}{dr} = \frac{r\Delta P}{2L} \tag{21}$$

Therefore, the maximum shear stress is along the tube wall and is equal to

$$\tau_s = \frac{R\,\Delta P}{2L} = \frac{4Q\mu}{\pi R^3} = \frac{4\bar{V}\mu}{R} \tag{22}$$

2. Flow Characteristics

The flow is parabolic (Poiseuille flow), and the shear stress varies linearly with the radial distance from the tube axis. The shear stress is zero along the centerline of the tube and maximum along the tube wall, $r = R = D/2$. The Reynolds number in a cylindrical tube or pipe is defined as

$$\text{Re} = \frac{\rho \bar{V} D}{\mu} = \frac{2\rho Q}{\pi R \mu} = \frac{\rho\,\Delta PR^3}{4\mu^2 L} = \frac{\rho \tau_s R^2}{2\mu^2} \tag{23}$$

The critical Reynolds number above which turbulent flow occurs in a tube is 2300. The critical entrance length L_e for pipe flow is (Hornbeck, 1964; Friedman et al., 1968)

$$L_e = 0.113\,R\,\text{Re} \tag{24}$$

This critical length again corresponds to the length at which the velocity reaches 99% of its undisturbed value. Unlike the parallel-plate flow channel in which the channel width can affect the flow field, it should be noted that for $L > L_e$, one-dimensional flow is guaranteed in a cylindrical tube. For a given fluid and tube diameter, there is a maximum shear stress that can be produced at the tube wall in the laminar domain. For example, since the critical Reynolds number is 2300, that maximum shear stress is 4600 dyn/cm² with $\mu = 1$ cP, $2R = 220$ μm, and $\rho = 1$ g/cm³ [Eq. (23)]. The tube diameter is chosen such that it is equal to the dimension of the channel gap used to calculate the maximum shear stress in a flow channel. The corresponding entrance length is found to be about 2.9 cm. In the laminar domain, higher shear stresses can be produced in a cylindrical tube than in a parallel plate, but the tube must be longer than the plates to provide an identical test section length.

3. Experimental Problems

In both tube flow and flow between parallel plates, the shear stress and shear rate vary over the cross section. It is, therefore, difficult to interpret tube flow data for suspended cells and for non-Newtonian suspensions. In the tube flow instrument, as in the case of the parallel-plate flow device, the nonuniform flow regions are near the ends of the capillary where the fluid flows from a reservoir into the tube, and out of the tube into a second reservoir. The shear stress at the tube wall can be deduced from various measurements [Eq. (22)], but if the calculation makes use of the pressure difference between locations in the two reservoirs, these effects must be considered. However, the effect of the downstream end is probably not important, and again the entrance length for shear stress development is less than for velocity development.

To compensate for these effects, one can again obtain flow data with tubes of the same diameter but different lengths, while keeping the upstream and downstream geometries constant. A simple subtraction will then give the wall shear stress free of end effects

$$\tau_s = \frac{D(\Delta P_1 - \Delta P_s)}{4(L_1 - L_s)} \tag{25}$$

provided that the shorter tube has a section of appreciable length where the flow is simple. The subscripts l and s refer to the long and short tubes, respectively.

In general, the experimental problems mentioned in Section II,B,3,b for the parallel-plate device are also relevant to the cylindrical-tube device. As a note of caution, wall effects in an enclosed circular boundary are more important than for infinite parallel plates, and may be important if the tube diameter is very small.

4. Discussion

Cylindrical-tube devices have been used extensively in the studies of blood rheology. Most of these studies deal with the bulk properties of blood. The main disadvantages of the cylindrical-tube devices are that microscopic observation in these instruments is difficult because of optical distortion caused by the curved surfaces and that they offer a smaller area of observation than through a flat plate. The problem of optical refraction can be circumvented by embedding the tube in a water-filled chamber. However, because of this inconvenience and the disadvantage of having a smaller area of observation, very few studies have been done on the effects of shear on the properties of individual or anchorage-dependent cells with these instruments. Studies that aim to follow individual cells are obviously very difficult, because flows of interest are relatively fast and rapidly take cells past a static observer. This experimental difficulty has, however, been overcome with the travelling microtube technique of Goldsmith and Mason (1975) for studying individual cells. In this technique, a thin glass tube, containing the cell suspension, is embedded in a surrounding fluid with a refractive index identical to that of the cell suspension medium and maintained vertically on a motor-driven stage. The tube is examined with a microscope mounted horizontally while a slow flow is generated in the tube and the stage is moved vertically in order to follow selected individual cells.

In summary, cylindrical-tube devices have advantages and disadvantages similar to those in parallel-plate systems. The main difference is that microscopic observation through a cylinder is more difficult and is limited to a smaller area of observation than through a flat plate.

C. Concentric Cylinder Device

1. Governing Equations

The motion (velocity) of a fluid between two infinite coaxial cylinders (Fig. 3) with radii R_1, R_2 ($R_2 > R_1$), rotating about their axes with angular velocities ω_1, ω_2, is of the form $v_z = v_r = 0$, $v_\theta = v(r)$, and $p = p(r)$.

The only nonzero component of the velocity distribution is obtained from the r-component of the Navier–Stokes equation

$$v_\theta = \frac{\omega_2 R_2^2 - \omega_1 R_1^2}{R_2^2 - R_1^2} r - \frac{(\omega_2 - \omega_1) R_2^2 R_1^2}{R_2^2 - R_1^2} \frac{1}{r} \tag{26}$$

and the pressure distribution is found from the θ component of the Navier–Stokes equation by straightforward integration

$$p = p_1 + \frac{\rho}{\left(R_2^2 - R_1^2\right)^2} \left[\frac{\left(\omega_2 R_2^2 - \omega_1 R_1^2\right)^2 \left(r^2 - R_1^2\right)}{2} \right.$$

$$- 2 R_1^2 R_2^2 (\omega_2 - \omega_1)(\omega_2 R_2^2 - \omega_1 R_1^2) \ln \frac{r}{R_1}$$

$$\left. - \frac{R_1^4 R_2^4 (\omega_2 - \omega_1)^2}{2} \left(\frac{1}{R_1^2} - \frac{1}{r^2} \right) \right] \tag{27}$$

where p_1 is the pressure at $r = R_1$. The magnitude of the shear rate between the two cylinders is equal to

$$\dot{\gamma}_{r\theta} = -r \frac{d}{dr} \left(\frac{v_\theta}{r} \right)$$

$$= \frac{2|\omega_1 - \omega_2| R_2^2 R_1^2}{R_2^2 - R_1^2} \frac{1}{r^2} \tag{28}$$

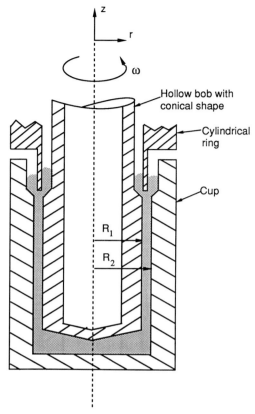

FIGURE 3   Concentric cylinder device. The ring is used to prevent fluid interfacial effect. The flow field between coaxial cylinders where the gap/cylinder radius is $\ll 1$ (Couette flow) can be approximated by the flow between parallel plates or by a linear velocity profile. Parameters are as follows: velocity field, $V_\theta = V_\theta(r)$, $V_r = 0$, $V_z = 0$; shear rate, $\dot{\gamma} = -r(d/dr)(V_\theta)/r$; lines of shear, circles of constant r and z; shearing surfaces, concentric cylinders.

and the torque T acting on the inner or outer cylinder of length L is

$$T = \frac{4\pi\mu|\omega_1 - \omega_2|R_2^2 R_1^2 L}{R_2^2 - R_1^2} \tag{29}$$

Although only one cylinder is rotating in most practical cases, these formulas were derived for the general case of two rotating cylinders.

In the case of a rotating outer cylinder (the cup)—$\omega_1 = 0$, the shear stress is maximum along the surface of the inner cylinder (the bob) and minimum at the surface of the cup $(r = R_2)$, with respective magnitudes

$$\tau_{r=R_1} = \frac{2\mu\omega_2 R_2^2}{R_2^2 - R_1^2}, \qquad \tau_{r=R_2} = \frac{2\mu\omega_2 R_1^2}{R_2^2 - R_1^2} \qquad (30)$$

The relationships between the torque and the shear stresses are

$$\tau_{r=R_1} = \frac{T}{2\pi R_1^2 L}, \qquad \tau_{r=R_2} = \frac{T}{2\pi R_2^2 L} \qquad (31)$$

The shear stress $\tau_{r\theta} = \mu\dot{\gamma}_{r\theta}$ in the case of a rotating bob can be found from Eq. (28) with $\omega_2 = 0$.

2. Flow Characteristics

A characteristic feature of this instrument under steady, laminar flow, is that the torque is independent of r. Another feature is that the shear stress varies in the gap as a function of $1/r^2$. Because of the assumptions made, it is essential that turbulence is not present and that Taylor vortices are not formed. These vortices can occur only when the inner cylinder (bob) is rotated. Rotation of the inner cylinder causes the inside layers of fluid to have a larger centrifugal force than the outside layers, setting up a centrifugal force field that produces an unstable equilibrium. The limit of stability for laminar motion is determined by the Taylor number Ta, which is related to the Reynolds number

$$\text{Ta} = \frac{v_1 h}{\nu}\sqrt{\frac{h}{R_1}} = \text{Re}\sqrt{\frac{h}{R_1}} \qquad (32)$$

where v_1 and R_1 are the tangential velocity and radius of the bob, respectively, h is the gap width, and ν is the fluid kinematic viscosity $(\nu = \mu/\rho)$. For a rotating inner cylinder, depending on the value of the Taylor number, three flow patterns can be produced:

Ta < 41.3	laminar flow;
41.3 < Ta < 400	laminar flow with Taylor vortices;
Ta > 400	turbulent flow.

3. Experimental Problems

In addition to the formation of Taylor vortices when the bob is rotating, other problems could occur involving torque readings and excluded volume effects next to the wall.

   a. Torque reading   Errors in the torque readings from a coaxial-cylinder-type device can originate in three spatial regions of the instrument: (1) the fluid in the liquid space above the bob, (2) the fluid between the bottom of the bob and the bottom of the cup, (3) the fluid near the edge of the gap formed by the cylindrical surfaces. Computational and experimental methods have been developed to correct for these effects. A common computational method assigns a fictitious length to the bob (greater than its real length), which is experimentally determined with Newtonian fluids of known viscosity. The assumption of tangential annular flow used to derive Eqs. (18)–(22) does not hold near the bottom of the inner cylinder. In order to compensate for this end effect, the bob is usually constructed with a conical shape. The flow at the bottom is thus like a cone-and-plate flow (Fig. 3). If the gap is narrow, that is, $(R_2 - R_1)/R_1 \ll 1$, and the cone angle is small, then the shear rate is uniform throughout the fluid.

   Alternatively, a simple experimental method exists that eliminates end and edge effects. This technique uses two bobs of identical shape and diameter but different lengths. Torque determinations are made with each bob at a given rotational speed with the same fluid space, geometry, and flow above and below the bob. The shear stress at a radial position r, free of end and edge effects is then calculated as

$$\tau = \frac{(T_1 - T_s)}{2\pi r^2 (L_1 - L_s)} \tag{33}$$

where the subscripts l and s refer to the long and short bobs, respectively.

   When there is interfacial film formation, as at the interface of a suspension of blood and air, a correction of the torque reading has to be made. This source of error can be removed by mechanically preventing the transmission of a torque through the film to the torque-sensing surface of the viscometer. One method is to place a cylindrical ring through the liquid–gas interface (Fig. 3). The ring is mounted independent of the torque measuring surface, but is capable of rotating with that surface as it moves in response to different torques. Contrary to some belief, the guard ring does not prevent the formation of the interfacial film; it only eliminates its effect. The use of the two bob technique described above for end and edge effects should also allow compensation for the film effect.

b. Wall effects   When a smooth wall forms an interface with a suspension of particles, the presence of the solid wall physically prevents the particles from occupying the fluid space next to the wall. This results in the particle being excluded from a certain volume of the fluid adjacent to the wall. For example, in blood, the centers of red cells can never be found in the fluid space immediately adjacent to the wall. Vand (1948) studied this wall effect with suspensions of glass spheres in one concentric cylinder viscometer and two different capillary viscometers. He determined that this wall effect was measurable in both types of instruments. Cokelet et al. (1963) reported a similar wall effect with blood in a concentric-cylinder viscometer when data were obtained from experiments with grooved and smooth surfaces. Their data indicated that the wall effect is negligible for blood with hematocrit (cell concentration) below 40% at high shear rates, but is substantial at very low shear rates (e.g., calculated plasma layer thicknesses at the smooth walls are 1 μm at a shear rate of 25 s^{-1} and 3 μm at 1 s^{-1}).

4. Discussion

It is useful to note that the flow of an incompressible fluid caused by the relative rotation of two concentric cylinders forming a small gap is known as Couette flow. The interest in Couette flow is that all cells are subjected to the same shear rate.

Most of the reported studies on mammalian cells using coaxial-cylinder devices have been on the rheological properties of blood cell suspensions (Chien et al., 1982; Tran-Son-Tay et al., 1986; Drasler et al., 1987) or on the complex responses of platelets to shear stress (Wurzinger et al., 1985a, 1985b; Sutera et al., 1988). However, very few studies have been performed on hybridoma cells (Smith et al., 1987; Petersen et al., 1988; Abu-Reesh and Kargi, 1989).

Concentric-cylinder devices do not allow direct study of individual cells. Direct observations of cell shape or determinations of mechanical properties are very difficult to accomplish with coaxial-cylinder devices. In general, only bulk properties are obtained and little information is given about the properties of individual cells. This may result in some difficulty in interpreting experimental data, due to the extensive structural and functional heterogeneity usually displayed from cell to cell even for a given cell line. The changes in cell morphology as a result of fluid flow, even in complex flows, may be studied indirectly by fixing the cells with glutaraldehyde (Sutera and Mehrjardi, 1975; Sutera et al., 1975). However, for this technique to be useful, it is necessary that the deformation rate of the cell be much less than the rate of the fixation process or that the cell be in a steady state of deformation.

Accurate controls of parameters such as temperature, pH, and dissolved oxygen are very difficult or impossible in the coaxial devices as currently designed. This is a major issue for cultured cells.

D. Cone-and-Plate Device

1. Governing Equations

This is another popular instrument because the shear rate is fairly constant throughout the gap for small cone angles (commercial cone–plate viscometers usually range from 0.5° to 8°).

A schematic of the cone-and-plate device is shown in Figure 4. The simplest analysis of this instrument makes use of the fact that the cone angle α_0 (angle between the cone and the plate) is so small that local flow can be regarded as essentially the same as that between parallel plates. For a cone rotating at an angular speed ω, the velocity profile in the gap at a distance r from the cone apex and at an angle α ($\alpha = \pi/2 - \theta$) is

$$v_\phi = \frac{r\omega\alpha}{\alpha_0} \tag{34}$$

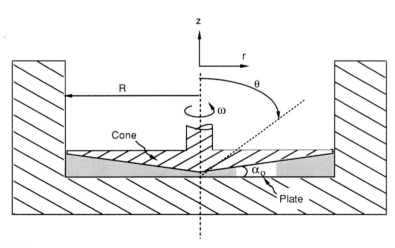

FIGURE 4 Cone-and-plate device. For small cone angles, the flow between cone and the plate is equivalent to the one between parallel plates. Parameters are as follows: velocity field, $V_\phi = V_\phi(r, \theta)$, $V_\theta = 0$, $V_r = 0$; shear rate, $\dot\gamma = -(1/r)(\partial V_\phi/\partial\theta)$; lines of shear, circles of constant r and z; shearing surfaces, cones of constant θ.

and the pressure distribution reduces to

$$p = p_a + \frac{1}{2}\rho r \omega^2 R^2 \left(\frac{\alpha}{\alpha_0}\right)^2 \left(\frac{r^2}{R^2} - 1\right) \tag{35}$$

where R is the cone radius, p_a is the pressure at the position $r = R$, and ρ is the fluid density.

The shear stress τ and rate $\dot{\gamma}$ are given by

$$\tau = \frac{3T}{2\pi R^3} = \frac{\mu\omega}{\alpha_0} \tag{36}$$

$$\dot{\gamma} = -\frac{\sin\theta}{r}\frac{d}{d\theta}\left(\frac{v_\phi}{\sin\theta}\right) \cong -\frac{1}{r}\frac{dv_\phi}{d\theta} = \frac{r\omega}{d} = \frac{\omega}{\alpha_0} \tag{37}$$

where T is the applied torque, d is the vertical thickness of the gap at the position r, and μ is the fluid viscosity.

2. Flow Characteristics

In the cone-and-plate instrument, the shear stress and shear rate are virtually constant throughout the gap for a given speed of rotation of the cone. However, the effects of secondary flow on the rate of deformation cap become important at high Reynolds number. In a cone-and-plate viscometer, this number is defined as

$$\text{Re} = \frac{R^2\omega}{\nu} \tag{38}$$

where ν is the kinematic viscosity ($\nu = \mu/\rho$). The onset of these effects have been studied by numerical integration (Fewell and Hellums, 1977). For example, with a 2° cone, secondary flow effects need to be taken into consideration if Re > 1000. For these values, the effects of the secondary flow on the rate of deformation is about 10% of the primary rate of deformation.

To characterize flow between a shallow cone and a flat plate, Bussolari et al. (1982) use the method of asymptotic expansion and find that the appropriate expansion parameter is a modified Reynolds number,

$$\widetilde{Re} = \frac{(r\alpha_0)^2 \omega}{12\nu} \tag{39}$$

where r is the radial distance from the cone apex. They report that secondary flow appears when $\widetilde{Re} \geq 1$ and that a gradual transition to turbulent flow occurs with increasing $\widetilde{Re}$. It is important to note that $\widetilde{Re}$ is a local Reynolds number that ranges from zero at the cone apex to a maximum at the outer boundary. Therefore, it is possible, under steady rotation, to have laminar, transitional, and turbulent flows simultaneously in the gap. Turbulence appears first at the outer boundary and spreads inward with increasing rotational speed. It is also of fundamental importance to note that in the turbulent region the shear stress is no longer uniform across the gap.

With regard to secondary flow effects, it can be shown that the results reported by Bussolari et al. (1982) and Fewell and Hellums (1977) do not agree. For example, at the edge of a cone ($r = R$) with a 2° angle, the theory due to Bussolari et al. predicts that secondary flow effects appear at a speed of rotation that is 10 times larger than that given by the analysis of Fewell and Hellums. This discrepancy comes from the level at which secondary flow is considered significant. Whereas the significance of the secondary flow is well documented in the work by Fewell and Hellums, it receives limited attention by Bussolari et al. However, a complete analysis given by Sdougos et al. (1984) shows that the analysis of Bussolari et al. (1982) is valid only if $\widetilde{Re} < 0.5$.

3. Experimental Problems

In addition to the requirement of a small cone angle, it is essential that the cone axis be closely aligned so that it is perpendicular to the plate. Eccentricity in the orientation of the cone or in the drive is a more serious problem than with concentric cylinders, which, to some extent, are self-centering.

If the fluid fills more than just the space between the cone and the plate and forms a meniscus, an additional torque from fluid shearing in this additional space is developed. The best experimental procedure to ensure that these extraneous effects are negligible is to take measurements with two cones of equal radius but different cone angles. If both sets of data give the same results, there is strong evidence that the method of analysis is correct. Of course, in varying the cone angle we must keep in mind that the theory is valid only for small cone angles.

Other experimental problems that can occur are the formation of a film at the surface of the medium and the sedimentation of the suspended cells. If there is formation of an interfacial film, the two cone technique suggested above for end-effect correction will not eliminate the film effect. With its essentially horizontal surfaces, the cone-and-plate device seems especially susceptible to any problems that might arise from cell sedimentation. At low nominal shear rates, below $1–2$ s^{-1}, a layer of essentially cell-free fluid could form at the upper surface, resulting in erroneous torque measurements (which would be lower than they should be with a homogeneous cell distribution because of the cell-free layer). Apparently this problem has not been investigated by users of the cone-and-plate viscometers. Also, no studies have been done on the effect of smooth versus rough walls in a cone-and-plate viscometer.

Finally, it should be remembered that the instrument has a nonideal flow pattern near the cone edge.

4. Discussion

The cone-and-plate geometry became popular in the last two decades for the measurements of rheological properties of suspended cells. More recently the cone-and-plate system was used to study flow effects on the properties of both suspended and anchorage-dependent cultured cells. Shear sensitivity of tumor cells was studied in a cone-and-plate viscometer (Brooks, 1984). Sutera et al. (1988) found that shear-induced platelet aggregation was significantly greater in response to pulsatile versus continuous shearing, except at the lowest applied stress of 10 dyn/cm^2. The level of aggregation was dependent on both stress amplitude and number of pulses. The maximum shear stress used in these experiments was 50 dyn/cm^2. When postconfluent endothelial cultures grown under static conditions were continuously exposed to a shear stress 1 or 5 dyn/cm^2 for up to 8 days, their configuration remained unchanged. In contrast, endothelial cultures (either postconfluent or subconfluent) exposed to 8 dyn/cm^2 exhibited dramatic time-dependent morphological changes (Dewey, 1984). Franke et al. (1984) reported that endothelial stress fibers can be induced by exposure for 3 h to a fluid shear stress of 2 dyn/cm^2 without affecting cellular shape and orientation. Bussolari et al. (1982) studied the effects of fluid shear stress on endothelial cell structure and function. They found that the flow induced by the rotating cone has three regimes defined by the dimensionless parameter $\widetilde{Re}$. It is important to note, as mentioned earlier, that the theory of Bussolari et al. (1982) is valid only for $\widetilde{Re} \leq 0.5$. This criterion seems to be quite restrictive. The position of the test specimen in their apparatus is not specified but assuming a nominal

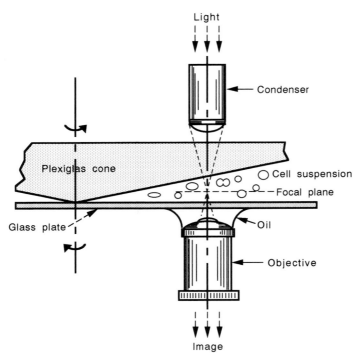

FIGURE 5   A rheoscope. The main feature of this instrument is the counter rotation of the cone and plate. This gives the advantage that a particle midway between the cone and plate, subjected to a well-defined shear stress can be studied without the aid of a high speed cinematography. At a distance r from the axis of rotation, the shear rate is equal to $\dot{\gamma} = 2r\omega/d$, where d is the local gap width and w is the angular velocity of the cone and plate. Along the midplane (plane midway between the cone and plate), the shear rate and shear stress are equal to zero.

radial distance of 3 cm, a fluid kinematic viscosity of 0.01 cm^2/s, $\widetilde{Re} = 0.5$, and a system with a 2° cone angle, Eq. (37) gives a limiting angular speed of 5.5 rad/s. The corresponding limiting shear stress is 1.6 dyn/cm^2.

To allow direct observation of suspended cells, a modified cone-and-plate viscometer, called a rheoscope (Fig. 5), has been developed (Schmid-Schonbein et al., 1973) in which the cone and plate counterrotate. This gives the advantage that a particle midway between the cone and plate, is subjected to a well-defined shear stress field and remains nearly stationary in the laboratory frame of reference so that it can be studied without the aid of high-speed cinematography. It is important to remark, for an identical speed of rotation, that the shear rate generated in the rheoscope is twice that in the conventional cone–plate device because of the counter rotation of the cone and the plate. Effects of shear and mechanical properties of individual cells (Tran-Son-Tay et al., 1984, 1987; Sutera et al., 1989) and anchorage-dependent cells (Franke et al., 1984) have been determined with the rheoscope.

In addition to the fact that the shear rate is fairly constant throughout the gap, the advantage of the cone-and-plate system is that it requires a smaller fluid volume than other geometries (channel, cylindrical-tube, concentric cylinders). The apparatus does allow optical microscopic observation during shear stress application. Further, if desired, the cells may be fixed in situ, under shear, by infusing fixative into the fluid volume as in the coaxial-cylinder device. However, at high-Reynolds-number values secondary flow effects are not negligible and the equations given above need to be corrected for these effects. As currently designed, the cone-and-plate device has a small cell/volume ratio, does not permit continuous sampling of the cell-incubating medium, and has the same environmental control problems associated with the incubating medium as the coaxial-cylinder device.

E. Parallel-Disk System

1. Governing Equations

The parallel-disk system is very similar in operation to the cone-and-plate device. A schematic of the parallel-disk apparatus is shown in Fig. 6.

In addition to assumptions (1)–(6) specified at the beginning of the section, it is generally assumed that inertia is negligible, and as a consequence the velocity and pressure distributions have the form $v_r = 0$, $v_z = 0$, $v_\theta = rf(z)$, and $p = p(r, z)$. When the upper disk is rotated with angular velocity ω, the velocity distribution in the gap of thickness h reduces to one component in the tangential direction θ

$$v_\theta = \frac{r\omega z}{h} \tag{40}$$

where r is the distance from the center of the disk, and z is the axial distance from the bottom plate.

The shear rate is a function of r alone, and for a Newtonian fluid the shear stress in the device varies with the radial distance r

$$\tau = \mu\dot{\gamma} = \mu\frac{dv_\theta}{dz} = \mu\frac{r\omega}{h} \tag{41}$$

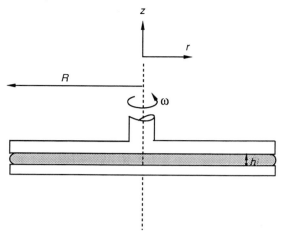

FIGURE 6   A parallel-disk apparatus. When the upper disk is placed in a container or cylinder, the instrument is called a *disk-and-cylinder device*. Parameters are as follows: velocity field, $V_\theta = (rz\omega/h)$, $V_r = 0$, $V_z = 0$; shear rate, $\dot\gamma = (r\omega/h)$; lines of shear, circles of constant r and z; shearing surfaces, parallel disks.

2. Flow Characteristics

Along the axis $r = 0$, the shear rate is zero, and therefore the fluid located on the z axis must be stress-free. At the rim of the device $r = R$, the shear stress is maximum. The Reynolds number in a parallel-disk apparatus is defined as

$$\mathrm{Re} = \rho\frac{\omega Rh}{\mu} = \frac{\rho\tau_s h^2}{\mu^2} \tag{42}$$

where τ_s is the maximum shear stress at the rim. The flow produced by a rotating disk is laminar when $\mathrm{Re} < 10^5$.

   The torque required to turn the upper disk can be shown to be

$$T = \frac{\pi\mu\omega R^4}{2h} \tag{43}$$

3. Experimental Problems

Disk viscometers have experimental problems that are similar to the cone-and-plate viscometer. The fluid between the disks is assumed to form hemispherical caps at the edges of the disks, but this is not observed experimentally. Also, if the fluid fills more than just the space between the disks, an additional torque from the fluid is developed. End and sedimentation effects, as in the cone-and-plate device, have not been fully investigated.

Very often, the rotating or upper disk is placed in a cylindrical container. The system is then referred to as a disk-and-cylinder viscometer. The solution of the primary velocity field in this device is more complicated than the one for the parallel-disk device.

Secondary flow effects, as in the cone-and-plate viscometer, also occur in the disk devices at low Reynolds numbers. Secondary flow patterns in the two-disk devices mentioned above are very similar. The primary fluid motion is the tangential direction because of the rotation of the disk. In addition, the fluid at the top of the device rotates with a larger, angular velocity than that at the bottom; consequently, the fluid nearer the upper disk experiences a larger outward centrifugal force. For Newtonian fluids there are no forces to counter this centrifugal force, and a weak secondary flow forms perpendicular to the primary flow. Secondary flow is directed in a circular pattern that moves radially outward near the upper disk, down the gap edge or down the side of the cylinder, inward along the bottom, and finally back up near the center or axis of the disk device. The magnitudes of velocities in the secondary flow are roughly 10% those in the primary flow. It is interesting to note that a polymer fluid placed in the disk-and-cylinder system exhibits a secondary flow in the opposite direction (Hill, 1972).

4. Discussion

Disk instruments are not widely used because the shear rate is not constant throughout the gap, making interpretation of the data on flow effects difficult. On the other hand, if the flow is well characterized, this can be an advantage for studies on anchorage-dependent cells since a range of shear stresses can be examined in one experiment. Using a rotating-disk system, Ando et al. (1987) found that shear stress can stimulate the migration and proliferation of endothelial cells. However, the cell layer exposed to shear stresses ranging from 0.3 to 1.7 dyn/cm^2 showed no remarkable changes in cell morphology. Effects of pulsatile shear stress have been investigated by Nomura et al. (1988). Cell adhesion, cell attachment, and cell morphology as a function of shear stress for the chosen time of rotation before fixation have been studied with a spinning disk in an infinite medium (Hochmuth et al., 1972; Horbett et al., 1988). The analysis used is based on the boundary-layer theory. In the study by Hochmuth et al., glutaraldehyde is injected into the chamber to fix the cells with the disk rotating.

In a disk apparatus system, the disk does not have to rotate. Cozens-Roberts et al. (1990) used a radial-flow device to study cell–surface interaction. Because the flow conditions in this apparatus are different from those in a rotating disk, the governing equations for the radial-flow device are different from those given in this section.

Disk devices have about the same advantages and disadvantages as cone-and-plate devices. However, analysis of experimental results are more difficult in the former instruments.

In general, by reducing the gap or radius size in all these devices it is possible to elevate the stress to a higher level without increasing the Reynolds number. However, when the gap or radius is too small, that is when the medium depth is too narrow, nutrient supply is limited and could reduce viability of the culture cells in the medium. Therefore, the method of increasing the shear stress at the wall by reducing the gap or tube radius has limitations in an actual culture system.

To conclude the section on well-defined flows, the question on the potential effects of pressure level on cell morphology and metabolism should be addressed. Tikunaga and Watanabe (1987), using a constant-pressure chamber, report that umbilical-vein endothelial cells are affected by pressure level whereas smooth-muscle cells are not. Endothelial cell growth is shown to vary with pressure level. The maximum gauge pressure (i.e., the pressure level is measured with respect to atmospheric pressure) used in their experiment is 160 mm Hg. Growth is minimum at atmospheric pressure (0 mm Hg gauge pressure) and maximum at 80 mm Hg for the first 2 days of incubation. However, the growth rates are reversed after the second day. A degeneration of the cells is seen at high pressures (above 80 mm Hg). In contrast to the cell growth, production of prostacyclin is maximum at 0 mm Hg and minimum at 80 mm Hg. These findings are particularly important in the case of a system in which the flow is produced by a hydrostatic-pressure head, since the pressure head increases with the level of shear stress used and thus with the pressure drop across the chamber. For this reason, it is difficult to differentiate the effects of shear stress and pressure in this type of system. As an example, let us consider the case of a parallel-plate flow channel with the dimensions $h = 0.025$ cm and $L = 6$ cm and a Reynolds number Re = 150. It can be shown that, for the values given above, the pressure drop across the chamber is about 52 mm Hg [Eq. (14)]. This change in pressure may have an effect on endothelial cell morphology and metabolism, but this remains to be tested since the results of Tokunaga and Watanabe (1987) have not yet been confirmed.

Finally, as a transition to the turbulent-flow section, it is worthy to note the system of Cooke et al. (1990) for studying flow-mediated endothelium-dependent vasodilation. This system is essentially a bioreactor. The flow chamber contains a stir bar and is mounted on a magnetic stirrer to induce vortical flow. A bioassay ring and a suspension of endothelial cells on beads are immersed in the flow chamber. The flow pattern is characterized by laser Doppler measurements and calculations. Shear stresses are also calculated. This approach is very useful since the velocity and shear stress distributions in a bioreactor are not well defined. However, one should keep in mind that the results of this approach are semiquantitative only.

III. TURBULENT FLOW

The effects of turbulent flow are of major concern in bioreactor culture of cells. At present it is difficult to assess the effects of hydrodynamic forces on cell damage from the published literature because the results are not related to an intrinsic cell damage parameter but to an applied external force. This difficulty is made worse by the diversity of bioreactors, instruments, agitation conditions, and cell lines used.

A number of cell damage mechanisms have been proposed for mammalian cells on microcarriers in a stirred tank (Cherry and Papoutsakis, 1986, 1988, 1989; Croughan et al., 1987, 1988, 1989; Croughan and Wang, 1989). These mechanisms are based on theories of particles in turbulent flow, but it is still not clear what types of hydrodynamic forces cause cell death. The theory developed for microcarriers assumes, among other things, that damage occurs only if a microcarrier encounters an eddy smaller than a critical length. For a typical bioreactor this critical length is on the order of 100–250 μm (Cherry and Papoutsakis, 1986, 1988, 1989; Croughan et al., 1989) and is much larger than the 10–20 μm diameter of an individual suspended cell. Thus, the assumptions from which these analyses are derived are not adequate to describe fluid effects in suspended cell cultures. Shear or elongational stresses within eddies and transient stresses as the cells move from one eddy to another must be postulated as possible causes of suspended cell damage. This theory is not adequate, either, for cells attached to a fixed substratum membrane. Comparisons between supported and suspension cultures have been carried out on hybridoma cells in the growth and metabolism studies of Murdin et al. (1989). They report that specific antibody productivity is higher in a packed-bed reactor than in suspension cultures. They also address the problem of mass transfer within aggregates of immobilized cells.

Bioreactors can be characterized either as homogeneous or heterogenous systems. Stirred tanks, airlift reactors, and hybrid systems that combine the airlift principle with mechanical agitation are the most common homogeneous systems. The major disadvantage of the conventional stirred-tank reactors is oxygen-transfer limitation at large-scale operation. This limitation is governed by the maximum rate of agitation that can be generated before cell damage occurs, which in turn must be matched with the level of shear sensitivity of the cell line used. The effects of hydrodynamic forces in a stirred-tank reactor have been addressed for anchorage-independent cells through studies on the effects of well-defined hydrodynamic forces in viscometers and capillary tubes (Augenstein et al., 1971; Tramper et al., 1986; Smith et al., 1987; McQueen et al., 1987; Petersen et al., 1988; McQueen and Bailey, 1989). Cell death has been found to occur beyond some threshold level of shear stress due to the external fluid. However, as mentioned earlier, it is difficult to assess these effects since different cell lines were used and different surface loadings were imposed on the suspended cells. Viscometric studies have indicated that hybridoma cells in lag and stationary phases are more sensitive to death by shear than are cells in logarithmic growth phase. Airlift reactors have been designed for shear and oxygen-supply considerations but, in general, direct sparging of air into serum-containing medium causes undesirable foaming (Handa et al., 1987; Handa-Corrigan et al., 1989; Passini and Goochee, 1989). The proposed mechanisms of cell damage due to sparging are (1) the formation of a foam in which the cells are entrained and die and (2) the formation and/or breakage of bubbles. It has been found that the surfactant Pluronic F-68 protects mammalian cells from the lethal effects of sparging (Handa et al., 1987; Handa-Corrigan et al., 1989). Interfacial tensions between air and water are large (72 dyn/cm) and are reduced on absorption of surface-active agents. It has been suggested that the protective mechanism of Pluronic F-68 relies on its ability to decrease plasma membrane fluidity through direct interaction with the plasma membrane, since it was found that increasing membrane fluidity correlates with increasing shear sensitivity (Ramirez and Mutharasan, 1990).

In heterogenous systems, cells are trapped or immobilized inside hollow fibers, in ceramic matrices, between flat membranes, or in beads. Heterogenous unit process systems seem to produce the highest cell densities and antibody concentrations. However, these factors do not provide flexibility in the event of mechanical failure or contamination; a longer time is required to restore a heterogenous reactor back to full production rates after stoppage. In addition, heterogeneous unit process systems do not offer the best option for direct monitoring and control of process parameters. In contrast, measurements done on a sample of a homogeneous system reflect the state of the system.

The differences in reported shear sensitivity are sometimes confusing but are not surprising. Hybridoma cells are morphologically and physiologically a diverse group. Needham et al. (1991) have shown that these diversities also exist in cells that are from a given cell line but at different phases in cell growth and in the cell cycle. Studies of the effect of agitation on hybridoma cells show variable results, ranging from decreased growth rate at relatively low agitation rates (Fazekas de St. Groth, 1983) to insensitivity at high agitation rates (Oh et al., 1989). It is also found that hybridoma cells in stationary and decline phases are more sensitive to elevated agitation rates than are cells in log-phase growth (Dodge and Hu, 1986; Lee et al., 1988; Petersen et al., 1990).

Reports on the influence of turbulent flow on mammalian cells, besides those in bioreactors, are very limited. Most of these reports are for suspended cells. Using a coaxial-cylinder viscometer, Abu-Reesh and Kargi (1989) find that turbulent shear causes a higher degree of damage than laminar shear for the same mean shear level and exposure time. Hybridoma cells are damaged above a shear stress level of 50 dyn/cm^2 in the turbulent regime. Using a mouse myeloma cell line growing in a suspension subjected to flow through a capillary tube with a sudden contraction in tube diameter, McQueen et al. (1987) observe that significant lysis occurs above a mean wall shear stress of 1800 dyn/cm^2. The only study on turbulence effects, to my knowledge, on anchorage-dependent cells is that by Davies et al. (1986), with a cone-and-plate viscometer. They report that it is the unsteady flow characteristics, rather than the magnitude of wall shear stress per se, that are responsible for enhanced endothelial turnover in vitro. The highest shear stress used was 15 dyn/cm^2 in laminar and 14 dyn/cm^2 in turbulent flow for a 24-h period.

IV. SURFACE STRETCH DEFORMATION

There is an increasing interest in using in vitro culture models under stretch to study cellular properties and activities. Stretching devices have been developed in an effort to investigate the mechanical properties of tissues and to explore the biosynthetic responses of arterial smooth-muscle cells to a series of mechanical stresses under carefully controlled conditions (Lanir and Fung, 1974a, 1974b; Leung et al., 1977). It has long been known that physical forces may influence cell proliferation and differentiation directed at maintenance of connective tissues, particularly in tissues that subserve a mechanical function. Whereas flow deformation devices allow a direct evaluation of the effects of *shear stress* on cell structure and function, stretching devices provide tools for directly assessing the influence of *strain*. A variety of biological tissues, such as the underlying smooth-muscle cells, are subjected to strain due to stretching of the

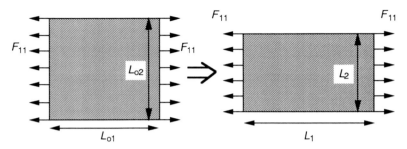

FIGURE 7   Deformation of a rectangular membrane. The membrane is subjected to tensile force in the x direction. There is no transversal force (shear force) acting on the edges of the membrane.

compliant tissue wall because of hydrostatic pressure loading, and do not experience shear stress due to blood flow. There are many types of stretching devices, and they are usually divided into two main groups: those producing uniaxial loading and those producing biaxial loading.

The purpose of the stretching devices is to subject the specimen under study to quantifiable levels of deformation. The major problem with analyzing the results with cells growing on a compliant substrate is that this quantification is given in terms of the deformation of the substrate and not of the cells. Cells cultured on a substrate or biomaterial membrane may not completely adhere to it, and may be subjected to a lower level of deformation. Therefore, in this particular instance, the stress and strain distributions on the cells cannot be quantitated and experiments performed on identical cells, but cultured on different substrates, may provide different results.

Deformation of a material that can be related to stresses is described by strain. Some investigators relate their results to a strain measure, but, very often, their selected strain is not correctly formulated or the assumptions made are not well specified. To illustrate the analysis of deformation, let us consider the extension of a plane element along the x axis with no change in the y direction as shown in Figure 7. Let the original (unstressed) length of the element be L_{o1} and the thickness of the original and deformed plate be h_o and h, respectively. After the imposition of the force F_{11} the length of the plate becomes L_1, and the corresponding nonzero stresses are

$$\sigma_{11} = \frac{F_{11}}{L_2 h}, \qquad T_{11} = \frac{F_{11}}{L_{o2} h_o}, \qquad S_{11} = \left(\frac{L_{o1}}{L_1}\right) T_{11} = \frac{\rho}{\rho_o}\left(\frac{L_{o1}}{L_1}\right)^2 \sigma_{11}$$

$$(44)$$

where ρ_0 and ρ are the density before and after deformation, respectively. Eulerian or Cauchy stresses (e.g., σ_{11}) are used in equations of equilibrium or motion. Lagrangian stresses (e.g., T_{11}) are the most convenient for the reduction of laboratory experimental data. Kirchoff stresses (e.g., S_{11}) are directly related to the strain energy function or strain potential.

For the description of deformation, the stretch ratio in the x axis, λ_1, is used

$$\lambda_1 = \frac{L_1}{L_{o1}} \tag{45}$$

where L_1 and L_{o1} are the length and the reference length, respectively. The reference state can be arbitrarily chosen but is commonly taken as the natural state or resting state of the material or specimen. The strain defined as

$$E_1 = \frac{1}{2}\left(\lambda_1^2 - 1\right) \tag{46}$$

was introduced by Green and is called Lagrangian or Green's strain, and as

$$e_1 = \frac{1}{2}\left(1 - \frac{1}{\lambda_1^2}\right) \tag{47}$$

by Almansi and is called Eulerian or Almansi's strain. The strain defined as

$$\varepsilon_1 = \frac{L_1 - L_{o1}}{L_{o1}} = \lambda_1 - 1 \tag{48}$$

was introduced by Cauchy and is called infinitesimal strain. Again, use of any of these strain measures is sufficient, but it is important to know their difference and that they are different numerically. For example, if $L_1 = 3$ cm and $L_{o1} = 1$ cm ($\lambda = 3$), then $E_1 = 4.0$, $e_1 = 0.44$ and $\varepsilon_1 = 2$. But if the extension ratio is close to unity (i.e., small deformation), the nonzero strain components (E_1, e_1, and ε_1) are approximately the same. For example, if $L_1 = 1.01$ cm and $L_{o1} = 1.00$ cm ($\lambda = 1.01$), then $E_1 \approx e_1 \approx \varepsilon_1 \approx 0.01$. The selection of proper strain measures is dictated basically by the stress–strain relations (i.e., the constitutive equation of the material). In the description of a large deformation, it turns out that the most convenient quantity to consider is the square of the distance between any two points.

Deformation of most materials in nature are much more complex than the example given above, and a more general method of analysis is needed. This method, in which the deformation is described by the displacement field, involves differential geometry. For example, in the case of infinitesimal displacement (small deformation), the distinction between the Lagrangian and Eulerian variables disappears and the infinitesimal strain tensor is defined in terms of the displacement vector u_i as

$$\varepsilon_{ij} = \frac{1}{2} \left(\frac{\partial u_j}{\partial x_i} + \frac{\partial u_i}{\partial x_j} \right) \tag{49}$$

The definitions and example given in this section are to familiarize the reader with the terminologies used, and to make them aware that the choice of a strain measure is not unique. For more detailed information, the interested reader is referred to classical continuum or solid mechanics handbooks (e.g., Fung, 1965).

A. Uniaxial Loading

The simplest stretch experiment that can be done on a biomaterial is the uniaxial tension test. A schematic of a typical uniaxial stretching device is shown in Figure 8. Except near the edge where the membrane is clamped, the strain experienced by the cells is uniform. The proper strain to be used should be prescribed by the type of experiment performed—large or small deformation—as mentioned earlier.

The effect of unidirectional stretching of the substratum on cell orientation has been investigated by many research groups. They found that endothelial and muscle cells undergoing continuous *stretch–relaxation cycling* elongate and orient perpendicular to the direction of the stretch (Buck, 1980; Ives et al., 1986; Terracio et al., 1988; Vandenburgh, 1988; Shirinsky et al., 1989). In contrast, growing proliferating muscle cells on a substratum with slow, continuous *unidirectional stretching* causes the cells to orient and elongate parallel to the direction of movement (Vandenburgh, 1988). Biochemical effects of physical stimuli have also been studied. Mechanical stretching has been found to initiate specific biochemical changes in the cell (Harell et al., 1977; Yeh and Rodan; 1984; Binderman et al., 1984) and to increase the number of cells synthesizing DNA (Binderman et al., 1984; Hasegawa et al., 1985).

The first report on the effects of mechanical strain on cultured cells was that of Harell et al. (1977), who used an orthodontic screw jack spanning two resin blocks glued to the bottom of a polystyrene culture

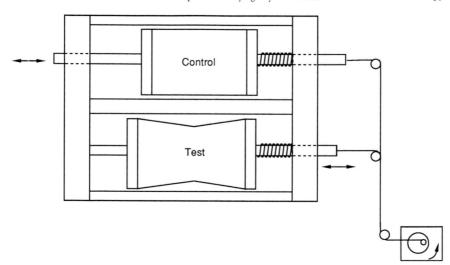

FIGURE 8   Uniaxial loading; strain field uniform except near clamps. This is the stretch device of Ives et al. (1986). In the top or control chamber the membrane is unstretched; in the bottom or test chamber the membrane is stretched.

dish. Osteoblasts were grown on the internal surface of the culture plate. To apply strain to the cells, the screw jack was turned, driving the two pieces of acrylic resin apart. An interesting uniaxial loading device for cultured cells is the one of Vandenburgh (1988). This device is basically a circular membrane forming the bottom of a well on which cells grow and is similar to devices using a vacuum or pressure to produce biaxial loading. Cell deformation in the device of Vandenburgh is performed by pushing a prong located at the center of the well upward. The motion of the prong is controlled by a computer interface system. Two main activity patterns can be produced: (1) a slow continuous motion at a specified speed and (2) a cyclic upward–downward motion at a specified frequency and time period. The membrane strain during an experiment is not defined. The membrane is shown to stretch uniformly across the membrane surface for an increase of up to 54% in the membrane length. However, this stretch characterization is performed on a rectangular piece of membrane under unidirectional loading as shown in Figure 7. In the deformation of a clamped circular membrane by a concentrated force, the strain is not expected to be uniformly distributed across the membrane surface. Of the many methods that use uniaxial loading to mechanically deform a circular elastic diaphragm, the one used by Hasegawa et al. (1985) is also worth noting. In this device, a Petriperm dish with a thin flexible Teflon base is placed over a convex template to deform the base. A lead weight is then placed on top of the dish to hold the dish to the template and stretch the base. The load in this device is not, strictly speaking, uniaxial since the weight of the lead is distributed over the convex surface. It is reported that this set up increases the surface area of the Teflon base by 4%. A potential problem with this procedure is that the Teflon may stick to the template, generating friction forces.

The uniaxial loading experiments discussed so far can yield some information on the mechanical properties of the cells but cannot provide the full relationship between all stress and strain components. To obtain the tensorial relationship, it is necessary to perform biaxial and triaxial loading tests.

B. Biaxial Loading

It is generally accepted that biological materials are incompressible. In incompressible materials, one can obtain three-dimensional mechanical properties from two-dimensional tests. For this reason, triaxial loading tests will not be discussed in this chapter.

Various devices have been developed to subject cultured cells to biaxial loading. However, the most popular is one that applies a positive or negative pressure to a flexible-bottomed culture plate. A schematic of a biaxial stretching device using uniform pressure load is shown in Figure 9. Results similar to unidirectional stretching, concerning factors such as cell orientation and levels of proteins produced by the stretched cells, have been reported for biaxial strain devices (Kanayama and Fukamizu, 1989; Gorfien et al., 1989; Banes et al., 1990). However, it is very difficult to

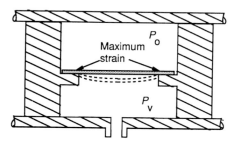

FIGURE 9   Biaxial loading; strain field nonuniform, maximum near support edge. As negative pressure is applied, the membrane deforms downward, yielding a region of maximum strain at the edge of the support and of minimum strain at the center of the culture well.

interpret the effects of strain or rate of strain on the morphology and function of cells in the biaxial, as well as in the uniaxial, stretching devices since the strain and rate of strain distributions *on the cells* are generally not well characterized.

The major problem that one faces when working with cells growing on a substrate is that the characterization of the strain is made from the deformation of the substrate and not of the cells. The substrate is generally a linear elastic material, but this is not necessarily true of the cells. Of course, this problem is irrelevant if the deformation and measurements can directly be done on the specimen [e.g., a piece of muscle or skin (Lanir and Fung, 1974a)].

It is only recently that investigators working on cultured cells are attempting to define the strain field generated in their devices. These recent publications concern stretching devices in which a circular elastic diaphragm, clamped at the edge, is deformed under the action of a uniform pressure (Thibault and Fry, 1983; Winston et al., 1989; Gilbert et al., 1990). However, the results of these reports appear to be inconsistent, even though the authors are using similar devices. Thibault and Fry (1983) and Winston et al. (1989) find that the strain in an inflated circular membrane is uniform, while Gilbert et al. (1990) report that the strain is nonuniform.

To understand this discrepancy, let us consider the bending or deflection, $w(r)$ of a circular plate under uniform symmetrical pressure loading p_o, and clamped at the edge [Fig. 10(a)]. It is assumed that the plate deflections are small in comparison with the thickness h, of the plate, and that the middle plane of the plate remains a neutral plane during bending, that is, is stress-free. The plate equation for small deformation is a fourth-order linear partial differential equation linking the displacement of the middle surface $w(r)$ to the load p_0. In polar coordinates, this equation is

$$\nabla_r^2 \nabla_r^2 w(r) = \frac{p_0}{D} \tag{50}$$

where

$$\nabla_r^2 = \frac{d^2}{dr^2} + \frac{1}{r}\frac{d}{dr}$$

$$D = \frac{Eh^3}{12(1 - \nu^2)}$$

where D Is the bending rigidity, E is the modulus of elasticity, and ν is the

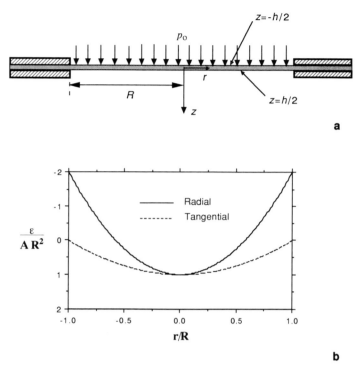

FIGURE 10   (a) Schematic of the circular clamped plate under uniform pressure loading. (b) Strain distribution on the surface, $z = h/2$, of the plate. Along the surface $z = h/2$, the maximum strain is in the radial direction and is located near the clamped edge. The surface is compressed near the edge and stretched around the plate center. The tangential strain on that surface is always positive; that is, the plate is in tension in that direction. When one side of the plate (e.g., $z = h/2$) is in compression, the other side (e.g., $z = -h/2$) is in tension, and vice versa.

Poisson ratio (for an incompressible material, $\nu = 0.5$). It can be shown that the solution of Eq. (50) has the form

$$w(r) = \frac{p_0}{64D} \left(R^2 - r^2 \right)^2 \tag{51}$$

where R is the plate radius. The maximum deflection is at the center of the plate ($r = 0$),

$$w_{max} = \frac{p_0 R^4}{64D} \tag{52}$$

The strain and stress distributions can be derived from Eq. (51) and the strain–stress equations. These derivations are beyond the scope of the chapter, but the solutions are given here to show that the strain and stress distributions are not uniform:

$$\varepsilon_r = A(R^2 - 3r^2)$$

$$\varepsilon_\theta = A(R^2 - r^2)$$

$$\sigma_r = \frac{EA}{(1 - \nu^2)}\left[(1 + \nu)R^2 - (3 + \nu)r^2\right] \qquad (53)$$

$$\sigma_\theta = \frac{EA}{(1 - \nu^2)}\left[(1 + \nu)R^2 - (3\nu + 1)r^2\right]$$

where ε and σ represent the strain and stress, respectively, whereas the subscripts r and θ denote the radial and tangential direction. The parameter A is defined as

$$A = \frac{p_0 z}{16D} \qquad (54)$$

where z denotes the normal distance from the neutral plane, and $z = h/2$ and $z = -h/2$ represent the surfaces of the flat plate. A plot of the strain distribution is shown in Fig. 10(b). It can be seen that the radial and tangential strain are not equal, and that the maximum strain is in the radial direction and occurs near the clamped edge.

Now, let us define the average elongating strain ε as

$$\varepsilon = \frac{\Delta l}{R} \qquad (55)$$

where Δl is the elongation of a radius caused by the deformation

$$\Delta l = \int ds - \int dr = \int \sqrt{dw^2 + dr^2} - \int dr$$

For small deformations, the condition $dw/dr \ll 1$ implies that

$$\sqrt{1 + (dw/dr)^2} \approx 1 + \tfrac{1}{2}(dw/dr)^2.$$

Therefore, the above equation simplifies to

$$\Delta l = \frac{1}{2} \int_0^R \left(\frac{dw}{dr} \right)^2 dr \tag{56}$$

Substitution of Eqs. (51), (52), and (56) into Eq. (55) yields

$$\varepsilon = \frac{64}{105} \frac{w_{max}^2}{R^2} \approx 0.61 \left(\frac{w_{max}}{R} \right)^2 \tag{57}$$

This average strain is the same in all directions (i.e., in the tangential and meridional directions; the tangential direction here is not the one denoted by θ, but the one given by the tangential radius of curvature). It is important to emphasize again that many strain measures can be used to characterize the deformation of a material. The choice matters only when a strain–stress relationship (constitutive equation) for the material is desired.

If the plate deflections are very large, $w_{max} > 3h$ (which can occur without yield stress only in very thin plates), the tensile stresses in the neutral plane are large in comparison with the bending stresses. As a consequence, bending stresses can be neglected and the plate can be treated as a "membrane" (Brodland, 1986, 1988). Unlike plates that are capable of taking bending moments and shear forces perpendicular to their own plane as well as shear forces lying in their own plane, a membrane is incapable of transmitting transverse bending moments or shear forces. A flat membrane can carry no load, and its load-carrying capacity develops only with the deflection. The load in a membrane, instead of being proportional to the deflection as in a plate [Eq. (52)], varies with $(w_{max})^3$ (Den Hartog, 1952)

$$p_0 = \frac{8}{3} \frac{E}{1 - \nu} \frac{h}{R} \left(\frac{w_{max}}{R} \right)^3 \tag{58}$$

From the equation of equilibrium, the deflection of the membrane is found to be

$$w(r) = \frac{w_{max} r^2}{R^2} + \text{const} \tag{59}$$

where

$$w_{max} = \frac{p_0 R^2}{4h\sigma}$$

and

$$\sigma = \frac{\varepsilon E}{(1 - \nu)} \tag{60}$$

The stress σ is the same in all directions (two-dimensional hydrostatic tension), and is related to the uniform strain ε defined in Eqs. (55) and (56). Combination of these equations and Eq. (59) yields

$$\varepsilon = \frac{\Delta l}{R} = \frac{2}{3} \left(\frac{w_{max}}{R} \right)^2 \tag{61}$$

The biaxial strain can be measured from the maximum, or center, deflection [Eq. (61)], or the applied pressure. This strain–pressure relationship is obtained by substituting Eq. (58) into Eq. (61)

$$\varepsilon = k_1 p_0^{2/3} \tag{62}$$

where the constant k_1 is equal to

$$k_1 = 0.218 \left(\frac{R}{Eh} \right)^{2/3} \tag{63}$$

There is a factor of $(\frac{2}{3})$ missing in Eq. (9) of Winston et al. (1989), since they list the coefficient in k_1 as equal to 0.327 instead of 0.218. Winston et al., find experimentally that ε varies linearly with $p^{2/3}$ and that the slope has a value of 0.0351. They also give a value of 0.968 for the slope in the figure legend of their Figure 5, but from their results, this must be a typographical error. Since Winston et al. did not provide the size of the well radius, it is not possible to know how closely their k_1 value of 0.0351 compares with the value predicted by Eq. (63) here. A value for the radius can be obtained by using Eq. (58), their calculated Young's modulus and experimental data of w^3 versus p_0. However, this calculated value gives a value for k_1 that does not agree with the value of their experimental slope. Surprisingly, the authors did not address this issue.

For the intermediate case in which tensile and bending stresses are of the same order of magnitude, a simple approximation solution exists that agrees well with the exact nonlinear theory for the circular plate (Den Hartog, 1952). This simple solution for large deflection states that the load carried by the plate is equal to the sum of the two partial loads, carried membranewise and bendingwise, respectively:

$$p_0 = \frac{64D}{R^3}\left(\frac{w_{max}}{R}\right) + \frac{8}{3}\frac{E}{1-\nu}\frac{h}{R}\left(\frac{w_{max}}{R}\right)^3 \tag{64}$$

This expression, relating the load to the ratio of the maximum deflection to plate radius, can be rewritten in terms of the plate thickness:

$$\frac{3}{16}(1 - \nu^2)\frac{p_0 R^4}{Eh^4} = \frac{w_{max}}{h} + \frac{1}{2}(1 + \nu)\left(\frac{w_{max}}{h}\right)^3 \tag{65}$$

This equation is equivalent to Eq. (7) of Winston et al. (1989), although there is a typographical error in their equation. The first term on the right-hand side of Eq. (65), which is linear in w_{max}, represents the contribution due to bending stresses; the second term, which is cubic in w_{max}, represents the contribution due to membrane stresses. However, if a characterization of the strain field is desired, it is necessary to derive it from the general equations for a clamped circular membrane inflated under a uniform pressure (Adkins and Rivlin, 1952; Hart-Smith and Crisp, 1967). If the center deflection/thickness ratio (w/h) is not too large, then the nonlinear plate theory for large deflection can be used (Way, 1934). The plate theory assumes that the applied load always acts in the axial direction (uniform load), whereas the membrane theory assumes that the applied load is always normal to the deformed diaphragm (pressure). It is interesting to note that in all these general theories the focus has been on the determination of the stress field. Solutions for the strain field have not been given, but can be derived from the solutions for the stress field.

In summary, it is clear that whether the strain in a circular stretching device is found to be uniform depends both on the definition used and on the experimental conditions (i.e., small or large deformation). Thibault and Fry (1983) and Winston et al. (1989) use the membrane theory to characterize the deformation in their devices [Eqs. (58)–(63)]. In their experiments, the ratio of maximum, or center, deflection-to-thickness is about four. In the work of Winston et al., this ratio ranges from 10 to 50. They find that the experimental data follow the trend predicted by the theory, although it is not clear if the experimental and theoretical results agree quantitatively. Gilbert et al. (1990), on the other hand, use a finite-element approach to solve the plate problem, and find that the strain distribution is not uniform and that the maximum strain is located just above the edge of the cell support (Fig. 9). The shape of the strain distribution is similar to the one for small deflection [Fig. 10(b)]. The ratio of maximum deflection to thickness is 2.9. This is the domain where bending and membrane stresses can still be equally important. It is therefore not surprising that their results differ from those of Thibault and Fry (1983) and Winston et al. (1989). Gilbert et al. define the strain as the radial strain, but, properly speaking, it is not. The strain calculated in their finite-element analysis is a local radial strain. The analysis of Gilbert et al. appears to be correct, but is incomplete since the tangential strain is not characterized.

C. Discussion

Although uniaxial stretch is theoretically a special case of biaxial stretch, in reality it is a separate case. In uniaxial deformations the lateral dimensions (the lateral sides are left free) may gradually develop arbitrary changes that are irreversible. In biaxial deformations the lateral dimensions are preserved.

A major concern with stretching devices is that substantial cell detachment and replacement of the lost cells by a stretch-induced increase in the rate of proliferation of the remaining cells might be a complicating factor in interpreting experimental data. Cell proliferation rate is known to increase under mechanical stimulation (Vandenburgh and Kaufman, 1979), and theories assume that the cells are completely attached to the membrane. This assumption is open to discussion since it is found that cell elongation is about 60% of the strain in the substrate (Winston et al., 1989). This value appears to be dependent on both cell age and morphology.

Other concerns with stretching devices are that they require an appropriate substratum material that must be biocompatible with cultured cells and yet has the elastic characteristics required for long-term physical stimulation. For example, for cyclic stretch experiments, the membrane must withstand continuous stretch–relaxation cycling without permanent membrane distortions. For direct microscopic observation the membrane must also be transparent; otherwise fixation and staining of the cells are necessary.

In general, the theoretical analyses developed for calculating the strain distributions experienced by the cells under biaxial stretch are either inadequate or incomplete. Most of the methods reported in the literature are lacking information needed to fully characterize the strain field in the substrate, or are described only in terms of overall increase in length or area and give no indication if a large strain variation due to end effects occurs or if the strain distribution is nonuniform. In addition, since the substrate is a purely elastic material, the strain is usually related to the maximum deflection only. However, if the cells under study exhibit viscoelastic behavior, such a characterization cannot provide the stress–strain relationship of the cells.

As a note of caution on the repeatability of results in experiments involving cycles of stretching and unstretching, it has been found that successive cyclic uniaxial tests of tissue yield unequal results that converge as the number of cycles increases (Fung, 1971). Repeatability of the results was ensured by preconditioning the tissue. (A test procedure was carried out several times until the results converge.)

Finally, the mechanisms that produce orientation among endothelial cells in response to mechanical stimulation remain obscure. Actin bundles orient perpendicularly to the direction of strain, while in shear-stressed endothelial cells they align parallel to the shear force. Franke et al. (1984), using a cone–plate device, found that microfilament bundle alignment is not directly related to the action of shear stress and takes place after endothelial orientation. On the other hand, Shirinsky et al. (1989) suggest that, in the case of endothelial stretching, actin bundle alignment may precede the cell body orientation. They also conclude that stress fibers are essential for the process of endothelial orientation induced by repeated strain, but not for the maintenance of endothelial orientation. Interpretation of the data in terms of strain effects is very difficult for lack of appropriate measurement techniques and mathematical models. In addition, interpretation in terms of strain effects only may not be sufficient.

V. TECHNIQUES FOR MEASURING CELL MECHANICAL PROPERTIES

In order to fully understand shear stress or strain effects on cells, it is necessary to have a quantitative assessment of the rheological properties of the cells. This section gives a brief review of the most popular techniques that have been developed to determine the mechanical properties of individual cells. The determination of material properties of biological membranes (e.g., elastic and viscous moduli) is conceptually simple. Cell deformation is produced by an applied force, and the deformation is related to the force as a function of time. Because of its molecular thinness, biological membrane material can be treated as a continuum only in the two dimensions that characterize the membrane surface: the surface plane. In the third dimension—thickness, the membrane exhibits anisotropy of molecular structure that prevents its treatment as a continuum. Consequently, membranes are represented as two-dimensional continua. Historically, many of the membrane experiments originated in the studies of blood rheology.

The most common techniques for measuring cell mechanical properties are those using area dilation produced by isotropic tension, those using fluid shear flow, and those studying the recovery of a deformed cell to its initial shape. More recently, laser trap methodology has been employed for similar purposes.

A. Area Dilation

There are three main instruments in this category. They are the micropipette, devices involving compression of the cell between two flat surfaces, and deflection of the cell surface by a rigid spherical particle [for review, see Evans and Skalak (1980)]. The most popular instrument among these is certainly the micropipette. The micropipette aspiration technique was first developed by Mitchison and Swann (1954) for studying the mechanical properties of sea-urchin eggs. This technique has been extensively used to characterize the mechanical properties of red and to a lesser degree white blood cells. Only very few studies have been performed on cultured cells (Sato et al., 1987a, 1987b; Needham et al., in press). Area dilation methods are best suited for spherical cells (e.g., sphered red cells or white blood cells). They can provide information on membrane shear elastic modulus, surface shear elasticity, surface tension, cell surface area, and volume.

An interesting area dilation technique is the deflection of the cell surface by a rigid material. Daily et al. (1984) and Duszyk et al. (1989) have used a cell poker (deflection of the surface by an indenter) to characterize red blood cell mechanical properties. This technique is interesting since it can provide information on the mechanical properties of the membrane locally.

B. Fluid Shear

Fluid shear instruments or techniques include the flow of extracellular fluid to deform a cell that is point-attached to a substrate (Hochmuth, 1980), the rheoscope (Tran-Son-Tay et al., 1985; Sutera et al., 1989), and more recently the flow into and down a tapered or straight pipette (Bagge et al., 1971; Dong et al., 1988; Yeung and Evans, 1989; Needham and Hochmuth, 1990; Needham et al., 1991). The latter methods, namely flow into and down a tapered or straight pipette, have been developed specifically for analyzing the mechanical properties of white blood cells. These cells are spherical and are so viscous [the apparent viscosity of a neutrophil is about 1000 poise (P)] that high surface shear stresses have to be applied to deform them. For this reason, in a simple shear flow, these cells may not deform at all and may just rotate like rigid spheres.

In general, the mathematical treatments to obtain the mechanical properties of the cells in these instruments are more complex than those in the previous section. However, these instruments permit the characterization of cell properties under dynamic conditions.

Needham et al. (1991) apply the theory developed for flow of a liquid drop into a pipette to determine the mechanical properties of hybridoma cells. They find that these cells flow into a capillary tube like Newtonian liquid drops with an extremly high internal viscosity. They also point out an important feature of proliferating cells; that a proliferating cell such as the hybridoma cannot be characterized by a single value for any one property (morphological, geometric, or mechanical) since these parameters depend on batch growth phase and cell cycle. It is this dependence that they consider of major importance if proliferating cells are to be fully characterized from a physical standpoint and if the mechanisms and structures involved in cell damage caused by hydrodynamic (and other) forces are to be identified.

C. Recovery

Methods that monitor the recovery of cell shape after small or large deformation can be performed with most of the instruments described above. Recovery of red blood cells (Hochmuth et al., 1979) and white blood cells (Dong et al., 1988; Tran-Son-Tay et al., 1991) have been studied with the micropipette. Other instruments such as the rheoscope (Tran-Son-Tay et al., 1985; Sutera et al., 1989) have also been used for that purpose.

The recovery of red cells was investigated by pulling a flaccid red cell disk at diametrically opposite locations on the rim of the cell: the cell is attached at one end to a surface and pulled with a micropipette from the opposite end. The cell is then released and the recovery to the familiar discoid shape is observed. The recovery experiment in the rheoscope is simple. When subjected to a shear flow such as in the rheoscope and under certain experimental conditions, the red blood cell takes the form of an ellipsoid. After the rheoscope is stopped abruptly, the cell recovers its resting shape and the recovery process is monitored. Since the membrane of a red blood cell is a linear viscoelastic material, the recovery time yields a time constant that is equal to the ratio of the membrane shear elastic module over the surface shear viscosity.

The white blood cell is a more complicated cell than the red blood cell in both structure and function. In the red blood cell, a hierarchy of elastic properties (area expansion, shear, and bending) are exhibited and are associated mainly with the cell membrane. In the white blood cell, and specifically the neutrophil, resistance to applied stresses appears to reside predominantly in the bulk cytoplasm. Not surprisingly, recovery experiments for white blood cells are different from those for red blood cells and analyses are more involved. The white blood cell is modeled as a shell-like body (cortical shell) under a prestressed tension $\bar{T}_0$ with a Newtonian or Maxwell fluid inside. The undeformed, or reference state, of the shell-like body is assumed to be a sphere since white blood cells have a spherical shape in the resting state. As a consequence, whole-cell recovery is performed by completely aspirating a white blood cell and holding it in the deformed cigar shape inside a pipette for few seconds. The cell is then expelled from the pipette and starts its recovery immediately. The deformed shell undergoes a relaxation displacement with convergence of its major and minor axes until the drop gradually returns to its spherical shape. The underlying mechanism that drives cell recovery and the general problems associated with deformation of white blood cells are quite involved and will not be discussed here. The interested reader should consult the works of Dong et al. (1988) Yeung and Evans (1989), and Tran-Son-Tay et al. (1991). To my knowledge, no recovery experiments on cultured cells have yet been reported.

D. Optical Tweezers (Laser Traps)

Infrared laser traps have recently been developed using the forces of radiation pressure to micromanipulate living cells (Ashkin and Dziedzic, 1987). One of the unique features of the optical tweezers or laser-trap technique is the ability to apply controlled local forces inside of cells while leaving the cell wall intact. This technique has been used to study the mechanical properties of living cytoplasm with minimal damage (Ashkin and Dziedzic, 1989).

As opposed to the techniques described above (with the exception of the cell poker, which are used to give the overall mechanical properties of the cell), the optical-tweezers technique can be used to probe the mechanical properties of the cell locally. This is an important option since cell properties could vary from point to point.

The principal limitation of the technique for internal cell manipulation is the maximum force that can be applied without causing optical absorption damage to the cell. This innovative technique is fairly new and has not yet been applied to cultured cells.

E. Discussion

In general, to assess whole-cell deformability, the particular method of micropipette manipulation seems to be preferable, although it lacks the simplicity and convenience of the flow channel experiment. Other measurement techniques, such as the rheoscope, involve complicated kinematics of cell deformations but are very useful for evaluating the dynamic deformability of cells.

To a first approximation, biological materials can be considered as elastic solids or viscous liquids, but usually exhibit viscoelastic behavior that must be determined experimentally. This determination is crucial since, without the constitutive laws, no meaningful analysis of how externally applied fluid forces might deform the cells can be done. The properties of materials are specified by constitutive equations. Parameterized mechanical models, which involve combinations of linear springs and dashpots, are often used to discuss the viscoelastic behavior of materials. It is important to note that the theories developed for the techniques described in this section are for phenomena that occur at the continuum level; that is, the grainness of the material must not be apparent.

VI. CONCLUDING REMARKS

In this chapter, I have attempted to provide the necessary background to better understand the problems and assumptions associated with several instruments and approaches to the study of the effects of fluid forces on mammalian cells. The reader should not consider it a complete study, but rather a review of some of the more popular and/or innovative devices or techniques used in the field.

It is evident that only a small portion of the problems having direct bearing on hydrodynamic and stretch effects on mammalian cells have been studied quantitatively. Much work remains to be done both to clear up the discrepancies in the reported experimental results and to develop more appropriate and complete theories. Most of the flow devices described in the text have been developed for studying red blood cells or adapted from them. However, with the increasing interest in studies of cultured cells, these devices need to be modified or redesigned to allow continuous infusion of fresh medium for experiments lasting several hours and even days. Better control of the cell incubating medium environment and continuous sampling of that medium are also necessary.

The study of physical forces on mammalian cells is open to much further investigation. The future holds promise for significant progress in the understanding of hydrodynamic forces on the structure and function of mammalian cells, and also in the development of better techniques and more complete theories for the analysis of the effects of these forces.

REFERENCES

Abu-Reesh, I., and Kargi, F. (1989). Biological responses of hybridoma cells to defined hydrodynamic shear stress. *J. Biotechnol.* **9**, 167–178.

Adkins, J. E., and Rivlin, R. S. (1952). Large elastic deformations of isotropic materials IX. The deformation of thin shells. *Phil. Trans. Royal Soc.* **224**, 505–531.

Ando, J., Nomura, H., and Kamiya, A. (1987). The effect of fluid shear stress on the migration and proliferation of cultured endothelial cells. *Microvasc. Res.* **33**, 62–70.

Ashkin, A., and Dziedzic, J. M. (1987). Optical trapping and manipulation of viruses and bacteria. *Science* **235**, 1517–1520.

Ashkin, A., and Dziedzic, J. M. (1989). Interna; cell manipulation using infrared laser traps. *Proc. Natl. Acad. Sci. USA* **86**, 7914–7918.

Augenstein, D. C., Sinskey, A. J., and Wang, D. I. C. (1971). Effect of shear on the death of two strains of mammalian tissue cells. *Biotechnol. Bioeng.* **13**, 409–418.

Bagge, U., Skalak, R., and Attefors, R. (1971). Granulocyte rheology. Experimental studies in an in vitro micro-flow system. *Adv. Microcirc.* **7**, 29–48.

Banes, A. J., Gilbert, J., Taylor, D., and Monbureau, O. (1985). A new vacuum-operated stress providing instrument that applies static or variable duration cyclic tension or compression to cells in vitro. *J. Cell Sci.* **75**, 35–42.

Banes, A. J., Link, G. W., Jr., Gilbert, J. W., Tran-Son-Tay, R., and Monbureau, O. (1990). Culturing cells in a mechanically active environment. *Amer. Biotechnol. Lab.* (May), 12–22.

Binderman, I., Shimshoni, and Somjen, D. (1984). Biochemical pathways involved in the translation of physical stimulus into biological message. *Calcif. Tiss. Int.* **36**, S82–S85.

Bodoia, J. R., and Osterle, J. F. (1961). Finite difference analysis of plane Poiseuille and Couette flow developments. *Appl. Sci. Res.* (sect. A), **10**, 265–276.

Brodland, G. W. (1986). Nonlinear deformation of uniformly loaded circular plates. *Solid Mechanics Arch.* **11**, 219–256.

Brodland, G. W. (1988). Highly non-linear deformation of uniformly-loaded circular plates. *Int. J. Solids Struct.* **24**, 351–362.

Brooks, D. E. (1984). The biorheology of tumor cells. *Biorheology* **21**, 85–91.

Buck, R. C. (1980). Reorientation response of cells to repeated stretch and recoil of the substratum. *Exp. Cell Res.* **127**, 470–474.

Bussolari, S. R., Dewey, C. F., Jr., and Gimbrone, M. A., Jr. (1982). Apparatus for subjecting living cells to fluid shear stress. *Rev. Sci. Instrum.* **53**, 1851–1854.

Cherry, R. S., and Papoutsakis, E. T. (1986). Hydrodynamic effects on cells in agitated tissue culture reactors. *Bioproc. Eng.* **1**, 29–41.

Cherry, R. S., and Papoutsakis, E. T. (1988). Physical mechanisms of cell damage in microcarrier cell culture bioreactors. *Biotechnol. Bioeng.* **32**, 1001–1014.

Cherry, R. S., and Papoutsakis, E. T. (1989). Growth and death rates of bovine embryonic kidney cells in turbulent microcarrier bioreactors. *Bioproc. Eng.* **4**, 81–89.

Chien, S., King, R. G., Kaperonis, A. A., and Usami, S. (1982). Viscoelastic properties of sickle cells and hemoglobin. *Blood Cells* **8**, 53–64.

Cokelet, G. R., Merrill, E. W., Gilliland, E. R., Shin, H., Britten, A., and Wells, R. M., Jr., (1963). The rheology of human blood—measurement near and at zero shear rate. *Trans. Soc. Rheol.* **VII**, 303–317.

Colton, C. K. (1969). Permeability and transport studies in batch and flow dialyzers with applications to hemodialysis. Ph.D. thesis, MIT (Massachusetts Institute of Technology), Department of Chemical Engineering, Cambridge, MA.

Cooke, J. P., Stamler, J., Andon, N., Davies, P. F., McKinley, G., and Loscalzo, J. (1990). Flow stimulates endothelial cells to release a nitrovasodilator that is potentiated by reduced thiol. *Am. J. Physiol.* **259**, H804–H812.

Cozens-Roberts, C., Quinn, J. A., and Lauffenburger, D. A. (1990). Receptor-mediated adhesion phenomena. Model studies with the radial-flow detachment assay. *Biophys. J.* **58**, 107–125.

Croughan, M. S., Hamel, J.-F., and Wang, D. I. C. (1987). Hydrodynamic effects on animal cells grown in microcarrier cultures. *Biotechnol. Bioeng.* **29**, 130–141.

Croughan, M. S., Hamel, J.-F., and Wang, D. I. C. (1988). Effects of microcarrier concentration in animal cell culture. *Biotechnol. Bioeng.* **32**, 975–982.

Croughan, M. S., Sayre, E. S., and Wang, D. I. C. (1989). Viscous reduction of turbulent damage in animal cell culture. *Biotechnol. Bioeng.* **33**, 862–872.

Croughan, M. S., and Wang, D. I. C. (1989). Growth and death in overagitated microcarrier cell cultures. *Biotechnol. Bioeng.* **33**, 731–744.

Daily, B., Elson, E. L., and Zahalak, G. I. (1984). Cell poking: Determination of the elastic area compressibility modulus of the erythrocyte membrane. *Biophys. J.* **45**, 671–682.

Davies, P. F., Remuzzi, A., Gordon, E. J., Dewey, C. F., Jr., and Gimbrone, M. A., Jr. (1986). Turbulent fluid shear stress induces vascular endothelial cell turnover *in vitro*. *Proc. Natl. Acad. Sci. USA* **83**, 2114–2117.

Den Hartog, J. P. (1952). *Advanced Strength of Materials*. McGraw-Hill, New York.

Desai, N. P., and Hubbell, J. A. (1989). The short-term blood biocompatibility of poly(hydroxyethyl methacrylate-co-methyl methacrylate) in an *in vitro* flow system measured by digital videomicroscopy. *J. Biomater. Sci. Polym. Edn.* **1**(2), 123–146.

Dewey, C. F., Jr. (1984). Effects of fluid flow on living vascular cells. *J. Biomech. Eng.* **106**, 31–35.

Diamond, S. L., Eskin, S. G., and McIntire, L. V. (1989). Fluid flow stimulates tissue plasminogen activator secretion by cultured human endothelial cells. *Science* **243**, 1483–1485.

Dodge, C. T., and Hu, S. (1986). Growth of hybridoma cells udner different agitation conditions. *Biotechnol. Lett.* **8**, 683–686.

Dong, C., Skalak, R., Sung, K.-L. P., Schmid-Shconbein, G. W., and Chien, S. (1988). Passive deformation analysis of human leukocytes. *J. Biomech. Eng.* **110**, 27–36.

Drasler, W. J., Smith, C. M., II, and Keller, K. H. (1987). Viscoelasticity of packed erythrocyte suspensions subjected to low amplitude oscillatory deformation. *Biophys. J.* **52**, 357–365.

Duszyk, M., Schwab, B., III, Zahalak, G. I., Qian, H., and Elson, E. L. (1989). Cell poking: Quantitative analysis of indentation of thick viscoelastic layers. *Biophys. J.* **55**, 683–690.

Eskin, S. G., Ives, C. L., McIntire, L. V., and Navarro, L. T. (1984). Response of cultured endothelial cells to steady flow. *Microvasc. Res.* **28**, 87–94.

Evans, E. A., and Skalak, R. (1980). *Mechanics and Thermodynamics of Biomembranes*. CRC Press, Boca Raton, FL.

Fazekas de St. Groth, S. (1983). Automated production of monoclonal antibodies in a cytostat. *J. Immun. Methods* **57**, 121–136.

Fewell, M. E., and Hellums, J. D. (1977). The secondary flow of Newtonian fluids in cone-and-plate viscometers. *Trans. Soc. Rheol.* **21**(4), 535–565.

Frangos, J. A., Eskin, S. G., McIntire, L. V., and Ives, C. L. (1985). Flow effects on prostacyclin production by cultured human endothelial cells. *Science* 227, 1477–1479.

Frangos, J. A., McIntire, L. V., and Eskin, S. G. (1988). Shear stress induced stimulation of mammalian cell metabolism. *Biotechnol. Bioeng.* 32, 1053–1060.

Franke, R. P., Gräfe, M., Schnittler, H., Seiffge, D., and Drenckhahn, D. (1984). Induction of human vascular endothelial stress fibres by fluid shear stress. *Nature* 307, 648–649.

Friedmann, M., Gillis, J., and Liron, N. (1968). Laminar flow in a pipe at low and moderate Reynolds numbers. *Appl. Sci. Res.* 19, 426–438.

Fung, Y. C. (1971). *Biomechanics. Its Foundation and Objectives.* Y. C. Fung, N. Perrone, and M. Anliker (eds.). Prentice-Hall, New Jersey.

Fung, Y. C. (1965). *Foundations of Solid Mechanics.* Prentice-Hall, Englewood Cliffs, NJ.

Gilbert, J. A., Banes, A. J., Link, G. W., and Jones, G. L. (1990). Video analysis of membrane strain: An application in cell stretching. *Exp. Tech.* (Sept./Oct.), 43–45.

Goldsmith, H. L., and Mason, S. G. (1975). Some model experiments in hemodynamics. V. Microrheological techniques. *Biorheology* 12, 181–192.

Gorfien, S. F., Winston, F. K., Thibault, L. E., and Macarak, E. J. (1989). Effects of biaxial deformation on pulmonary artery endothelial cells. *J. Cell. Physiol.* 139, 492–500.

Grapa, E., Truskey, G. A., and Reichert, W. M. (1990). Digitized total internal reflection video microscopy: Analysis of scatter and cell-glass contacts. *Trans. Soc. Biomat.* 13, 160–161.

Grimm, J., Keller, R., and de Groot, P. G. (1988). Laminar flow induces cell polarity and leads to rearrangement of proteoglycan metabolism in endothelial cells. *Thromb. Haemostas.* 60, 437–441.

Handa, A., Emery, A. N., and Spier, R. E. (1987). On the evaluation of gas-liquid interfacial effects on hybridoma viability in bubble column bioreactors. *Devel. Biol. Stand.* 66, 241–253.

Handa-Corrigan, A., Emery, A. N., and Spier, R. E. (1989). Effect of gas-liquid interfaces on the growth of suspended mammalian cells: Mechanisms of cell damage by bubbles. *Enzyme Microb. Technol.* 11, 230–235.

Happel, J., and Brenner, H. (1986). *Low Reynolds Number Hydrodynamics.* Martinus Nijhoff Publishers, Boston, MA.

Harell, A., Dekel, S., and Binderman, I. (1977). Biochemical effect of mechanical stress on cultured bone cells. *Calcif. Tiss. Res.* 22 (suppl.), 202–209.

Hart-Smith, L. J., and Crisp, J. D. C. (1967). Large elastic deformations of thin rubber membranes. *Int. J. Eng. Sci.* 5, 1–24.

Hasegawa, S., Sato, S., Saito, S., Suzuki, Y., and Brunette, D. M. (1985). Mechanical stretching increases the number of cultured bone cells synthesizing DNA and alters their pattern of protein synthesis. *Calcif. Tiss. Int.* 37, 431–436.

Hill, C. T. (1972). Nearly viscometric flow of viscoelastic fluids in the disk and cylinder system. II: Experimental. *Trans. Soc. Rheol.* 16, 213–245.

Hochmuth, R. M. (1980). Viscoelastic solid behavior of red cell membrane. In *Erythrocyte Mechanics and Blood Flow*, G. R. Cokelet, H. J. Meiselman, and D. E. Brooks (eds.). Kroc Foundation Series, Vol. 13, Alan R. Liss, New York.

Hochmuth, R. M., and Mohandas, N. (1972). Uniaxial loading of the red-cell membrane. *J. Biomech.* 5, 501–509.

Hochmuth, R. M., Mohandas, N., Spaeth, E. E., Williamson, J. R., Blackshear, P. L., Jr., and Johnson, D. W. (1972). Surface adhesion, deformation and detachment at low shear of red cells and white cells. *Trans. Am. Soc. Artif. Int. Organs* 18, 325–332.

Hochmuth, R. M., Worthy, P. R., and Evans, E. A. (1979). Red cell extensional recovery and the determination of membrane viscosity. *Biophys. J.* 26, 101–114.

Horbett, T. A., Waldburger, J. J., Ratner, B. D., and Hoffman, A. S. (1988). Cell adhesion to a series of hydrophilic-hydrophobic copolymer studied with a spinning disc apparatus. *J. Biomed. Mater. Res.* **22**, 383–404.

Hornbeck, R. W. (1964). Laminar flow in the entrance region of a pipe. *Appl. Sci. Res.* (Sect. A), **13**, 224–232.

Ives, C. L., Eskin, S. G., and McIntire, L. V. (1986). Mechanical effects on endothelial cell morphology: *in vitro* assessment. *In Vitro Cell Devel. Biol.* **22**, 500–507.

Ives, C. L., Eskin, S. G., McIntire, L. V., and DeBakey, M. E. (1983). The importance of cell origin and substrate in the kinetics of endothelial cell alignment in response to steady flow. *Trans. Am. Soc. Artif. Int. Organs* **29**, 269–274.

Jo, H., Dull, R. O., Hollis, T. M., and Tarbell, J. M. (in press). Endothelial albumin permeability is shear-dependent, time dependent, and reversible. *Am. J. Physiol.* **260**, 1992–1996.

Kanayama, N., and Fukamizu, H. (1989). Mechanical stretching increases prostaglandin E_2 in cultured human amnion cells. *Gyn. Obstet. Invest.* **28**, 123–126.

Koslow, A. R., Stromberg, R. R., Friedman, L. I., Lutz, R. J., Hilbert, S. L., and Schuster, P. (1986). A flow system for the study of shear forces upon cultured endothelial cells. *J. Biomech. Eng.* **108**, 338–341.

Krueger, J. W., Young, D. F., and Cholvin, N. R. (1971). An *in vitro* study of flow response by cells. *J. Biomech.* **4**, 31–36.

Landau, L. D., and Lifshitz, E. M. (1987). *Fluid Mechanics*, 2nd ed. In *Course of Theoretical Physics*, (Translated from the Russian by J. B. Sykes and W. H. Reid). Vol. 6. Pergamon Press, New York.

Lanir, V., and Fung, Y. C. (1974a). Two-dimensional mechanical properties of rabbit skin—I. Experimental system. *J. Biomech.* **7**, 29–34.

Lanir, V., and Fung, Y. C. (1974b). Two-dimensional mechanical properties of rabbit skin—II. Experimental results. *J. Biomech.* **7**, 171–182.

Lansman, J. B. (1988). Going with the flow. *Nature* **331**, 481–482.

Lansman, J. B., Hallam, T. J., and Rink, T. J. (1987). Single stretch-activated ion channels in vascular endothelial cells as mechanotransducers? *Nature* **325**, 811–813.

Lee, G. M., Huard, T. K., Kaminski, M. S., and Palsson, B. O. (1988). Effect of mechanical agitation on hybridoma cell growth. *Biotechnol. Lett.* **10**, 625–628.

Leung, D. Y. M., Glagov, S., and Mathews, M. B. (1977). A new in vitro system for studying cell response to mechanical stimulation. *Exp. Cell Res.* **109**, 285–298.

Levesque, M. J., and Nerem, R. M. (1985). The elongation and orientation of cultured endothelial cells in response to shear stress. *J. Biomech. Eng.* **107**, 341–347.

Levesque, M. J., and Nerem, R. M. (1989). The study of rheological effects on vascular endothelial cells in culture. *Biorheology* **26**, 345–357.

McIntire, L. V., Frangos, J. A., Rhee, B. G., Eskin, S. G., and Hall, E. R. (1987). The effects of fluid mechanical stress on cellular arachidonic acid metabolism. *Ann. N.Y. Acad. Sci.* **516**, 513–524.

McQueen, A., and Bailey, J. E. (1989). Influence of serum level, cell line, flow type and viscosity on flow-induced lysis of suspended mammalian cells. *Biotech. Lett.* **11**, 531–536.

McQueen, A., Meilhoc, E., and Bailey, J. E. (1987). Flow effects on the viability and lysis of suspended mammalian cells. *Biotech. Lett.* **9**, 831–836.

Mitchison, J. M., and Swann, M. M. (1954). The mechanical properties of the cell surface. I. The cell elastimeter. *J. Exp. Biol.* **31**, 443–460.

Murdin, A. D., Thorpe, J. S., Groves, D. J., and Spier, E. E. (1989). Growth and metabolism of hybridomas immobilized in packed beds: Comparison with static and suspension cultures. *Enzyme Microb. Technol.* 11, 341–346.

Nakache, H., and Gaub, H. E. (1988). Hydrodynamic hyperpolarization of endothelial cells. *Proc. Natl. Acad. Sci. USA* 85, 1841–1843.

Needham, D., and Hochmuth, R. M. (1990). Rapid flow of passive neutrophils into a 4 μm pipet and measurement of cytoplasmic viscosity. *J. Biomech. Eng.* 112, 269–276.

Needham, D., Ting-Beall, H. P., and Tran-Son-Tay, R. (1991). A physical characterization of GAP A3 hybridoma cells: Morphology, geometry, and mechanical properties. *Biotechnol. Bioeng.* 38, 838–852.

Nomura, H., Chiharu, I., Komatsuda, T., Ando, J., and Sapporo, A. (1988). A disk-type apparatus for applying fluid shear stress on cultured endothelial cell. *Biorheology* 25, 461–470.

Oh, S. K. W., Al-Rubeai, M., Emery, A. N., and Nienow, A. W. (1989). The effect of agiation on growth and antibody production of hybridoma cells—preliminary results. In *Advances in Animal Cell Biology and Technology for Bioprocesses*, pp. 221–223, Spier, R. E., Griffiths, J. B., Stephenne, J., and Crooy, P. J. (eds.). European Society for Animal Cell Technology 9th Meeting.

Olesen, S. P., Clapham, D. E., and Davies, P. F. (1988). Haemodynamic shear stress activates a K^+ current in vascular endothelial cells. *Nature* 331, 168–170.

Passini, C. A., and Goochee, C. F. (1989). Response of a mouse hybridoma cell line to heat shock, agitation, and sparging. *Biotechnol. Prog.* 5, 175–188.

Petersen, J. F., McIntire, L. V., and Papoutsakis, E. T. (1988). Shear sensitivity of cultured hybridoma cells (CRL-8018) depends on mode of growth, culture age and metabolite concentration. *J. Biotech.* 7, 229–246.

Petersen, J. F., McIntire, L. V., and Papoutsakis, E. T. (1990). Shear sensitivity of hybridoma cells in batch, fed-batch, and continuous cultures. *Biotechnol. Prog.* 6, 114–120.

Ramirez, O. T. and Mutharasan, R. (1990). The role of the plasma membrane fluidity on the shear sensitivity of hybridomas grown under hydrodynamic stress. *Biotechnol. Bioeng.* 36, 911–920.

Sato, M., Levesque, M. J., and Nerem, R. M. (1987a). An application of the micropipette technique to the measurement of the mechanical properties of cultured bovine aortic endothelial cells. *J. Biomech. Eng.* 109, 27–34.

Sato, M., Levesque, M. J., and Nerem, R. M. (1987b). Micropipette aspiration of cultured bovine aortic endothelial cells exposed to shear stress. *Arteriosclerosis* 7, 276–286.

Schlichting, H. (1979). *Boundary-Layer Theory*, 7th ed. McGraw-Hill, New York.

Schmid-Schonbein, Gosen, H. J. V., Heinich, L., Klose, H. J., and Volger, E. (1973). A counter-rotating "rheoscope chamber" for the study of the microrheology of blood cell aggregation by microscopic observation and microphotometry. *Microvasc. Res.* 6, 366–376.

Sdougos, H. P., Bussolari, S. R., and Dewey, C. F. (1984). Secondary flow and turbulence in a cone-plate device. *J. Fluid Mech.* 139, 379–404.

Shiga, T., Sekiya, M., Maeda, N., and Oka, S. (1985). Statistical determination of red cell adhesion to material surface. *J. Colloid Interface Sci.* 107, 194–198.

Shirinsky, V. P., Antonv, A. S., Birukov, K. G., Sobolevsky, A. V., Romanov, Y. A., Kabaeva, N. V., Antonova, G. N., and Smirnov, V. N. (1989). Mechano-chemical control of human endothelium orientation and size. *J. Cell Biol.* 109, 331–339.

Smith, C. G., Greenfield, P. F., and Randerson, D. H. (1987). A technique for determining the shear sensitivity of mammalian cells in suspension culture. *Biotech. Tech.* 1, 39–44.

Stathopoulos, N. A., and Hellums, J. D. (1985). Shear stress effects on human embryonic kidney cells *in vitro*. *Biotech. Bioeng.* **27**, 1021–1026.

Sutera, S. P., and Mehrjardi, M. H. (1975). Deformation and fragmentation of human red blood cells in turbulent shear flow. *Biophys. J.* **15**, 1–10.

Sutera, S. P., Mehrjardi, M. H., and Mohandas, N. (1975). Deformation of erythrocytes under shear. *Blood Cells* **1**, 369–374.

Sutera, S. P., Nowak, M. D., Joist, J. H., Zeffren, D. J., and Bauman, J. E. (1988). A programmable, computer-controlled cone-plate viscometer for the application of pulsatile shear stress to platelet suspensions. *Biorheology* **25**, 449–459.

Sutera, S. P., Pierre, P. R., and Zahalak, G. I. (1989). Deduction of intrinsic mechanical properties of the erythrocyte membrane from observations of tank-treading in the rheoscope. *Biorheology* **26**, 177–197.

Terracio, L., Miller, B., and Borg, T. K. (1988). Effects of cyclic mechanical stimulation of the cellular components of the heart: *in vitro*. *In Vitro Cel. Devel. Biol.* **24**, 53–58.

Thibault, L. E., and Fry, D. L. (1983). Mechanical characterization of membranelike biological tissue. *J. Biomech. Eng.* **105**, 31–38.

Tokunaga, O., and Watanabe, T. (1987). Properties of endothelial cell and smooth muscle cell cultured in ambient pressure. *In Vitro Cell. Devel. Biol.* **23**, 528–534.

Tramper, J., Williams, J. B., Joustra, D., and Vlak, J. M. (1986). Shear sensitivity of insect cells in suspension. *Enzyme Microb. Technol.* **8**, 33–36.

Tran-Son-Tay, R., Coffey, B. E., and Hochmuth, R. M. (1990). The motion of a ball oscillating in a bounded fluid: Inertial and wall effects. *J. Rheology* **34**(2), 169–191.

Tran-Son-Tay, R., Nash, G. B., and Meiselman, H. J. (1985). Effects of dextran and membrane shear rate on red cell membrane viscosity. *Biorheology* **22**, 335–440.

Tran-Son-Tay, R., Nash, G. B., and Meiselman, H. J. (1986). Oscillatory viscometry of red blood cell suspensions: Relations to cellular viscoelastic properties. *J. Rheol.* **30**, 231–249.

Tran-Son-Tay, R., Sutera, S. P., and Rao, P. R. (1984). Determination of RBC membrane viscosity from rheoscopic observations of tank-treading motion. *Biophys. J.* **46**, 65–72.

Tran-Son-Tay, R., Sutera, S. P., Zahalak, G. I., and Rao, P. R. (1987). Membrane stress and internal pressure in red blood cells freely suspended in shear flow. *Biophys. J.* **51**, 915–924.

Tran-Son-Tay, R., Needham, D., Yeung, A., and Hochmuth, R. M. (1991). Time dependent recovery of passive neutrophils after large deformation. *Biophys. J.* **60**, 856–866.

Truskey, G. A., and Pirone, J. S. (1990). The effect of fluid shear stress upon cell adhesion to fibronectin-treated surfaces. *J. Biomed. Mat. Res.* **24**, 1333–1353.

Vand, V. (1948). Viscosity of solutions and suspensions. II. Experimental determination of the viscosity-concentration function of spherical suspensions. *J. Phys. Colloid. Chem.* **52**, 300–321.

Vandenburgh, H. H. (1988). A computerized mechanical cell stimulator for tissue culture: Effects of skeletal muscle organogenesis. *In Vitro Cell Devel. Biol.* **24**, 609–618.

Vandenburgh, H. H., and Kaufman, S. (1979), In vitro model for stretch-induced hypertrophy of skeletal muscle. *Science* **203**, 265–268.

Van Dyke, M. (1970). Entry flow in a channel. *J. Fluid Mech.* **44** (part 4), 813–823.

Viggers, R. F., Wechezak, A. R., and Sauvage, L. R. (1986). An apparatus to study the response of cultured endothelium to shear stress. *Trans. ASME* **108**, 332–337.

Way, S. (1934). Bending of circular plates with large deflection. *Trans. ASME.* **56**, 627–636.

Wechezak, A. R., Viggers, R. F., and Sauvage, L. R. (1985). Fibronectin and F-actin redistribution in cultured endothelial cells exposed to shear stress. *Lab. Invest.* **53**, 639–647.

Winston, F. K., Macarak, E. J., Gorfien, S. F., and Thibault, L. E. (1989). A system to reproduce and quantify the biomechanical environment of the cell. *J. Appl. Physiol.* **67**, 397–405.

Wurzinger, L. J., Opitz, R., Wolf, M., and Schmid-Schönbein, H. (1985a). Shear induced platelet activation—a critical reappraisal. *Biorheology* **22**, 399–413.

Wurzinger, L. J., Opitz, R., Blasberg, P., and Schmid-Schönbein, H. (1985b). Platelet and coagulation parameters following millisecond exposure to laminar shear stress. *Thromb. Haemostas.* **54**, 381–386.

Yeh, C. K., and Rodan, G. A. (1984). Tensile forces enhance prostaglandin E synthesis in osteoblastic cells grown on collagen ribbons. *Calcif. Tiss. Int.* **36**, S67–S71.

Yeung, A., and Evans, E. (1989). Cortical shell-liquid core model for passive flow of liquid-like spherical cells into micropipets. *Biophys. J.* **56**, 139–149.

CHAPTER 2

■ ■ ■ ■ ■

Mechanochemical Transduction across Extracellular Matrix and through the Cytoskeleton

■ ■ ■ ■ ■

Donald Ingber, Seth Karp, George Plopper, Linda Hansen, and David Mooney

I. INTRODUCTION

The importance of mechanical forces as regulators of three dimensional tissue form has been recognized for many years (Koch, 1917; Thompson, 1977; Russell, 1982). Tissues and organs (e.g., bone, muscle, skin, vessels) remodel in response to applied mechanical stress and undergo changes in form and function until that stress is minimized. It is likely that individual cells are responsible for these physiological changes. However, little is known about the cellular basis of mechanochemical transduction.

Many investigators study the effects of large-scale forces (e.g., shear stress, blood pressure, gravity) on cell form and function. However, when you consider the amount of force that is actually exerted on each individual cell, these "large-scale" forces may become relatively minor compared to endogenous cell-generated forces that hold tissues together. For example, the force exerted by gravity on a single cell has been estimated to be less than 1 μdyn/cell (Albrecht-Buehler, 1990). In contrast, tensile forces that are generated within the intracellular cytoskeleton and exerted on extracellular matrix (ECM) attachment points can be 50–10,000 times greater (James and Taylor, 1969; Dennerll et al., 1988).

Thus, in this chapter, we focus on the role of cell-generated forces as biological regulators. We first review work that demonstrates that ECM molecules regulate cell growth and differentiation based on their ability to physically resist cell tension. We then describe what is known about the molecular basis of force transmission in cells and explain how intracellular cytoskeletal filaments function both as force generators and as load-supporting elements. Studies with three-dimensional cell models that are constructed out of sticks and elastic string according to the rules of tensegrity (tensional integrity) architecture are also presented. The importance of these modelling studies is that they suggest that utilization of a tensegrity arrangement for cytoskeletal organization may provide an efficient mechanism for direct transduction of mechanical forces into changes of biomolecular organization inside the cell and nucleus. We suggest that cells may also use a tensegrity mechanism for mechanochemical transduction of externally applied forces.

II. CONTROL OF GROWTH AND DIFFERENTIATION BY CELL TENSION

It is relatively easy to envision that tissues and organs that serve a mechanical function might be sensitive to physical forces. The pattern of bone deposition with the femur, for example, corresponds precisely to engineering lines of tension and compression for any structure of that size and form (Koch, 1917). However, what is more difficult to visualize is that any three-dimensional structure, whether it is a building, a living organ, or a cell, is subject to a pattern of inherent structural forces that vary depending on the size and configuration of that structure. It is for this reason that the form of any tissue may be thought to represent a "diagram of underlying forces" (Thompson, 1977).

The forces that hold cells and tissues together are produced by cells. All cells generate tensile forces within their cytoskeleton (see Section IV) and exert these forces on their ECM attachments as well as on neighboring cells (Emerman and Pitelka, 1977; Harris et al., 1980; Ingber and Jamieson, 1985; Li et al., 1987). As described above, cell tension is a major force acting on cells, and all additional mechanical loads (e.g., shear stress) are imposed on an already existing force equilibrium.

On the basis of this concept of tissue architecture, we set out to devise an in vitro system in which we could change the cellular force equilibrium by altering the ability of the ECM to resist cell-generated tensile forces. We did this by precoating nonadhesive bacteriological dishes with increasing densities of purified ECM molecules, such as fibronectin (FN). Cells were plated on these dishes in chemically defined, serum-free medium containing a constant, saturating amount of soluble mitogen [e.g., fibroblast or epidermal growth factor]. Serum had to be excluded because it contains soluble ECM molecules (e.g., FN and vitronectin) as well as a variety of different growth factors.

Using this system, we were able to demonstrate that capillary endothelial cells can be switched between growth and differentiation (i.e., capillary tube formation) in the presence of mitogen simply by varying ECM coating densities (Fig. 1). On dishes coated with high FN densities, cells attached, extended long processes, and formed many cell–cell contacts, but no tubes were observed. When plated at similar densities on a low FN density, cells attached but they could not extend because of the low mechanical resistance of the substratum. However, formation of extensive branching, tubular networks was observed within 24–48 h when capillary cells were plated on substrata of moderate adhesivity (100–500 ng FN/cm^3). Cells first attached, spread, and formed multiple cell–cell contacts over the first 6 h after plating, then multicellular retraction was observed. Retraction resulted in elevation of cellular cords above the surface of the dish and formation of stable tubular structures.

Interestingly, we also found that we could overcome the inhibitory effects of a highly adhesive substratum by increasing cell plating number (Fig. 1). Under these conditions, adherent cell multilayers formed in which many cells only contacted and exerted tractional forces on other cells. This configuration resulted in amplification of cell-generated forces while the mechanical resistance of the substratum remained constant. Once again, both multicellular retraction and tube formation were observed. Finally, similar results were also obtained using different types of ECM molecules (FN or type IV collagen), suggesting that the mechanical integrity of the ECM may be more important than its chemical specificity, at least in terms of regulation of capillary morphogenesis.

Using the same ECM control system with sparsely plated cells, we have shown that capillary cell growth can be similarly controlled by changing the mechanical resistance of the ECM (Ingber, 1990). Highly adhesive ECM configurations that supported cell spreading also promoted entry into S phase and stimulated cell proliferation. In contrast, cells which appeared round on ECM configurations with low mechanical integrity remained quiescent, even in the presence of saturating amounts of soluble growth factors. These findings are consistent with the concept that cell shape is critical for growth control (Folkman and Moscona, 1978).

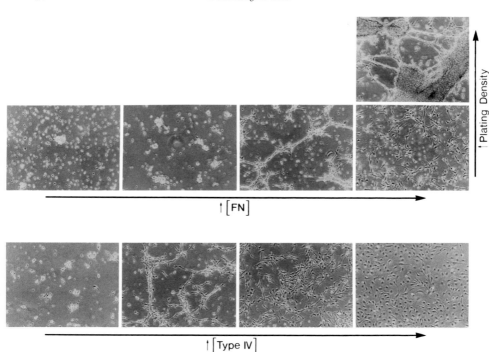

FIGURE 1   Mechanochemical switching between growth and differentiation in cultured capillary endothelial cells. Bacteriological dishes were precoated with 10, 50, 100, or 2500 ng/cm^2 (from left to right) of fibronectin (FN) or type IV collagen (type IV), as described in Ingber and Folkman (1989a). These phase-contrast views show that cell attachment and spreading increased in parallel with the number of ECM molecules available to resist cell-generated tensile forces. Tube formation was observed only on dishes of intermediate adhesivity when cells were plated at a moderate density (4×10^4 cells cm^{-2}). Endothelial cells formed extensive capillary networks on the highest FN density with the greatest mechanical resistance when higher cell numbers were plated (2×10^5 cells cm^{-2}; top right). Tube formation was observed on lower coating densities on type IV than on FN; type IV promoted more extensive cell attachment and spreading at all coating concentrations (magnification $\times 50$). From Ingber and Folkman, 1989a. Printed with the permission of the Rockefeller University Press.

More recently, we have found that primary rat hepatocytes use the same "mechanochemical" mechanism for switching between growth and differentiation and that this control is exerted at the level of gene expression (Mooney et al., 1992). Both hepatocyte spreading and growth increased in parallel when cells were plated on dishes coated with increasing densities of laminin (Fig. 2). Stimulation of growth by high ECM densities

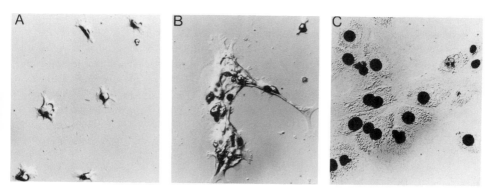

FIGURE 2   Control of hepatocyte shape and growth by varying the mechanical resistance of the ECM. ^{3}H-Thymidine autoradiography of hepatocytes cultured on dishes coated with 1 (A), 50 (B), and 1000 (C) ng/cm^2 of laminin (LM), as described in Mooney et al. (1992). Cell spreading and nuclear labeling increased in parallel as the ECM coating density was raised. The dark shadows on the upper right of cells on the low LM coating density (A) result from the use of Hoffman optics; none of these cells exhibited nuclear grains.

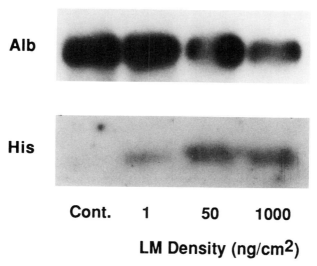

FIGURE 3   Northern blot analysis of albumin (Alb) and histone 3.2 (His) mRNA in freshly isolated hepatocytes used as a control (Cont.) and cells cultured for 48 h on increasing densities of LM (from left to right).

was accompanied by an increase in expression of the growth-related gene, histone 3.2, and concomitant down regulation of expression of differentiation-specific genes, such as albumin (Fig. 3). Thus, ECM coatings with high mechanical resistance turned on growth-associated genes and, at the same time, switched off genes necessary for differentiation. Again, similar results were obtained using a variety of different ECM molecules, derived from both basement membrane (laminin, type IV collagen) and interstitial matrix (type I collagen, FN). These data strongly suggest that ECM molecules regulate cell function based, to a large degree, on their ability to resist cell-generated forces.

III. TRANSMISSION OF PHYSICAL FORCES ACROSS EXTRACELLULAR MATRIX RECEPTORS

In any structure, mechanical stresses are distributed only across structural elements that are physically interconnected. Thus, to understand how individual cells experience changes in tension, we must first identify the path by which these forces are transmitted across the cell surface. As described above, cells exert tensile forces on their ECM anchors, which, in

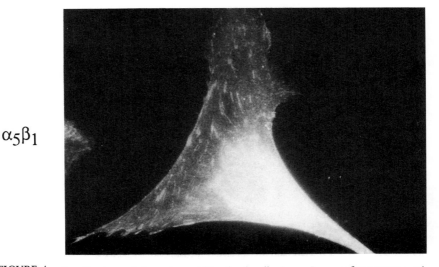

$\alpha_5\beta_1$

FIGURE 4   Localization of integrins within focal adhesions. Immunofluorescence microscopy was carried out using antibodies directed against the integrin $\alpha_5\beta_1$ complex within capillary cells cultured on FN-coated dishes.

Act

Tln

Vnc

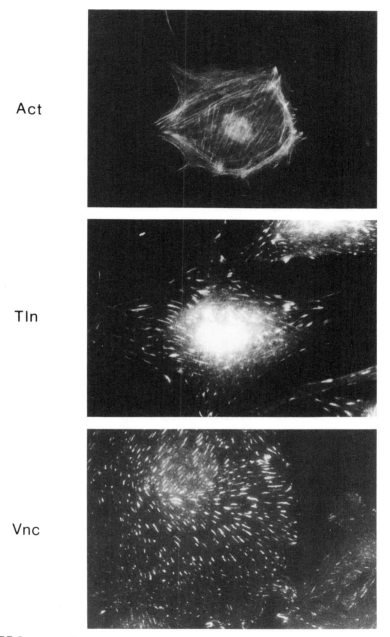

FIGURE 5 Actin filaments interconnect with integrins in focal adhesions via binding to bridging proteins, such as talin and vinculin. F-Actin (Act) was localized using rhodaminated phalloidin; talin (Tln) and vinculin (Vnc) were localized using specific antibodies in conjunction with immunofluorescence microscopy.

turn, apparently feed back and regulate cell function based on their ability to physically resist these forces. But how are these forces transmitted to and from the ECM?

Over the past 5 years, great advances have been made in the search for cell-surface receptors that physically link cells to ECM. Perhaps the best characterized are members of the "integrin" family of transmembrane receptors that recognize a common sequence, arg-gly-asp (RGD), that is found within many different ECM proteins (Ruoslahti and Pierschbacher, 1987; Hynes, 1987). Integrins are heterodimers comprised of one α and one β subunit, both of which span the plasma membrane. Several different α subunits are capable of association with each type of β subunit. However, it is the α subunit that determines the binding specificity of the receptor complex (e.g., $\alpha_5\beta_1$ binds FN whereas $\alpha_2\beta_1$ binds vitronectin).

Cell-surface integrin receptors become clustered within defined plaques or "focal adhesions" along the basal cell surface (Burridge, 1986) when cells attach to ECM-coated surfaces (Fig. 4). Interestingly, intracellular actin bundles or "stress fibers" also appear to end in the same type of adhesion complex (Fig. 5). A number of cytoskeletal-associated proteins (e.g., talin, vinculin) both localize within focal adhesions (Fig. 5) and physically interact with the cytoplasmic portion of integrins (Burridge, 1986), suggesting that they may form a molecular bridge linking integrins to actin filaments. Thus, it is thought that focal adhesions act as anchors

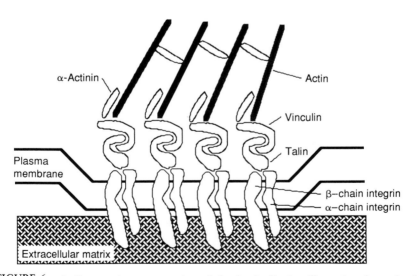

FIGURE 6   A diagramatic representation of the focal adhesion illustrating the molecular basis of force transmission across the cell surface.

that hold the actin filaments in place and thereby provide a direct linkage between ECM and the cytoskeleton (Fig. 6).

In summary, cells experience forces via transmission across specific transmembrane receptors, such as integrins, which physically interconnect cytoskeletal filaments with immobilized ECM molecules. In the next section, we will examine how cells generate tension and discuss the role of internal support elements during the establishment of a cellular force equilibrium.

IV. CYTOSKELETAL FILAMENTS AS FORCE GENERATORS AND LOAD-BEARING ELEMENTS

It is commonly assumed that all cells generate tension based on actomyosin interactions. However, this has never been demonstrated directly in nonmuscle cells because experimental perturbation of actomyosin interactions commonly results in dissociation of actin and myosin filaments. Using membrane-permeabilized endothelial cells, we have been able to directly demonstrate that cells utilize an actomyosin filament sliding mechanism to generate tension (Sims et al., 1992). Previous investigators have shown that saponin-treated cell monolayers retract their cell borders when incubated with calcium and adenosine triphosphate (ATP) in a cytoskeletal stabilization buffer that also supports tension generation (Wysolmerski and Lagunoff, 1990). Using the same membrane-permeabilization system with sparsely plated cells, we found that both the cell and its nucleus spontaneously round on addition of calcium and ATP. Furthermore, this retraction could be prevented using a synthetic myosin heptapeptide that was previously shown to inhibit actomyosin rigor complex formation in muscle (Suzuki et al., 1987). Thus, the mechanical forces that are responsible for changes of cell and nuclear shape are not generated by membrane surface tension; rather, they result from the same energy-dependent molecular binding interactions that mediate tension generation in muscle.

The intracellular tension that is generated within contractile microfilaments is then distributed across transmembrane integrin receptors and exerted on the ECM. Cells exert a resting tension (Harris et al., 1980; Dennerll et al., 1988) and spontaneously retract flexible substrata (Emerman and Pitelka, 1977; Harris et al., 1980; Ingber and Jamieson, 1985; Li et al., 1987). Conversely, cell shape can be controlled by varying the number of ECM resistance sites on the surface of an otherwise nonadhesive culture dish, and thus controlling the ability of the ECM to support these forces, as shown above.

However, cytoskeletal tension also may be resisted by internal support elements. For example, intracellular struts may be required to resist inward-directed tension during the initial stages of cell spreading in order to permit outward extension of the cell and nucleus (i.e., before ECM contacts have formed). Microfilaments apparently act as force-bearing elements when bundled into larger stress fibers. Disruption of microfilaments using cytochalasin D prevents initial cell spreading and causes dramatic changes in the morphology of adherent cells (Miranda et al., 1974). An increase in actin bundles is also observed in cells exposed to exogenous forces (Wong et al., 1983; Dewey et al., 1983; Sumpio et al., 1988; Terracio et al., 1988) and in cells cultured on ECM densities that exhibit high mechanical resistance (Mochitate et al., 1991).

Microtubules and intermediate filaments, which constitute the other two major cytoskeletal filament systems, do not appear to be involved in force generation. However, they apparently can act as force-supporting and transducing elements. Microtubules resist compressive forces when organized within bundles (Joshi et al., 1985; Dennerll et al., 1988) and physically interconnect with actin filaments (Schliwa and van Blerkom, 1981). Destabilization of microtubules results in altered morphology of fibroblasts (Middleton et al., 1988; Danowski, 1989) and the bundled microtubules present in neurite processes are necessary for their extension (Joshi et al., 1985; Dennerll et al., 1988). The intermediate filament array also interconnects with both microfilaments and microtubules and its distribution, stretching from the nucleus to the cell surface, is suggestive of a role as a cellular integrator. Perhaps the best example of the mechanical role played by intermediate filaments is in skin, where keratin filament networks are known to provides great tensile strength.

The different cytoskeletal filament systems and ECM anchors may provide redundant load-bearing functions. For example, induction of retraction or disruption of either intermediate filaments (Lin and Feramisco, 1981; Hollenbeck et al., 1989), microfilaments (Miranda et al., 1974), or microtubules (Middleton, 1988; Danowski, 1991) alone does not produce complete cell retraction or rounding. Furthermore, intact microtubules are required for cell extension in neurites when nonadhesive dishes are utilized for cell culture, yet microtubules are not required if the proper ECM is provided (Dennerll et al., 1988). Similarly, microtubule depolymerization has little effect during the initial stages of cell spreading on ECM unless the continuity of the actin network is compromised; then total inhibition of cell extension is observed (unpublished data). Redundancy of load-bearing elements is a critical feature of cell architecture since it provides greater structural stability and allows for local remodeling without causing a major change of cell morphology.

V. A TENSEGRITY MECHANISM FOR MECHANOCHEMICAL TRANSDUCTION

Past studies have shown that the cell and its nucleus and surrounding ECM form a structural continuum (Wolosewick and Porter, 1979; Schliwa and Van Berkblom, 1981; Fey et al., 1984; Burridge, 1986). This "solid state" quality of cells and tissues makes it difficult to explain how cells are able to undergo dynamic and coordinated deformations in response to changes in the ECM or mechanical stress. Most man-made structures (e.g., a house) depend on compressive forces (i.e., gravity) for their stability. These structures are usually rigid and often break in response to local force application (e.g., when exposed to high winds) and thus are unlikely to be used by cells. However, tension-dependent structures also exist that gain stability from a balance of internal forces. Cells may be thought of as tension-dependent structures since they generate tensile forces in their cytoskeleton and change their shape by exerting tension on their ECM anchors (Ingber et al., 1981; Ingber and Jamieson, 1985; Ingber and Folkman, 1989b).

The most efficient type of architectural system is based on maintenance of tensional integrity and is known as tensegrity (Fuller, 1975;

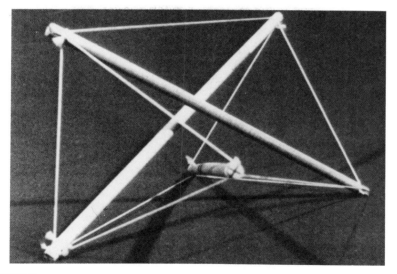

FIGURE 7   A basic tensegrity system. A tensegrity network is composed of a discontinuous series of compression-resistant struts that are pulled up and open via interconnection with a continuous series of tensile elements, as described in Ingber and Jamieson (1985).

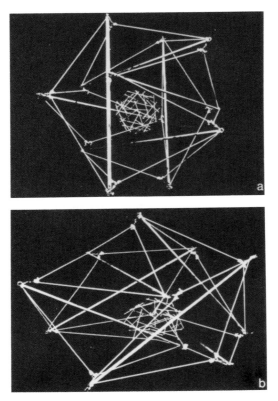

FIGURE 8   Studies with nucleated tensegrity cell models. (a) When unattached, both the tensegrity cell and its nucleus take on a round form. (b) The same model takes on a flattened or extended form when allowed to attach to a rigid foundation that can resist cell-generated forces. Note that the nucleus also extends in a coordinated fashion due to the presence of elastic elements that physically interconnect the nucleus with the cell surface. These black filaments cannot be seen because of the black background in these photographs. Additional descriptions of these models can be found in Ingber and Jamieson (1985).

Edmondson, 1987). In rigid structures, compression is a continuous element that connects all structural members. In contrast, tension is the continuous element that interconnects local compression-resistant islands in tensegrity structures (Fig. 7). We have previously shown that cell models can be built using tensegrity architecture that predict much of cell behavior (Ingber and Jamieson, 1985; Ingber and Folkman, 1989b). For example, when a tensegrity cell is unanchored, it tends to take on a round form because this is the minimal energy form in a symmetrical structure that exerts centripetal tension [Fig. 8(a)]. However, when the same structure is attached to a rigid foundation, it spontaneously changes shape and takes on a more flattened and extended form [Fig. 8(b)]. Furthermore, if the tensegrity cell attaches to a malleable foundation, it pulls in, wrinkles the substrate, and takes on a more rounded shape (Ingber and Jamieson, 1985). Similar behavior is exhibited by living cells cultured on silastic rubber or on malleable ECM gels (Emerman and Pitelka, 1977; Harris et al., 1980; Ingber and Jamieson, 1985; Li et al., 1987). Thus, studies with tensegrity cell models predict that changes of cell shape will depend on the ability of the ECM substratum to resist cell-generated tensile forces, as demonstrated above. Furthermore, studies with nucleated tensegrity models also suggest that both the cell and the nucleus will extend in a coordinated fashion when the cell spreads across a rigid substratum and that both should retract when cell–ECM contacts are dislodged (Fig. 8). Again, this is exactly what happens in living cells (Ingber et al., 1987; Ingber and Folkman, 1989b; Sims et al., 1992).

Use of a tensegrity arrangement by cells would permit forces to be directly transmitted from the cell surface, through the cytoskeleton, and to the nucleus. More importantly, this architectural arrangement would also provide a rapid and efficient mechanism for transduction of mechanical stresses into changes of cytoskeletal filament polymerization and organization inside the cell. Key to this mechanochemical transduction mechanism are two major concepts: (1) mechanical stresses are distributed in specific patterns and only across structural elements that are physically interconnected and (2) physical forces can alter chemical reactions, protein–protein binding, and polymerization processes as a result of local changes of thermodynamic parameters. Structural molecules that physical interconnect with cell–ECM anchors (e.g., integrins, cytoskeletal filaments) may therefore experience mechanical forces that nearby soluble molecules cannot recognize. In this manner, thermodynamic parameters may vary over nanometer distances inside the cell and thus result in distinct patterns of cytoskeletal organization.

In fact, thermodynamic analysis of microtubule polymerization reveals that the free tubulin concentration (i.e., monomer that is available for polymerization) is a function of the net compressive or tensile forces on the microtubule (Hill and Kirschner, 1982; Buxbaum and Heidemann, 1988). Microtubules are very dynamic structures that are capable of rapid turnover ($7-17$ μm/min in living cells; Cassimeris et al., 1988). Changes in the mechanical load exerted on individual microtubules could therefore result in rapid alterations of microtubule polymerization. Studies with living cells confirm that transfer of mechanical loads from either actin filaments or ECM anchors onto microtubules results in microtubule depolymerization (Joshi et al., 1985; Dennerll et al., 1988).

While the assembly of intracellular actin is also sensitive to mechanical forces (Dewey et al., 1983; Terracio et al., 1988), its response may be based on a different thermodynamic mechanism. For example, cells can undergo large-scale changes of cell shape (e.g., during trypsinization) without altering the amount of total polymerized actin (Bereiter-Hahn et al., 1990). Mechanical tension apparently induces actin bundle assembly by promoting lateral filament–filament realignment within an already existing isotropic actin lattice, rather than by inducing changes in actin polymerization (Isenberg and Wohlfarth-Bottermann, 1976; Bereiter-Hahn, 1987; Opas, 1987).

These changes of cytoskeletal organization may mediate the effects of force on cell metabolism. The distribution and function of a variety of intracellular molecules and organelles have been shown to depend on their association with cytoskeletal filaments [e.g., polyribosomes, mRNA, mitochondria, glycolytic enzymes; reviewed in Ingber and Folkman (1989b)]. Transmission of force across the cytoskeleton also may be responsible for modulating the function of receptors (Yahara and Edelman, 1975) and "stress-activated" ion channels (Sachs, 1989) on the cell surface. This is to be expected since it is unlikely that mechanical tension would be efficiently transmitted through the lipid bilayer. For example, even experiments that use micropipettes to stimulate stress-sensitive ion channels by applying suction directly to the cell surface require cytoskeletal interconnections for force transmission and channel activation (Sachs, 1989).

It is important to emphasize that the tensegrity models predict that application of mechanical force on the cell surface may be translated directly into changes of nuclear structure. This is important because the nuclear protein matrix is itself a structural entity that is involved in chromosome organization and contains fixed sites for regulation of DNA replication and transcription (Nelson et al., 1986). Thus, mechanically induced changes of nuclear structure may alter gene expression and DNA replication directly by changing the arrangement and function of associated DNA regulatory proteins (Nelson et al., 1986), by releasing mechanical restraints to DNA unfolding (Roberts and D'Urso, 1988) or, as recently demonstrated, by increasing nuclear pore size and thereby altering nucleocytoplasmic transport (Feldherr and Akin, 1990).

VI. IMPLICATIONS FOR EXTERNALLY APPLIED FORCES

Before one analyzes the effects of external forces on cells, it is helpful to consider the magnitude of these forces. As described above, the force of gravity on a single cell is on the order of 1 μdyn, while the resting tension generated intracellularly can be in the range of 10,000 μdyn. It is only when the force of gravity acts on macroscopic structures and becomes concentrated, such as in bone, that it becomes an important regulator of cell function. Furthermore, forces are distributed to individual cells via transmission across ECM that links all adherent cells within living tissues and organs. Thus, these externally applied forces are imposed on a pre-existing force balance that is present whenever cells are attached to ECM. A simple way to visualize this is to think of a violin string. Generation of sound (harmonic oscillations) on stroking with a bow requires that the string be under tension. Similarly, cells have to be attached and under tension in order to experience externally applied forces. For example, shear stresses have little effect on unattached cells because these cells rotate in response to flow. In contrast, application of shear along the apical surfaces of adherent cells both activates ion channels (Lansman et al., 1987; Olesen et al., 1988) and promotes actin bundle assembly inside the cell (Dewey et al., 1983).

The tensegrity model was developed to explain how changing the mechanical resistance of the ECM results in coordinated changes of cell, cytoskeletal, and nuclear forms. However, the way in which a continuous tensegrity structure distributes forces and deforms in response to stress does not depend on the origin of the forces. All forces, whether large or small, would be expected to be transmitted over the same series of molecular bridges that physically interlink the cell with its ECM. Thus, the effects of externally applied forces can be easily incorporated into this model. This possibility is supported by studies demonstrating that varying the mechanical stress applied to substrata coated with a constant ECM density results in changes in cell form and function similar to those that we obtained by altering the mechanical resistance of the ECM (Vandenburgh, 1983; Sumpio et al., 1988; Terracio et al., 1988). Furthermore, shear stress also may be transduced via a tensegrity mechanism since, as described above, its effects on both chemical and cystoskeletal signaling events appear to be mediated by pulling adherent cells away from their basal ECM anchors.

VII. DO CELLS UTILIZE TENSEGRITY ARCHITECTURE?

Whether cells utilize tensegrity architecture for their organization remains to be demonstrated directly. However, the importance of a balance between opposing forces for the maintenance of cell shape has been recognized in many cells, from plants as well as animals (Hahne and Hoffman, 1984; Joshi et al., 1985; Madrepelia and Adler, 1989; Danowski, 1989; Vasiliev, 1987). We and others have been able to show that the structural stability of the cell and nucleus depends on a dynamic balance of mechanical forces that are generated within contractile microfilaments, transmitted across cell surface ECM receptors, and resisted by ECM binding sites (Emerman and Pitelka, 1977; Harris et al., 1980; Ingber and Jamieson, 1985; Li et al., 1987; Sims et al., 1992). It is known that both microtubules (Joshi et al., 1985; Danowski, 1989) and actin bundles (Mogensen and Tucker, 1988) can act as semirigid struts inside the cell, and there is strong evidence suggesting that all three filament systems are intimately associated (Wolosewick and Porter, 1979; Schliwa and van Blerkom, 1981; Fey et al., 1984; Vasiliev, 1987). It also has been possible to demonstrate that microfilaments, microtubules, and ECM anchors play complementary roles in terms of carrying loads and changing their assembly in response to force. For example, disruption of microtubules results in both increased actin bundle assembly and a rise in tension that is exerted on the ECM (Danowski, 1989). However, perhaps most telling is the finding that stress fibers appear as islands of relative rigidity surrounded by a highly elastic cytoplasm when viewed by acoustic microscopy (Hildebrand et al., 1981). This observation is not consistent with a compression-dependent architectural arrangement; rather, it demonstrates that tension is a continuous element inside the cell. In summary, studies from a variety of different laboratories support the concept that cells exhibit tensional integrity and show that cytoskeletal filaments interconnect with ECM anchors so as to form a structurally integrated system for transduction of mechanical forces. Hopefully, the tensegrity paradigm will assist bioengineers and structural biologists by facilitating analysis of the mechanical basis of both force transmission and transduction at the molecular level.

ACKNOWLEDGMENTS

This work was supported by grants from the National Aeronautics and Space Administration, National Cancer Institute, Johnson and Johnson Research Fund, Takeda Chemical Industries, Ltd., and Neomorphics, Inc.

REFERENCES

Albrecht-Buehler, G. (1990). In defense of "nonmolecular" cell biology. *Int. Rev. Cytol.* 120, 191–241.
Bereiter-Hahn, J. (1987). Mechanical principles of architecture of eurkaryotic cells. In *Cytomechanics: The Mechanical Basis of Cell Form and Structure*, Bereiter-Hahn, J., Anderson, O. R., and Reif, W.-E. (eds.), pp. 3–30. Springer-Verlag, Berlin.
Bereiter-Hahn, J., Luck, M., Miebach, T., Stelzer, H. K., and Voth, M. (1990). Spreading of trypsinized cells: cytoskeletal dynamics and energy requirements. *J. Cell Sci.* 96, 171–188.

Burridge, K. (1986). Substrate adhesion in normal and transformed fibroblasts: Organization and regulation of cytoskeletal, membrane, and extracellular matrix components at focal contacts. *Cancer Rev.* 4, 18–78.

Buxbaum, R. E., and Heidemann, S. R. (1988). A thermodynamic model for force integration and microtubule assembly during axonal elongation. *J. Theor. Biol.* 134, 379–390.

Cassimeris, L., Pryer, N. K., and Salmon, E. D. (1988). Real-time observations of microtubule dynamic instability in living cells. *J. Cell Biol.* 107, 2223–2231.

Danowski, B. A. (1989). Fibroblast contractility and actin organization are stimulated by microtubule inhibitors. *J. Cell Sci.* 93, 255–266.

Dennerll, T. J., Joshi, H. C., Steel, V. L., Buxbaum, R. E., and Heidemann, S. R. (1988). Tension and compression in the cytoskeleton of PC-12 Neurites. II: Quantitative measurements. *J. Cell Biol.* 107, 665–674.

Dewey, C. F., Jr., Bussolari, S. R., Gimbrone, M. A., Jr., and Davies, P. F. 1983. In *Fluid Dynamics as a Localizing Factor for Atherosclerosis*, Gimbrone, M. A., Jr., and Davies, P. F. (eds.), pp. 182–187. Springer-Verlag, Berlin.

Edmondson, A. C. (1987). *A Fuller Explanation: The Synergetic Geometry of R. Buckminster Fuller.* Birkhauser, Boston.

Emerman, J. T., and Pitelka, D. R. (1977). Maintenance and induction of morphological differentiation in dissociated mammary epithelium on floating collagen membranes. *In Vitro Cell. Devel. Biol.* 13, 316–328.

Feldherr, C. M., and Akin, D. (1990). The permeability of the nuclear envelope in dividing and nondividing cell cultures. *J. Cell Biol.* 111, 1–8.

Fey, E. G., Wan, K. M., and Penman, S. (1984). Epithelial cytoskeletal framework and nuclear matrix-intermediate filament scaffold: Three dimensional organization and protein composition. *J. Cell Biol.* 98, 1973–1984.

Folkman, J., and Moscona, A. (1978). Role of cell shape in growth control. *Nature* 273, 345–349.

Fuller, J. B. (1975). *Synergetics.* Macmillan, New York.

Hahne, G., and Hoffman, F. (1984). The effect of laser microsurgery on cytoplasmic strands and cytoplasmic streaming in isolated plant protoplasts. *J. Cell Biol.* 33, 175–179.

Harris, A. K., Wild, P., and Stopak, D. (1980). Silicone rubber substrata: a new wrinkle in the study of cell locomotion. *Science* 208, 177–180.

Hildebrand, J. A., Rugar, D., Johnston, R. N., and Quate, C. F. (1981). Acoustic microscopy of living cells. *Proc. Natl. Acad. Sci. USA* 78, 1656–1660.

Hill, T., and Kirschner, M. (1982). Bioenergetics and kinetics of microtubule and actin filament assembly-disassembly. *Int. Rev. Cytol.* 78, 1–178.

Hollenbeck, P. J., Bershadsky, A. D., Pletjushkina, O. Y., Tint, I. S., and Vasiliev, J. M. (1989). Intermediate filament collapse is an ATP-dependent and actin-dependent process. *J. Cell Sci.* 92, 621–631.

Hynes, R. O. (1987). Integrins: A Family of cell surface receptors. *Cell* 48, 549–554.

Ingber, D. E. (1990). Fibronectin controls capillary endothelial cell growth by modulating cell shape. *Proc. Natl. Acad. Sci. USA* 87, 3579–3583.

Ingber, D. E., and Folkman, J. (1989a). Mechanochemical switching between growth and differentiation during fibroblast growth factor-stimulated angiogenesis in vitro: Role of extracellular matrix. *J. Cell Biol.* 109, 317–330.

Ingber, D. E., and Folkman, J. (1989b). Tension and compression as basic determinants of cell form and function: Utilization of a cellular tensegrity mechanism. In *Cell Shape: Determinants Regulation and Regulatory Role*, pp. 1–32, Stein, W. D., and Bronner, F. (eds.), Academic Press, Orlando, FL.

Ingber, D. E., and Jamieson, J. D. (1985). Cells as tensegrity structures: Architectural regulation of histodifferentiation by physical forces transduced over basement membrane. In *Gene Expression during Normal and Malignant Differentiation*, pp. 13–32, Andersson, L. C., Gahmberg, C. G., and Ekblom, P. (eds.), Academic Press, Orlando, FL.

Ingber, D. E., Madri, J. A., and Folkman, J. (1987). Endothelial growth factors and extracellular matrix regulate DNA synthesis through modulation of cell and nuclear expansion. *In Vitro Cell. Devel. Biol.* **23**, 387–394.

Ingber, D. E., Madri, J. A., and Jamieson, J. D. (1981). Role of basal lamina in the neoplastic disorganization of tissue architecture. *Proc. Natl. Acad. Sci. USA* **78**, 3901–3905.

Isenberg, G., and Wohlfarth-Bottermann, K. E. (1976). Transformation of cytoplasmic actin: Importance for the organization of the contractile gel reticulum and the contraction-relaxation cycle of cytoplasmic actomyosin. *Cell Tiss. Res.* **173**, 495–528.

James, D. W., and Taylor, J. F. (1969). The stress developed by sheets of chick fibroblasts in vitro. *Exp. Cell Res.* **54**, 107–110.

Joshi, H. C., Chu, D., Buxbaum, R. E., and Heidemann, S. R. (1985). Tension and compression in the cytoskeleton of PC 12 neurites. *J. Cell Biol.* **101**, 697–705.

Koch, J. C. (1917). The laws of bone architecture. *Ann. J. Anat.* **21**, 177–298.

Lansman, J. B., Hallam, T. J., and Rink, T. J. (1987). Single stretch-activated ion channels in vascular endothelial cells as mechanotransducers? *Nature* **325**, 811–813.

Li, M., Ageller, J., Farson, D. A., Hatier, C., Hassell, J., and Bissell, M. J. (1987). Influence of a reconstituted basement membrane and its components on casein gene expression and secretion in mouse mammary epithelial cells. *Proc. Natl. Acad. Sci. USA* **84**, 136–140.

Lin, J.-C., and Feramisco, J. R. (1981). Disruption of the in vivo distribution of the intermediate filaments in fibroblasts through the microinjection of a specific monoclonal antibody. *Cell* **24**, 185–193.

Madreperla, S. A., and Adler, R. (1989). Opposing microtubule and actin-dependent forces in the development and maintenance of structural polarity in retinal photoreceptors. *Devel. Biol.* **131**, 149–160.

Mochitate, K., Pawelek, P., and Grinnel, F. (1991). Stress relaxation of contracted collagen gels: disruption of action filiment bundles, release of cell surface fibronectin, and down regulation of DNA and protein synthesis. *Exp. Cell Res.* **193**, 198–207.

Middleton, C. A., Brown, A. F., Brown, R. M., and Roberts, D. J. H. (1988). The shape of cultured epithelial cells does not depend on the integrity of their microtubules. *J. Cell Sci.* **91**, 337–345.

Miranda, A. F., Godman, G. C., Deitch, A. D., and Tanenbaum, S. W. (1974). Action of cytochalasin D on cells of established lines. I. Early events, *J. Cell Biol.* **61**, 481–500.

Mogensen, M. M., and Tucker, J. B. (1988). Intermicrotubular actin filaments in the transalar cytoskeletal arrays of Drosophila. *J. Cell Sci.* **91**, 431–438.

Mooney, D., Hansen, L., Vacanti, J., Langer, R., Farmer, S., and Ingber, D. (1992). Switching from differentiation to growth in hepatocytes: Control by extracellular matrix. *J. Cell Physiol.* **151**, 497–505.

Nelson, W. G., Pienta, K. J., Barrack, E. R., and Coffey, D. S. (1986). The role of the nuclear matrix in the organization and function of DNA. *Annu. Rev. Biophys. Chem.* **15**, 457–475.

Olesen, S.-P., Clapham, D. E., and Davies, P. F. (1988). Haemodynamic shear stress activates a K^+ current in vascular endothelial cells. *Nature* **331**, 168–170.

Opas, M. (1987). The transmission of forces between cells and their environment. In *Cytomechanics: The Mechanical Basis of Cell Form and Structure*, pp. 273–286, Bereiter-Hahn, J., Anderson, O. R., and Reif, W.-E. (eds.), Springer-Verlag, Berlin.

Roberts, J. M., and D'Urso, G. (1988). An origin unwinding activity regulates initiation of DNA replication during mammalian cell cycle. *Science* **241**, 1486–1489.

Ruoslahti, E., and Pierschbacher, M. D. (1987). New perspectives in cell adhesion: RGD and integrins. *Science* **238**, 491–497.

Russell, E. S. (1982). *Form and Function*. University of Chicago Press, Chicago.

Sachs, F. (1989). Ion channels as mechanical transducers. In *Cell Shape: Determinants, Regulation, and Regulatory Role*, pp. 63–92, Stein, W. D., and Bronner, F. (eds.), Academic Press, San Diego, CA.

Schliwa, M., and van Blerkom, J. (1981). Structural interaction of cytoskeletal components. *J. Cell Biol.* **90**, 222–235.

Sims, J. R., Karp, S., and Ingber, D. E. (1992). Altering the cellular force balance results in integrated changes in cell, cytoskeletal, and nuclear shape. *J. Cell Sci.* In press.

Sumpio, B. E., Banes, A. J., Buckley, M., and Johnson, G. (1988). Alterations in aortic endothelial cell morphology and cytoskeletal protein synthesis during cyclic tensional deformation. *J. Vasc. Surg.* **7**, 130–138.

Suzuki, R., Nishi, N., Tokura, S., and Morita, F. (1987). F-actin-binding synthetic heptapeptide having the amino acid sequence around the SH1 cysteinyl residue of myosin. *J. Biol. Chem.* **262**, 11410–11412.

Terracio, L., Miller, B., and Borg, T. K. (1988). Effects of cyclic mechanical stimulation of the cellular components of the heart: in vitro. *In Vitro Cell. Devel. Biol.* **24**, 53–58.

Thompson, D. W. (1977). *On Growth and Form*. Cambridge University Press, New York.

Vandenburgh, H. H. (1983). Cell shape and growth regulation in skeletal muscle: Exogenous versus endogenous factors. *J. Cell. Physiol.* **116**, 363–371.

Vasiliev, J. M. (1987). Actin cortex and microtubular system in morphogenesis: Cooperation and competition. *J. Cell Sci.* (Suppl.) **8**, 1–18.

Wolosewick, J. J., and Porter, K. (1979). Microtrabecular lattice of the cytoplasmic ground substance. *J. Cell. Biol.* **82**, 114–139.

Wong, A. J., Pollard, T. D., and Herman, I. M. (1983). Actin filament stress fibers in vascular endothelial cells in vivo. *Science* **219**, 867–869.

Wysolmerski, R. B., and Lagunoff, D. (1990). Involvement of myosin light-chain kinase in endothelial cell retraction. *Proc. Natl. Acad. Sci. USA* **87**, 16–20.

Yahara, I., and Edelman, G. M. (1975). Modulation of lymphocyte receptor mobility by locally bound concanavalin A. *Proc. Natl. Acad. Sci. USA* **72**, 1529–1583.

CHAPTER 3

■ ■ ■ ■ ■

Mechanical Strain
and the Mammalian Cell

■ ■ ■ ■ ■

Albert J. Banes

I. INTRODUCTION

A battery of techniques has been developed to apply strain to tissues and cells in vivo and in vitro. In vivo, strain is applied to tissue via prosthetic devices such as orthopedic or dental implants, orthotic devices, or therapeutically during physical therapy or exercise. Tissue generally, and muscle particularly, exhibit a training response; that is, they alter their metabolic rate, grow larger, and increase in strength in proportion to the type, amount, and frequency of applied load.

With increased interest in the molecular changes that accompany strain-induced cellular adaptation, scientists have turned to in vitro methods to investigate strain effects on individual cells. Devices to apply compression, tension, and shear have been developed to study cells from tissues that are derived from mechanically active environments in vivo. However, as biologists and biochemists interact with engineers to better define the problems of strain-induced cellular responses, it has become evident that tissues and cells experience more than one form of strain. The strain field may be some complex combination of tension, compression and

81

shear stress, have a biaxial or a simple uniaxial field, represent a complex rather than uniform gradient, and finally, be cyclic with intermixed static periods or have periods of high activity followed by lower activity and rest. Hence one must carefully study each model of strain application to tissue or cells to characterize the types of strain a cell actually experiences. Examples of biologic variation in strain fields include the aorta, which is subjected to flow-induced, largely unidirectional (with some eddy currents), shear stress to endothelial cells on the luminal side, as well as tension on endothelial cells and the underlying smooth-muscle cells. Striated skeletal muscle is subjected to tension as sarcomeres contract. The central bulk of muscle experiences compression as actin and myosin filaments slide over each other, creating shear stress and thus increasing the volume of the shortened muscle. The force of muscle contraction applies tension to attached tendons; however, as the tendon collagen and elastin resist force applied in a largely uniaxial direction, a force component normal to the long axis provides some compression as well. Likewise bone is subjected to cyclic loading during standing, walking, or running. Torque is applied to long bones and joints, compression and shear stress to joints, and tension and compression to long bones.

What is the hierarchy of signals delivered to matrix and cells? How are the signals interpreted, and what are the key cellular responses, both early (in milliseconds and seconds) and late (in minutes and hours; days and weeks), that constitute a single stretch response or a training response?

Section II will address specific findings concerning application of strain to tissues and cells. Section III will address general theories of how strain acts as well as provide a summary of mechanisms suggesting how strain is interpreted biochemically to effect a cellular response.

II. BIOCHEMICAL AND BIOLOGICAL RESPONSES OF SPECIFIC CELL TYPES SUBJECTED TO MECHANICAL DEFORMATION

The initial investigations involving testing of the effects of applied mechanical strain to biological tissues were performed using whole-tissue preparations or implanting mechanical devices in vivo to superimpose a strain field on living tissue. Beginning in the mid-1970s, researchers began to focus on effects of mechanical strain on isolated cells in culture. Additional work was performed on bone cells, tendon cells, chondrocytes, and muscle cells. The field of application of mechanical activity to cultured cells to simulate dynamic environments has grown now to include tissue studies by researchers in disciplines other than orthopedics or dentistry and has encompassed cells from aorta skin, lung, the brain, uterus, and other sources.

As the use of prosthetic devices increases in application in human medicine, understanding the mechanism of cell attachment and adhesion to surfaces of orthopedic and dental implants, vascular grafts, tendon, ligament and skin substitutes, synthetic knuckles, and breast implants also increases. In most cases, cells are challenged with a surface that has dynamic loads applied while they attempt to populate the surface and interior of the device. Moreover, motion must be considered to understand normal processes such as development and growth; skeletal muscle strength; training response; cardiac muscle hypertrophy; mineralization or pathologic processes such as atherosclerosis and osteoporosis; overuse syndromes for joint, ligament, tendons, and muscle; exercise-induced amenorhea; or how motion enhances healing in bone, joints, connective tissue, muscle, and skin.

If in vivo conditions can be simulated in vitro, perhaps the use of large animals for testing of cardioactive drugs, eye irritants, and other mediator testing can be minimized by testing cardiac myocytes, aortic smooth muscle or endothelial cells, or corneal epithelial cells, respectively, in a dynamic culture environment as an initial tier test.

The following section addresses studies performed in representative tissue and cell systems under tension or compression.

A. Endothelial Cells: Aortic, Venous, Capillary

Until recently, most of the studies involving the biochemistry of endothelial cells have been performed using whole-vessel preparations or isolated endothelial cells grown under static conditions. Clearly, aortic endothelial cells are constantly subjected to cyclic pulsations that create two types of strain on the cells: one strain is that of flow-induced shear stress; the second is the mechanical strain that results in radial tension on the endothelial cells and vessel wall. The initial work involving imposition of mechanical strain on cultured endothelial cells was performed using flow-induced shear stress. This subject will be addressed in Chapters 6 and 7. Only in the last 5 years have reports been published concerning the influence of mechanical strain on isolated endothelial cells in culture.

Ives and co-workers were the first to report on effects of mechanical strain on cultured endothelial cells (Ives et al., 1986). These workers utilized an apparatus designed after Leung and co-workers, specifically, a motor-driven cam shaft and cord clamped to a flexible polyetherurethane urea membrane on which human umbilical vein endothelial cells or bovine aortic endothelial cells were grown (Leung et al., 1977). The cells were isolated as primary cultures grown to confluence in 3–4 days on the urethane membranes, and were then subjected to 48 h of intermittent tension equivalent to 10% elongation at a frequency of 1 Hz. Experiments were performed to control for medium movement induced by the membrane cycling. Results indicated that both cell types aligned in a direction perpendicular to that of the applied strain. They also demonstrated actual elongation of the endothelial cells by 17.7%. Immunofluorescence studies revealed that microtubules aligned in a direction parallel with the long axis of the cells. They reported that highly confluent cells aligned less rapidly than did subconfluent cells since development of strong intercellular contacts resisted the movement of individual cells to the perpendicular position. The authors reasoned that flow-induced shear stress would cause parallel alignment of endothelial cells to reduce fluid drag. They suggested that alignment of the cells to a position perpendicular to the applied strain might be a mechanism by which cells reduced stress or strain energy exerted on the cell. Likewise, alignment to minimize the tension on the cell might minimize tension on microtubules or other cytoplasmic filament networks during stretching. The net effect could be to minimize cell movement during each stretch cycle to reduce mechanical strain on the cells.

Sumpio and co-workers, utilizing a computer-driven instrument developed by Banes et al., 1985, 1990, placed intermittent strain of 0.05 Hz with 10% elongation for 0–7 days on cultured endothelial cells (Sumpio et al., 1987). The strain regimen consisted of 10 s of strain followed by 10 s of relaxation, that is, a 3-cycle/min (cpm) regimen. The instrument was a pressure-operated, microprocessor-controlled unit that regulated pressure to the bottom of specially designed flexible-bottomed six-well culture plates. The culture surface was a hydrophilic, amino-derivatized, silicone elastomer membrane. The strain in the culture well was a gradient, maximal at the periphery and minimal at the central part of each culture dish. Primary, bovine, aortic endothelial cells were permitted to attach overnight to the flexible bottomed culture plates, then subjected to the 3-cpm strain regimen. Cultures sustained only 1.7 doublings from days 1 to 7, whereas cells subjected to cyclic strain achieved 2.6 doublings and had a significantly increased rate of division (Sumpio et al., 1987). The greatest difference in rate of division was identified between days 0 and 3. From days 3 to 7, the growth rates in the stretched and static cultures were the same. Increased cell counts paralleled increased DNA synthesis. The latter experiments were carried out in medium that contained 10% fetal calf serum. In general, fetal calf serum contains sufficient growth factor activity to maximally stimulate cell division. However, cell growth rate on the amino surface was much reduced compared to growth on standard positively or negatively charged polystyrene surfaces. Hence, the stretch-induced stimulation of division was detected because control cells were maintained at a basal rate. The authors emphasized that studying the nature of the cell biology and biochemistry of endothelial cells in a mechanically active environment was more relevant than using a static one. They also indicated that this instrument could be used to characterize the impact of altering the magnitude, duration, and frequency of the applied strain on endothelial cell function and morphology.

Sumpio and co-workers extended their original observations with bovine aortic endothelial cells to study protein synthetic changes. Cells were plated as above and subjected to a maximum of 24% elongation at 0.05 Hz for 5 days. For the final 24 h of cyclic strain, the cells were incubated with 500 μCi ^{35}S-methionine and then lysed and subjected to two-dimensional polyacrylamide gel electrophoresis (PAGE). In separate cultures, cells from stretched or control cultures were fixed then stained with rhodamine phalloidin, a fluorescent, bicyclic octapeptide that binds strongly to filamentous actin (Sumpio et al., 1988a). Results of the actin-staining experiments indicated that control cells cultured on the derivatized, flexible, rubber surface did not have florid actin stress cables as do cells cultured on polystyrene substrata, whereas cells subjected to cyclic strain on the rubber surfaces revealed well-developed actin stress cables. Actin cables were aligned in a direction parallel with the long axis of the cells. Approximately 1000 protein spots were detected and quantitated by computer analysis from autoradiograms from stretched and control endothelial cells labeled with ^{35}S-methoionine. Of these proteins, amounts of 14 were significantly increased in endothelial cells subjected to 5 days of 0.05-Hz cyclic strain. Synthesis of 20 other proteins was either decreased or totally inhibited by mechanical strain. Biochemical characterization of most of these proteins remains to be established. However, amounts of β-tubulin and the 100-kD (kilodalton) heat-shock protein were increased compared to normal, but not statistically so. The amount of G-actin in the cell was not increased by 5 days of cyclic stretch; however, actin polymerization occurred in a time-dependent fashion in response to strain. The latter data indicate that endothelial cells had a sufficient actin monomer pool to polymerize to F-actin, to contract or migrate in response to strain. This observation seems reasonable in that cells subjected to an insult such as heat shock or a mechanical challenge must be able to respond rapidly within seconds or minutes in order to survive. Hence, cells may be prepared to migrate away from unfavorable conditions as a protective response to minimize the strain to which they are subjected (Sumpio et al., 1987; Ives et al., 1986).

Sumpio and Banes extended their previous studies by exploring second-messenger activity that might be responsible for altering the rate of cell division and actin polymerization. Bovine aortic endothelial cells subjected to 0.05-Hz cyclic strain for 1, 3, or 5 days were transferred to serum-free medium for the final 24 h, with or without supplemental $20\text{-}\mu M$ arachidonic acid, 5 min prior to media collection (Sumpio et al., 1988a). Media from stretched and control cells were assayed for 6-keto $PGF_{1\alpha}$, a stable product of prostacyclin hydrolysis, and thromboxane B_2. The results indicated that basal production of PGI_2 in control and stretched cells did not vary until arachidonic acid was added. Arachidonic acid addition resulted in stimulation of PGI_2 production in cyclically strained endothelial cells. The degree of stimulation increased with increasing duration of mechanical stimulation. There were no significant differences in the amount of thromboxane A_2 produced by endothelial cells maintained in a static-versus-cyclic stretch environment with or without arachidonic acid. It is possible that long-term mechanical stimulation to endothelial cell membranes depleted the endogenous arachidonic acid stores. Addition of arachidonic acid to mechanically perturbed cells stimulates prostacyclin secretion. The latter result may be due to an increase in the amount or activity of phospholipase A_2, as proposed by Hong, Harell, and co-workers (Hong et al., 1976; Harell et al., 1977; Somjen et al., 1980; Binderman et al., 1988). Sumpio and Banes hypothesized that PGI_2 biosynthesis may be different at diverse intracellular sites regulated independently by physical deformation (Sumpio et al., 1988a; Sumpio and Banes, 1988a).

In a later study conducted in our laboratory, prostacyclin secretion was determined in both venous and arterial endothelial cells cultured in either a nontension or mechanically active environment (Upchurch et al., 1989). Primary cells isolated from bovine aorta or vena cave were subjected to 1 Hz deformation and up to 17% elongation in a Flexercell strain unit. Cells were cultured on six-well, flexible-bottomed Flex I culture plates covalently bonded with genetic type I collagen. Prostacyclin determinations were performed on control and experimental cultures on days 3, 4, and 5 using a radioimmunoassay for the stable hydrolysis product of prostacyclin, 6-keto $PGF_{1\alpha}$. The results indicated that venous endothelial cells maintained a higher endogenous level of prostacyclin production than did aortic endothelial cells, but that mechanical perturbation reduced the amount of prostacyclin produced by each cell type. Control cultures treated with $20\text{-}\mu M$ ibuprofen reduced prostacyclin secretion 27-fold in venous and aortic endothelial cells (ECs) and 22-fold in stretched venous ECs. The authors concluded that the reduction in prostacyclin production in mechanically deformed cells could be a result of reduced cycloxygenase activity. A second explanation may be that the true endogenous level of prostacyclin may be artificially high in statically cultured cells and that once restoration of cyclic mechanical deformation occurs (60 cpm), the amount of prostacyclin synthesis is down-regulated to a new, lower set point. Alternatively, the amount of arachidonic acid available to yield prostacyclin may have been depleted early during the mechanical activity regimen. Addition of arachidonate increases the amount of prostacyclin available to the cell (Sumpio and Banes, 1988a).

In a recent report, Sumpio and co-workers, using bovine aortic endothelial cells cultured in flexible-bottomed plates (Flex I amino-derivatized) subjected to 0.05 Hz, found that collagen production was reduced (Sumpio et al., 1990). Media and cells were collected separately on days 1, 3, and 5. Collagen and noncollagen protein syntheses were quantitated by analyzing the amount of hydroxyproline and proline present in hydrolyzed samples. Secretion of newly synthesized collagen and noncollagen proteins was decreased to 50%, 66%, and 30% of the control level and 80%, 75%, and 25% on days 1, 3, and 5, respectively, in cyclically stretched endothelial cells compared to the nonstretched counterparts (Sumpio et al., 1990). Protein synthesis in the cell sheets also decreased on days 3 and 5. The authors concluded that the degree of collagen expression may be inversely related to cell proliferation and that rapidly proliferating endothelial cells synthesized and secreted less collagen than did their static counterparts. However, the authors stated that the biological response of a cell to a given deformation regimen may vary with changes in the magnitude, frequency, and duration of the applied strain. Therefore, a variation in strain regimen may alter the degree to which the cell synthesizes and secretes or degrades collagen. Interestingly, the reverse relationship was found for aortic smooth-muscle cells (collagen synthesis was increased) whose DNA synthesis was decreased by strain (Sumpio et al., 1988b).

Shirinsky and co-workers utilized an apparatus similar to that of Leung to provide cyclic strain to cultured human umbilical-vein endothelial cells (HUVECs) grown on hydrophilic-treated silicone membranes (Shirinsky et al., 1989; Leung et al., 1976). The deformation regimen included a 20% amplitude in tension at 52 cpm. The authors indicated that the HUVEC cultured for 48 h in a cyclic strain regimen became elongated and oriented perpendicularly to the applied strain as was shown by Sumpio, Ives, and their co-workers (Sumpio et al., 1988a; Ives et al., 1986). They also reported that actin cytoskeletal filaments oriented parallel with the axis of the aligned cells. They indicated that the stretch-induced orientation response could be inhibited by the addition of 10-μM forskolin. The authors concluded that in the case of cyclically stretched endothelial cells, actin bundle alignment may precede cell body orientation. After 10–15 min of exposure to pulsatile strain, cells displayed transverse or oblique distribution of stress fibers with respect to the cell axis. The adenyl cyclase activator forskolin inhibited stretch-induced cell alignment by depolymerization of stress fibers. However, cells that had been oriented by cyclic strain but depleted of stress fibers did not lose their orientation for several hours. Thus, the authors concluded that actin was involved in stretch-induced orientation but that stress fibers were not required for the maintenance of the oriented state of EC.

Gorfien and co-workers, using a pressure-operated instrument providing biaxial deformation to polyurethane urea membranes, deformed cells with increasing strain at 1 Hz for 7 h (Gorfien et al., 1989). The degree of strain included 0.78%, 1.76%, 4.9%, or 12.5% tension. Fetal bovine pulmonary arterial endothelial cells were cultured on urethane membranes fitted into stainless-steel cylinders with a transparent base screwed into the cylinder to immobilize the membrane. Compressed air controlled with a solenoid valve was used to deform the urethane membranes. Replicate cultures were exposed to varying degrees of strain, or control cultures were incubated under similar conditions on a gyratory shaker platform to simulate the degree of medium movement experienced by the deformed cells. Fibronectin production was measured with an enzyme-linked immunoassay. The results indicated that cells subjected to 4.9% or 12.5% strain produced significantly less fibronectin than did nondeformed cells. The authors reported that depression in fibronectin synthesis was transient and after 7 h returned to control levels. The authors made an analogy between nonlethal cell injury induced by cyclic, biaxial deformation and wound healing in vivo. They conjectured that if cyclic biaxial strain constituted injury to the cells, then the cellular response was to deposit a matrix of fibronectin similar to that seen in wounded endothelium in vivo.

B. Muscle Cells: Smooth, Striated Skeletal, Cardiac

All cells are endowed to some extent with cytoplasmic filaments capable of providing contraction to the cell (Burridge, 1986). The latter function is necessary for cell movement, pinocytotic and phagocytotic activities, and intracellular movement of organelles. Muscle cells are particularly well endowed with actin and myosin contractile elements. Three species of muscle fiber types are present in mammalian white muscle—types A, B, and C—whereas in red muscle, there is essentially only one type, analogous to C (Pollack and Sugi, 1984). A fibers are fast-twitch fibers that metabolize glycogen anaerobically, having little capacity for oxidative metabolism. Type C fibers are tonically active fibers that metabolize oxidatively. Type B fibers have aspects of both A and C fibers and can utilize both pathways. During development, striated skeletal muscle hypertrophies, partly in response to long-bone growth. Hence, early in muscle development passive stretch may be the key stimulating mechanical aspect to muscle hypertrophy. Later in life, active tension is the prime stimulus for muscle hypertrophy, although passive stretch can still stimulate hypertrophy in mature muscle. The latter concept is a basis for physical therapy in patients who are immobilized. Muscles can be made to hypertrophy by enforced exercise or denervation, removal, or tenotomy of synergistic muscles (Binkhorst, 1969; Goldberg and Goodman, 1969; and Reitsma, 1969). Placing a load on muscle induces the stretch response resulting in a measurable heat rise and metabolism of the loaded muscle (Feng, 1932). Palmer and co-workers showed that intermittent changes in tension had more stimulating effects on protein, and specifically collagen synthesis, than did static strain regimens (Palmer et al., 1981). The instrument that they used was a modification of that originally described by Feng in 1932: an immobilized whole-muscle preparation inside a tube attached at the other end to an electromagnetic coil that was energized for 100 ms every 3 s. Force was applied to the muscle to increase its length to 110% of the resting length for 6 h. Muscle preparations were compared under static loads of 0–10,000 dyn and intermittent stretching with an 8000-dyn load. Both strain regimens increased rates of total protein and collagen synthesis compared to nonloaded control preparations. However, in muscles maintained in vitro under constant tension, total protein synthesis was 22% of the rate in vivo compared with muscles placed under intermittent stretch, which had 73% greater synthesis than that in nonstretched controls and were 38% of the value in vivo (Palmer et al., 1981). Concentrations of adenosine tri-, di-, monophosphate (ATP), (ADP), (AMP), and pyruvate were unchanged after 6 h of incubation in vitro in constant tension. However, tissue lactate and glycogen concentrations were increased twofold over the level observed in fresh muscle. In another report, Palmer and co-workers showed that $PGF_{2\alpha}$ release was twofold greater in intermittently stretched muscles than in statically stretched controls (Palmer et al., 1983). This in vitro finding substantiated the in vivo finding that exercise increased the release of prostaglandins from muscle reported by Young and Sparks (1980).

Summers and co-workers reported that soluble growth promoting factors extracted from homogenized chicken patagialis muscles induced to grow by passive stretch, stimulated chick myoblasts to divide in vitro (Summers et al., 1985). The optimum dose of extract was 1.5 mg/mL of culture medium for 48 h in vitro. Stimulation was 1.5-fold higher than in those cells grown in control extract or basal media. In other experiments, the authors showed that transferrin at 10 mg/mL could stimulate chick embryo skin fibroblasts to divide. They concluded that transferrin or transferrinlike factors in stretched-muscle extracts could stimulate muscle growth. Barnes and Worrell used an apparatus similar to that of Feng, in which paired sartori muscles of frogs were placed in oxygen gassed tubes in vitro, then were either not loaded or loaded with a constant 30,000-dyn force for 2, 4, or 24 h in vitro (Barnes and Worrell, 1985; Feng, 1932). Glycogen content of stretched muscles decreased 73% over the 24 h incubation period compared to a 23.1% reduction in nonstretched control muscles. However, stretched muscles retained 55% of their total protein content compared to 38% in the nonstretched control counterparts. In a later study with the same apparatus, Barnes showed rectus abdomini muscles of male frogs to have oxygen consumptions linearly related to the passive load, whereas secreted lactate was inversely related to load (Barnes, 1987). Barnes proposed that as stretch-related oxidative energy metabolism increased, dependence on anaerobiosis decreased.

Vandenburgh and Kaufman utilized a stretching frame to apply 10.8% static strain to isolated skeletal muscle myotubes from breast muscle of 12-day-old embryonic chick (Vandenburgh and Kaufman, 1979). Myotubes were cultured for 72 h on a silicone membrane; then the membrane was stretched to increase the length by 10.8% by displacing screws in a steel frame that held the membrane. The cells were labeled with radioactive α-aminoisobutyric acid (AIB) or radioactive leucine. Cells were labeled with no stretch or stretch for 18 h in vitro. Results indicated a 21–44% increase in AIB accumulation, indicating increased amino acid transport into stretched cells. The concentration of soluble ^{14}C-leucine in cell pools indicated only a 2.3–8% increase in radioactivity, whereas ^{14}C-leucine incorporation was increased an average of 14% in statically stretched cells. In particular, synthesis of myosin heavy chains was increased 14.7% after 18 h of constant stretch. In a later report, Vandenburgh utilized a microprocessor controlled device to cyclically stretch muscle cells in multiwell plates (Vandenburgh, 1987; 1988). Cells were grown on collagen-coated silastic membranes deformed from beneath by a pin centered under the bottom of each well. Cyclically deformed cells subjected to growth on a slowly, continuously stretching substratum oriented parallel to the strain field, whereas cells subjected to a 17% stretch–relaxation regimen oriented perpendicularly to the strain field. He reported no effect of mechanical activity on total cell density or myotube formation.

Wright reported that embryonic cardiac or striated skeletal myocytes subjected to cyclic stretch at 15 cpm with a 3-s duty cycle were stimulated to divide 2–20-fold (Wright et al., 1988). Analyses of DNA synthesis were carried out after 5 days of a 15-cpm strain regimen. DNA synthesis was monitored by incorporating ^{3}H-thymidine for the last 24 h of culture in control and stretched cells. Absolute activities of aldolase, lactate dehydrogenase, and malate dehydrogenase indicated no reproducible alterations in their activities in stretched versus control cells. However, in an attempt to test the hypothesis that the amplitude of deformation was biologically significant in stretched, chick, striated skeletal, or cardiac myocytes, the rate of cell alignment was measured at either 15% or 24% elongation. Cells were plated at subconfluency so that cell processes were not touching, and were then subjected to 15 cpm with 3-s duty cycle for up to 24 h. At time points, 1, 2, 3, 5, 7, 12, and 24 h, static and experimental cultures were removed from the incubator, fixed, and stained, and the cell angles were determined with respect to a radius. The data indicated that a 20% increase in the A term, or degree to which the cells were stretched, resulted in a 3.2-fold increase in the rate of cell alignment [see Section III.A, Eq. (2)]. The data indicated that the amplitude of the applied strain affects the rate at which muscle cells migrate, apparently to minimize the strain to which they are subjected (Wright et al., 1988).

Tidball utilized a microapparatus to test the effects of cyclic strain on single muscle cells (Tidball, 1986). His apparatus involved clamping a cell between the folds of bent, stainless steel wires on each end. Tidball had already reported that single-twitch muscle cells can generate a maximum isometric stress of 3×10^5 N/m^2. This represents a force of 2×10^{-3} N for a typical myocyte that is 80 μm in diameter. The load device could resolve a force 20 times greater than that force created by the twitch of a single cell. He subjected single muscle cells from frog semitendonosus muscle to sinusoidal oscillation to simulate strain placed on the cells at the close of passive extension and initiation of active contraction, simulating the action of swimming. A second group of cells had their basement membranes removed by both enzymatic and mechanical procedures. The results indicated that muscle cells with intact basement membranes had complex moduli, that is, dynamic stiffness and loss of tangent, greater than those of cells without basement membranes. The relative magnitude of energy loss to energy storage (i.e. specific loss) was 3 times greater for intact cells than for basement-membrane-depleted cells. He concluded that the basement membrane serves as a brake to retard passive muscle extension prior to contraction.

Leung, Glagov, and Matthews published two papers reporting their findings on matrix changes in cyclically stretched arterial smooth-muscle cells in vitro (Leung et al., 1976, 1977). The first largely outlined the mechanical system they utilized. The second publication concentrated on the findings that 52 cpm of 10% elongation, for up to 3 days, on rabbit aortic smooth-muscle cells increased total protein and collagen synthesis 3.0- and 3.7-fold respectively (Leung et al., 1976). DNA synthesis was unchanged. Both genetic types I and III collagens were increased 2.8- and 3.5-fold respectively. Hyaluronate synthesis was increased 3.3-fold, while synthesis of chondroitin 4-sulfate was unchanged and that of chondroitin 6-sulfate increased 2.9-fold. Dermatan sulfate synthesis increased 20%. This latter work was the most elegant one performed at the time and indicated beyond doubt that application of cyclic strain to cultured cells reproducibly altered their biochemical phenotype. A drawback to the system was the preparation of elastic membranes as sections of aorta and the difficulty in preparing and sterilizing the equipment for use. In a later publication from the same group, Sottiurai and co-workers utilized growth of smooth-muscle cells on elastin membranes that were subjected to either cyclic stretching at 10% elongation, 52 cpm, or were stretched and held stationary 8, 48, or 56 h (Sottiurai et al., 1983). A morphologic examination indicated that myofilaments were replaced by rough endoplasmic reticulum (RER) in cyclically stretched cells. Cells subjected to stationary tension had diminished myofilament contents but fewer RER profiles. There were approximately 4.6-fold more RER profiles in cyclically stretched cells than in cells subjected to stationary tension. The authors concluded that cyclic stretching resulted in RER formation and preservation of myofilaments, whereas immobilization in tension resulted in disappearance of myofilaments and cytoplasmic degradation.

Dartsch and Hammerle studied the responses of aortic smooth-muscle cells from rabbit media in a machine like that of Leung and co-workers, where the stretch amplitude was varied from 2% to 20% (Dartsch and Hammerle, 1986; Leung et al., 1976; 1977). They reported that cells became more oriented as the degree of strain increased. Cell orientation was approximately perpendicular to the direction of the strain field. They also indicated that in short-term experiments where cells were exposed to 3–12% cyclic strain with an amplitude of 10% elongation, actin filaments aligned in the direction of cell alignment.

In a later publication, Sumpio and Banes showed that porcine aortic smooth-muscle cells also aligned perpendicularly to the applied strain when subjected to a 3-cpm strain gradient of up to 24% maximum elongation (Sumpio and Banes, 1988a; 1988b). However, during the first 72 h after initiating the strain experiment, cell division was less than that in control cultures. The authors interpreted this finding that, initially, smooth-muscle cells may be retarded from dividing when subjected to the strain field. In another report, Sumpio and Banes showed that collagen synthesis increased in aortic, porcine, smooth-muscle cells cycled at 0.05 Hz (Sumpio and Banes, 1988b; Sumpio et al., 1988b). On days 3 and 5, control and stretched cells were separated into media and cell components and assayed for collagen and noncollagen protein synthesis. Amounts of collagen in the medium and cell layers in cells stretched for 3 days were not significantly different from control levels. However, after 5 days of stretching, the amount of collagen secreted into the medium or present in the cell layer was increased over that of the control nonstretched group. Results for noncollagen protein synthesis paralleled those for collagen synthesis.

It has been known that pressure overload in cardiac muscle for a period of weeks can increase ventricle weight and enzyme content (Dowell et al., 1983). Myocardial content increased at the rate of 9% compared with 5% per day for nonloaded rat hearts. Terracio and co-workers showed that cardiac myocytes on silastic membranes coated with laminin and subjected to 10% cyclic stretch at 10 cpm for 72 h, showed the orientation response to be perpendicular to that of the direction of strain as shown in other cell systems (Terracio et al., 1988).

Perhaps the most exciting new finding concerning the mechanism of cyclic strain and cardiac myocyte division rate is that of the Japanese group, which reported that the protooncogene c-*fos* was up-regulated in statically stretched cardiac muscle cells (Komuro et al., 1990; 1991). The c-*fos* gene codes for a nuclear transcription factor that is involved in the regulation of messenger ribonucleic acid (mRNA) synthesis (Sorrentino et al., 1986). Cardiac myocytes from 1-day-old rats were grown on a silicone membrane and subjected to a static stretch of approximately 10% in a screw-frame device (Mann et al., 1989). The c-*fos* mRNA was increased after as little as 15 min of static stretch, was maximal at 30 min and then returned to baseline after only 2 h. The c-*fos* gene expression was increased when cells were stretched as little as 5% and was maximal at 20% increase in cell length (Komuro et al., 1990). Relative c-*fos* mRNA levels estimated by Northern blot increased from 15% ± 5% at 5% stretch to 89% and 100% at 10% and 20% elongation, respectively. In neonatal rat myocytes transfected with chimeric c-*fos* genes ligated to the reporter gene, chloramphenicol acetyltransferase (CAT) (PSVO *fos* CAT) 10% passive stretch 48 h after transfection induced a seven-fold increase in CAT activity compared to nonstretched counterparts (100% activity vs. 12% ± 3% in controls with no stretch). Nonmyocyte, fibroblastic cells from the cardiac myocyte isolates did not demonstrate the stretch-activated c-*fos* induction.

In a second report, Komuro and co-workers reported that a passive stretch of 10% for 24 h increased rat myocyte total RNA by 45% and increased skeletal actin mRNA by 16% and that c-*fos* mRNA was increased 15-fold within 30 min (Komuro et al., 1991). To further understand how c-*fos* message induction was stimulated by stretch, constructs were made with successive deletions upstream from the 5′ flanking region of c-*fos* and its promoter. The 3′ borders of the deletion map were at positions −1450 base pairs (bp) (pFC_1), -404 (pFC_2) and -227 (pFC_3). Myocyte stretching induced a greater than seven-fold activity in CAT in the pFC_1 and pFC_2 constructs, whereas little activity was observed in the 3' end deletion at -227 pFC_3 construct missing a serum-response element. Results indicated that sequences between -227 and -404 were required for efficient c-*fos* transcription and that sequences 227 bp upstream of the mRNA start site may not be involved in stretch-induced c-*fos* induction.

Serum-response element transcription can be activated by cAMP and Ca^{2+} stimulation of a sequence, 60 bp upstream from the mRNA start site (Komuro et al., 1991). Protein kinase C and/or tyrosine kinase are two such elements that could be involved in stimulation of stretch-induced c-*fos* mRNA production. Phorbol esters activate protein kinase C activity. In myocytes pretreated for 4 h with 10 ng/mL 12-0-tetradecanoyl phorbol-13-acetate (TPA), c-*fos* mRNA could not be induced by additional TPA or stretch, but was responsive to epidermal growth factor (EGF) or forskolin. If cells were pretreated with EGF or forskolin, c-*fos* message accumulated in response to stretch or TPA. The authors concluded that in myocytes, TPA and tension induce c-*fos* gene expression through a common signal transduction pathway, protein kinase C. Experiments involving kinase inhibitors and down-regulation of protein kinase C indicated inhibition of c-*fos* transcription induced by stretching. Inositol monophosphate (IP_1) and inositol bisphosphate (IP_2) but not inositol triphosphate (IP_3) were increased 55% and 90%, respectively, within 1 min of initiating passive tension in myocytes. The authors concluded that mechanical strain might directly stimulate protein kinase C activity via phospholipase C activation to induce early and late genes yielding c-*fos*, c-*myc*, and β-type myosin heavy-chain and skeletal α-actin, respectively.

Although its actual role remains unknown, c-*fos* is involved in further messenger induction in the nucleus. Induction of other protooncogenes such as c-*myc* may be involved in stimulating cell division (Sorrentino et al., 1988).

C. Tendon and Tendon Cells

Tendons are highly fibrous structures that transmit the contractile force of muscle from the tendon origin or epimyceum, over the surface of and deep into muscle through its substance to the insertion into bone. Tendons are quite varied in their structures and compositions, particularly where tendons glide through sheaths or over bony prominences such as the calcaneus and talus. Tendons are composed of 70–80% genetic type I collagen, 10–40% elastin, large and small proteoglycans, fats, and minor noncollagen proteins including fibronectin (Oakes and Bialkower, 1977; Vogel and Evanko, 1988; Banes et al., 1988b; Amiel et al., 1989). The collagen appears to be present in prestressed undulations due to the imposed strain intrinsic in the elastin (Oakes and Bialkower, 1977). It is presumed that the internal tendon cells are responsible for prestressing collagenous components. Stopak and Harris have suggested that tendons form between bone and muscle by an action called tractional structuring, wherein leader cells lay down an initial matrix pathway and subsequently contract, laying down more collagen as the organ progresses in development (Stopak and Harris, 1982). Perhaps these sequential accretions of matrix and cellular structuring of tendon are involved in the prestressing phenomenon noted by Oakes and Bialkower. A load–strain hysteresis relationship has been demonstrated for the elastic wing tendon of the chick (Oakes et al., 1977). A large energy loss occurs during the first cycle of deformation. Moreover, elastase-treated tendons show a two- to threefold extension beyond the limit length of tendons with intact elastin. The latter experiment indicates the importance of elastin in limiting extension of the tendon and allowing rebound after deformation. Scott and Hughes studied the proteoglycan–collagen relationships in developing chick and bovine tendons (Scott and Hughes, 1985). They reported that in the chick, distal, flexor digitorum longus tendon, the collagen content increased 30-fold from days 11 to 22 in ovo, and increased 25% in mass up to maturation. Interestingly, collagen fibril diameter in ovo increased linearly up to the time of decreased frequency of fetal movement, at which time fibril diameter plateaued. After hatching, fibril diameter increased linearly three fold from the time of hatching at day 22 to day 68. An inverse relationship was found between proteoglycan and collagen contents of flexor tendon. Proteoglycan content, as measured by uronic acid content, declined almost 1000-fold from day 10 to hatching, and maintained a low level throughout life. A plot of collagen surface area versus interfibrillar volume indicated that glycosaminoglycans (GAGs) present in the tendon appear on the surface of fibrils not between them. The proteoglycan distributed between collagen fibrils was chondroitin sulfate-rich, whereas collagen-bound proteoglycans were dermatan sulfate-rich. They reasoned that since the ratio of collagen in the fibril cylinder to that of GAGs was proportional as the collagen volume to that of the collagen fibril surface area, the GAGs were entirely present at the periphery of collagen fibrils. A plot of ratio of hydroxyproline to GAG concentration versus fibril radius was a straight line. This was true for dermatan sulfate. They concluded that the transition in the embryo from a largely proteglycan matrix (phase I) to a largely collagen matrix (phase II) was associated with initial active muscle loading of tendon.

Vogel and Evanko reported that the predominant proteoglycans of proximal and distal flexor tendons of the cow are small pg_1 and pg_2 molecules (Vogel and Evanko, 1988). Only pg_2 was found in adult flexor tendon. The large proteoglycans, prominent in adult tendon were not found in proximal or distal regions of fetal tendon. The latter observation indicates that the differentiated proteoglycan phenotype in fibrocartilage in adult distal tendon develops as a direct response to compressive force once the animal begins to walk.

Gillard and co-workers utilized a rabbit flexor digitorum profundous tendon model wherein they translocated the portion of the tendon that glided over the talus and calcaneous bones (Gillard et al., 1979). The latter area contains an original pressure-bearing region compared with the tension-bearing region of the remainder of tendon. Reducing strain on the original pressure-bearing region for up to three and one-half months resulted in a rapid loss of GAGs, largely chondroitin sulfate, which was difficult to replace. There was an increase in GAG content in the original tension-transmitting region. When tension was restored to the translocated tendon by reoperation, the GAG content decreased to normal values, whereas the high overall GAG concentration was maintained by increased concentration of dermatan sulfate.

Merrilees and Flint published an ultrastructural study of the tension and pressure zones in rabbit flexor tendon (Merrilees and Flint, 1980). They showed that the pressure-bearing region was fibrocartilagelike and had a high concentration of GAGs. The pressure zone had linear clusters of fibrocartilage cells oriented normal to the direction of pressure. The cells were surrounded by dense arrays of 11-nm-diameter microfilaments and loosely packed collagen fibrils of 55-nm periodicity. Numerous lipid droplets were also observed in this region. In the zone of tension, cells were quite elongate with extensive cytoplasmic processes contacting neighboring cells. The cells showed a scalloped surface architecture due to the imprint of collagen fibrils in intimate association with the cell. The latter phenomenon was observed because the preparations were fixed in dorsiflexion, which placed tension on the collagen. The authors concluded that in the pressure-bearing area of the flexor digitorum profundus tendon, intermediate filaments, and lipid droplets may have an internal cytoskeletal function protecting cells from applied compression. They also suggested that changes in the availability of surface positive charge on collagen fibers in different tensional states might be the basis of transduction of physical information into chemical events.

Few reports have been published concerning the impact of cyclic or static tension or compression on cultured tendon cells. Banes and co-workers were the first group to utilize in vitro deformation to gauge the impact of cyclic compression on tendon cell biology (Banes et al., 1985). Avian tendon internal fibroblasts were isolated from flexor profundus tendons and cultured on hard-bottomed polystyrene culture plates subjected to a 0.13% compressive force, intermittently for 25 s, with 5-min intervening rest periods for 5 days. On the fifth day of the experiment, control or strained cells were labeled under no-load conditions with ^{35}S-methonine for 2 h. After incubation, samples were collected, proteins separated by PAGE, and the autoradiogram analyzed by densitometry. No significant changes in actin synthesis were detected. However, the amount of a 52-kD protein that comigrated with tubulin was decreased 18-fold in cells that underwent cyclic compression (3500 cycles, 0.13% compression over 5 days) compared with the static controls (Banes et al., 1985). The authors pointed out that the cells inside tendon matrix are like osteoblasts bound in bone subjected to both tension and compression in vivo. Tubulin is a major cytoskeletal element involved in maintenance of cell form (Dustin, 1980). In order for the cells to survive a decrease in length, the plasticity of the cell increased, which resulted in a reduction in the concentration of tubulin.

More recently, Banes and co-workers have studied the morphologic changes that accompany applied static and cyclic tension in vitro. Cyclic strain of up to 24% elongation at 15 cpm induced tendon internal fibroblasts (IFs) and synovial cells (SCs) from the surface of tendon to align in a direction perpendicular to that of the applied strain (Gilbert, 1989; Banes et al., 1990). In addition, internal tendon fibroblasts subjected to a 15-cpm regimen for 8 h and cycled with a 16-h rest period, aligned perpendicularly to the applied strain in the region of maximum stretch, but aligned parallel with applied strain at the 10% to minimal elongation in the same culture plate well. Moreover, in growth curve experiments where both tendon internal fibroblasts and synovial cells were grown in 10% serum-containing medium, no stimulatory effect on the rate of cell division was detected using either 3-, 10-, or 40-cpm regimens. However, if tendon cells were cultured in 1–2% calf serum, 1–2% horse serum, medium containing platelet-poor plasma or passaged in 5% fetal calf serum-containing medium and without changing the medium so that growth factor activity would be depleted, cells could be stimulated by cyclic strain (Banes et al., 1990). In experiments where cells were depleted of growth factors by culture for 5 days without medium change, tendon IF and SC responded to as little as one deformation event of two sec duration of 0.1% elongation by increasing the incorporation of ^{3}H-thymidine into TCA precipitable material. The maximum stimulatory event for DNA synthesis occurred between 10 and 100 stretches of a 15-cpm regimen. Addition of cytochalasin B or D to tendon cells inhibited the orientation response of the cells, but was unable to prevent the mitogenic effect. As an inhibitor of actin filament polymerization, cytochalasin D might be expected to decrease or ablate the stimulatory response of strain. However, when tendon cells become confluent without medium changes, sufficient matrix and mature cell–matrix contacts form such that addition of cytochalasin D alters cell shape to a more rounded phenotype, yet leaves enough cell contacts to produce intrinsic strain (Banes, personal observation). The latter result suggests that intrinsic strain may be sufficient to stimulate cell division maximally if cell substrate contours or cytoplasmic filament equilibrium are altered.

D. Osteoblasts: Whole Bone, Primary Calvarial, Ros 17/2.8, UMR-106; Response of Bone and Bone Cells to Mechanical Strain

Hayes and Carter and co-workers have come as close as any investigators in modern time to elaborating on the mathematics of bone remodeling in the tradition of Wolff (Carter and Hayes, 1977a; 1977b; Carter et al., 1980). From data involving strain-related fatigue failure of cortical bone, Carter has extrapolated the upper limit in cycles to failure for human activity in the range of 10^6–10^7 cycles for running or walking activity (Carter et al., 1980). However, rigorous exercise may lead to failure in the range of 100,000–1,000,000 cycles. In the latter publication, the authors noted that rigorous exercise involved strains of 200–400 microstrain, running involved 100–200 microstrain and walking generated strains of below 100 microstrain. He pointed out that military recruits may experience fatigue fracture in bone within 6 weeks after initiating a rigorous training regimen, aggregating 100–1000 miles of rigorous exercise, or 10^5–10^6 load cycles. From data collected on human bone, Carter estimated that repeated bending loads that are 25% of the ultimate bending strength will induce bone fracture after 12,400 load cycles (Carter and Hayes, 1977b). The prediction was made from the following expression:

$$N = 0.25\left(\frac{\sigma}{\sigma_{\text{ult}}}\right)^{-7.8} \tag{1}$$

where N represents the number of loading cycles to complete failure, σ (sigma) equals the nominal maximum fatigue bending stress, and σ_{ult} (sigma ultimate) equals nominal bending strength (modulus of rupture) (Carter and Hayes, 1977a).

Early work by Carter and Hayes indicated that repeated loading of bone caused progressive loss of stiffness and ultimate strength in monotonic loading (Carter and Hayes, 1977a; 1977b). Fatigue failure in bone was markedly different from that of metals that do not exhibit loss of stiffness or strength until a fatigue crack initiates. Fatigue failure in bone was induced by diffuse structural damage caused by microcracking and debonding of osteons (Carter and Hayes, 1977a).

Rubin and co-workers showed that under actual exercise conditions in athletes running the Boston Marathon, sound velocities across tibias were markedly higher in runners finishing the 26-mile course in under 3 h compared to those requiring longer than 3 h (Rubin et al., 1987). Velocities were higher in male than female runners, and the tibial velocity of wheelchair racers was 28% lower than combined velocities for all runners. Immediately after the race, velocities were higher across both tibia and fibula, suggesting that adaptive mechanisms exist in healthy bone to withstand microdamage and osteon debonding induced by repetitive cyclic loading. The relevancy of the measurement is that ultrasonic properties are directly related to elastic properties of bone (Ashman et al., 1984).

Booth and Gould, in a review on the effects of training and disuse on connective tissue, have cited the work of Price-Jones, and Saville and Whyte in results of exercise regimens with rats on bone growth (Booth and Gould, 1975; Price-Jones, 1927; Saville and Whyte, 1969). With wheel-running exercise regimens of 5 h/day for 4 months, or 6 h/day for 7 weeks, there were no differences in bone weights and lengths expressed as a ratio to body weight (Price-Jones, 1927; Saville and Whyte, 1969). But both groups reported that trained rats had heavier bones than did control rats. Booth and Gould pointed out that with animals exercised for nearly a quarter of a day, only a very low V_{O_2} maximum (low work intensity) was achieved. Therefore, low-intensity training had no negative effect and may have even stimulated bone growth. Later work substantiated the earlier work but included a higher-intensity training regimen (Kiiskinen and Heikkinen, 1973a, 1973b). Femurs were reported to be shorter and lighter than those in nonexercised control groups. Hence, high-intensity training may actually inhibit bone growth. The latter thought was substantiated by the study of Forwood and Parker, who showed that pubescent male rats subjected to an intensive one-month exercise program had significant reductions in tibial and femoral lengths and weights (Forwood and Parker, 1987).

Carter and co-workers indicated that bone had extremely poor resistance to fatigue failure (Carter et al., 1981). However, Burr and co-workers, working with a canine model, tested the hypothesis that osteonal remodeling is triggered by microdamage to bone (Burr et al., 1985). They used a

3-point bending model on the forelimb of the dog to determine that loads producing strains as little as 1500 microstrain on the radius and 1400 microstrain on the ulna for 10,000 cycles produced significant bone microdamage. There were 44 times as many microcracks associated with resorption spaces in limbs subjected to 10,000 cycles of 1500 microstrains in 1–4 days of loading. The authors concluded that fatigue microdamage may be a significant factor in intracortical bone remodeling. Part of the response to loading may be a vascular response. Forwood and Parker showed that the size of new Haversion canals increased following a one month-training program in male rats (Forwood and Parker, 1986).

Overall, although stress fractures in athletes resulted from overtraining (Matheson et al., 1987), the bone density of distance runners aged 50–72 years old was 40% greater than that of nonrunners (Lane et al., 1986). Skerry and co-workers reported that a single load cycle on bone could initiate cellular events that resulted in new-bone formation (Skerry et al., 1988). Working with the avian ulna, they utilized a load of 300 microstrain at a frequency of 1 Hz and a duration of 5 min. The latter regimen resulted in a substantial increase in bone mass over an 8-week period (Rubin and Lanyon, 1985). A single load period of 5 min increased ^{3}H-uridine incorporation six-fold in osteocytes in midshaft ulna. In addition, glucose-6 phosphate dehydrogenase activity was also increased. The authors suggested that the pathway toward pyrimidine synthesis was stimulated as an immediate response to applied strain.

Both Carter and co-workers and Lanyon substantiated the hypothesis that addition of mechanical strain applied to skeletal tissues significantly reduced bone degeneration and increased ossification in the appendicular skeleton. Carter emphasized that these affects can be accelerated by intermittently applied shear stresses and inhibited by intermittent hydrostatic pressure (Carter, 1982). Lanyon argued that in locations where resistance to repetitive loading is important, only the general form of bone will develop as a result of growth alone (Lanyon, 1987). Other major characteristics of bone such as localized bone density or periosteal thickening resulted from use; that is, functional adaptation occurred. This mechanism ensures that bone can withstand loads without failure during work or recreation in individuals who have exercised appropriately. This issue has even more relevance for the aged population (particularly females who sustain postmenopausal bone loss) where decreased bone density predisposes toward bone fracture.

Rodan and co-workers were among the first to subject living bone to strain in culture (Rodan et al., 1975a; 1975b). Chick tibias were isolated from 16-day-old embryos and subjected to compression at 80,000 dyn/cm^2 in a piston device. The authors calculated that the force exerted on the tibia of a standing, newly hatched chick was 100,000–200,000 dyn/cm^2.

To account for the static load that an embryonic bone might withstand, 50% or half the bird's weight (to account for weight borne on one leg and the reduced demand of an embryonic environment) at hatching was applied normally to the plane of the vertical bone. DNA synthesis was 1.16-fold greater in tibias subjected to constant compression for 4 days, whereas glucose consumption decreased to 0.4 mg glucose/day compared with 1 mg/day in nonstressed controls. Twenty-four hours after removal of pressure, glucose levels equaled those of the control cultures, indicating that tibias remained viable and that the reduction in metabolism was due to applied pressure.

Bone exhibits an electrical response to deformation or strain, and it is known that this effect can be attributed to streaming potentials and to the piezoelectric property of bone. Early research by investigators such as Fakuda and Yasuda (1957) and Bassett and Becker (1962) focused on the piezoelectric nature of bone and its role in generating electrical potentials in response to mechanical strain. Subsequent studies by Anderson and Eriksson (1968; 1970) identified the two distinct sources of the electrical signals in wet bone. Now it is generally accepted that streaming potentials, voltages generated by the flow of ion-rich fluid in response to deformation of bone, are the primary factor in generating the observed electrical effects in wet bone (Gross and Williams, 1982; Pienkowski and Pollack, 1983), but piezoelectric effect cannot be completely ruled out. Martin and Burr (1979) even suggested that there may be some locations in bone such as the lining of Haversian canals, where the piezoelectric effect may dominate. Research continues in the area of electrical potentials in bone (Guzelsu and Walsh, 1990; Scott and Korostoff, 1990) in order to better understand this phenomenon, which may play a role in the control of bone remodeling.

Davidovitch and co-workers utilized an orthodontic strain model in cats to gauge the effects on alveolar bone remodeling (Davidovitch et al., 1976). Springs attached to the right side canine teeth and third premolars exerted a tipping force to the canines of 80 g. Animals were treated for 1 to 4 weeks. Strain applied in the latter way resulted in two regions of compression on the periodontal ligament (PDL), one at the alveolar crest and the other at the tooth apex. Simultaneously, tension occurred at the opposite regions. The authors utilized an immunochemical stain to detect cyclic adenosine monophosphate (cAMP) in PDL and alveolar osteoblasts. Within hours after application of strain, the PDL on the compression side thinned in diameter and stained intensely for cAMP, unlike bone cells, which did not show these changes. After one week of strain, osteoblasts on the compression side stained positively for cAMP but the intensity of the PDL staining was less than it had been after 6 h of strain. On the tension side, the PDL stained only weakly for cAMP while the osteoblasts and surface fibrocytes stained more intensely. The tension sites were dominated by new-bone formation and cAMP-rich cells. It is likely that as time progressed, the tension on the tissues diminished as tooth movement ensued. Nevertheless, some cellular responses (rise in cAMP) occurred within hours on the compression side in PDL that required days for bone cells to follow, whereas on the tension side, bone cells responded first and resulted in bone deposition.

Harell and co-workers were the first to develop a system to apply strain to osteoblasts in culture (Harell et al., 1977). Using a screw device to stretch the plastic 0.05–0.1%, producing a force of 10 kg/cm^2 to polystyrene culture plates, they subjected avian calvarial periosteal cells to static strain. They found that PGE$_2$ increased after 5 min and achieved a plateau by 30 min (three-fold increase compared to control level). Strain-induced PGE$_2$ production could be suppressed by addition of in-domethacin, indicating that de novo synthesis occurred. Secondarily, cAMP production and calcium release peaked at 15 min. The latter results could be duplicated by exogenous application of PGE$_2$ without strain.

Somjen, in Binderman's group, extended that group's results with the same system, verifying the rapid production of PGE$_2$ followed by release of cAMP in response to strain in osteoblasts (Somjen et al., 1980). In an effort to understand the mechanism involved in the prostaglandin–cyclic nucleotide cascade, the authors reported that adenyl cyclase activity reached a maximum 5 min after strain application and remained at 30% above the control level for 60 min, whereas stimulation of adenyl cyclase activity by exogenous PGE$_2$ was only transient. Phosphodiesterase activity was also increased by strain and was maximal at 10 min. Exogenous PGE$_2$ also increased activity with similar kinetics as strain, but the effect diminished after 20 min and returned to baseline levels by 60 min. DNA synthesis was increased 1.5-fold 24 h after static strain was applied continuously to bone cells. Increased DNA synthesis was dependent on PGE$_2$ production.

Binderman and co-workers, using their mechanical strain model of static tension on bone cells grown in a polystyrene dish, extended their original results implicating PGE$_2$ as the initial chemical signal in the mediator response to strain (Binderman et al., 1988). They applied 0.1% tension to primary bone cells isolated from 19–20-day-old rat calvaria for 15 min and assayed for cAMP in cultures pretreated with gentamycin, an aminoglycoside antibiotic that binds to phospholipids, or antiserum to phospholipids from patients with systemic lupus erythematosus. In both cases, strain-stimulated production of cAMP was reduced approximately 50%. Exogenous PGE$_2$ plus the inhibitor returned cAMP levels to normal, indicating the primary effect was at the level of PGE$_2$ synthesis and that subsequent steps in the mediator cascade were unaffected at that level of inhibition. Addition of exogenous arachidonic acid also reversed the effects of antiphospholipid antiserum. Moreover, cAMP production in response to added phospholipase A_2 could be decreased by the antiserum, showing specificity of the antibody for phospholipids on the outer surface of the plasma membrane (Binderman et al., 1988). The authors concluded that strain affected only a fraction of the bone cells in culture because the effects of exogenous PGE_2 on isolated bone cells were much greater than that of mechanical strain alone. They proposed that the mediator cascade on bone cells involved (1) activation of phospholipase A_2 (PLA_2), (2) arachidonic acid release, (3) PGE_2 synthesis, and (4) increased cAMP production followed by DNA synthesis. Strain-induced membrane perturbation may expose phospholipids to phospholipase A_2 hydrolysis or directly activate PLA_2 and/or induce a Ca^{2+} concentration flux that itself could activate membrane-bound PLA_2.

Meikle and co-workers applied continuous tension with a 30-g force to newborn rabbit coronal suture explants (Meikle et al., 1979, 1980). Collagenase, gelatinase, and neutral metalloproteinase activities increased 34%, 95% and 36% respectively during a 4-day culture period. However, degradation of matrix was not increased due to coproduction of protease inhibitors. The authors also demonstrated an 87% increase in DNA synthesis in sutures subjected to strain.

Yeh and Rodan grew fetal rat calvarial osteoblast-like cells on collagen strips 5 cm $\times$ 0.45 μm prepared by Ethicon (Yeh and Rodan, 1984). These were clamped into a frame connected to transducers to report the degree of tension. The tension regimen applied with a piston of an infusion pump was composed of 8–10 cycles yielding a 2–3-mm displacement. Each cycle (rest to maximum amplitude back to rest) was of 15-min duration. PGE_2 was increased 3.7-fold after 2 h of applied strain. The latter report verified in a second system the earlier work of Harell and Somjen that PGE_2 was increased after limited stretching of bone cells in vitro (Harell et al., 1977; Somjen et al., 1980).

Hasegawa and co-workers utilized a flexible-bottomed culture plate placed over a convex hemispherical dome to distend rat calvarial osteoblasts by 4% (Hasegawa et al., 1985). The authors indicated that tension was uniform over the entire surface of the culture dish; however, no direct evidence was shown to substantiate this claim. Results indicated that DNA synthesis was stimulated up to 64% above control levels in cells that were stretched continuously or intermittently every 10 min. Both leucine and proline incorporation into bone cells was increased in stretched cultures, but the increased proline incorporation could not be attributed to collagen.

Buckley and co-workers were the first to utilize a computerized system to regulate strain applied to bone cells grown in flexible bottomed, silicone rubber culture plates (Buckley et al., 1988). They subjected avian calvarial osteoblasts to a 3-cpm, 0–24% elongation strain regimen for up to 7 days. The cyclically strained bone cells responded to deformation by (1) aligning in a direction perpendicular to the applied strain in the maximum deformation region, (2) polymerizing actin, (3) increasing DNA synthesis during the first 72 h after initiating strain, and (4) increasing cell numbers over the same time periods. In a second report, the authors showed that alkaline phosphatase activity was also increased by cyclic strain (Buckley et al., 1990). Moreover, actin polymerization was stimulated by cyclic strain and synthesis of specific proteins was altered after 5 days of strain (Buckley et al., 1988, 1990).

Sandy and co-workers, using flexible-bottomed culture plates and the model of Hasegawa, subjected murine calvarial osteoblasts to intermittent or continuous strain and induced increased synthesis and/or release of bone resorptive factors (Sandy et al., 1989). The results were most positive when cells were cultured in 10% serum. However, both high- and low-molecular-weight bone-resorbing factors were produced by mechanically deformed cells. Fractionation of the media from stimulated cells and assay against ^{45}Ca-labeled bone revealed three fractions of activity: I, 50–60 kD; II, 5–20 kD; and III, < 1 kD, none of which were attributable to interleukin-1 (IL_1) or PGE_2.

E. Chondrocytes: Whole Cartilage, Epiphyseal, Sternal, Temperomandibular Joint

One of the early studies addressing the nature of mechanical forces on cartilage was performed by Caterson and Lowther (Caterson and Lowther, 1978). A sheep model (1–2 years-old, mixed breed and either sex) was used in which one foreleg was immobilized by casting with plaster and wire mesh so that the animal could not bear weight on the limb. Changes in GAGs in the metacarpal and metatarsophalangeal ankle joints were quantitated after 4 weeks. Load on the contralateral foreleg was increased to compensate for the weight reduction on the cast leg while load on the hind legs was unchanged. Pressure transducer measurements on all legs before casting indicated that load distribution was shared equally among the four legs. After 4 weeks, samples of articular cartilage from the loaded, non-loaded, and control ankle joints from each animal were isolated and incubated with radiolabeled acetate in vitro, and proteoglycans were extracted and separated by column chromatography. Hexuronic acid and hydroxyproline values for control cartilage from hind and forelimb joints of 1–2-year-old sheep were approximately equivalent, indicating that the mixed-breed and age groups selected were valid. Hexuronic acid and galactosamine levels were 1.56- and 1.5-fold greater in load-bearing joints compared with values for joints from immobilized sheep. However, glucosamine levels were 70% of the immobilized level; hydroxyproline values were unchanged. Moreover, an important observation was that the molecular weight of the proteoglycans from the non-load-bearing joints was half that of the loaded or hind-leg control cartilage. The authors concluded that articular cartilage has the capacity to respond to mechanical strain and motion and that load is important in maintenance of cartilage constituents under normal conditions.

Vailas and co-workers showed that rats exercised on a treatmill with a workload of 1.8 km/h on a 12% grade for 1 h/day, 5 days/week for 12 weeks significantly increased the hydroxyproline, uronic acid, and calcium contents of the posterior knee meniscus 1.26-, 1.44-, and 1.42-fold, respectively (Vailas et al., 1986). Values for the anterior meniscus were unchanged because the posterior meniscus receives greater load-bearing forces than the anterior portion during exercise.

Bourret and Rodan showed in cultured epiphyseal cartilage cells from the proliferative or hypertrophic zones of chick tibia, that only the proliferative zone cells responded to hydrostatic pressure of 60,000 dyn/cm² by decreasing cAMP production by 20%, decreasing adenyl cyclase activity and increasing intracellular calcium 2.5-fold by a calcium dependent mechanism (Bourret and Rodan, 1976). The authors postulated that like differentiating cells that entered the proliferation phase on reduction of adenyl cyclase activity and lowering the [cAMP], proliferation-zone chondrocytes may be stimulated by mechanical strain to lower [cAMP] and divide as part of the remodeling process.

Klein-Nulend and co-workers applied compression to sections of growth plate cartilage in vitro and found that mineralization was increased in response to both intermittent and continuous compressive forces (Klein-Nulend et al., 1986). The investigators calculated the physiologic load on a mouse metatarsal rediment by calculating the forces exerted by flexor and extensor muscles divided by the cross-sectional area of the bone. The calculated force was 120,000 dyn/cm². The force applied to the cartilaginous metatarsal rudiments was 132,000 dyn/cm² in a pressure vessel, where effects of dissolved gases were accounted for. Intermittent force at 0.3 Hz resulted in 3.3- and 2.4-fold increases in ^{45}Ca and ^{32}P incorporation into the acid-soluble mineral phase, respectively, after 5 days of culture. Continuous compressive force increased the latter values on day 5, 2.2- and 1.5-fold, respectively. The authors concluded that intermittent stimulation induced a greater mineralization response than did continuous strain and that a cellular receptor may exist to transduce the strain.

De Witt and co-workers were among the first to apply strain to chondrocytes cultured in vitro (De Witt et al., 1984). Rather than stretch a substrate on which cells were attached, the investigators grew high-density, matrix-rich cultures of embryonic chick epiphyseal chondrocytes for 14 days. The multilayered cultures were removed from their substrata intact and placed in a motor and cam-driven stretching device wherein the cell multilayer itself was clamped and cyclically stretched 5.5% at 0.2 Hz for up to 2 weeks. Results indicated that DNA synthesis was increased 2.4-fold in the first 24 h after the strain regimen was begun. However, protein synthesis was unaffected during this time. Incorporation of sulfate and glucosamine were stimulated into the cyclically loaded chondrocytes, but neither the size nor the loss of proteoglycans from the matrix were altered by strain. Intracellular cAMP concentration increased 2.2-fold in cyclically loaded cultures. Moreover, addition of the cAMP analog N^6-monobutyryl cAMP to the medium stimulated incorporation of sulfate and glucosamine into GAGs. The authors concluded that short-term loading of chondrocytes in vitro can alter their metabolism in a similar fashion to that in vivo.

Recently, Inoue and co-workers utilized centrifugal force to apply compression to chondrocytes maintained as packed masses in centrifuge tubes (Inoue et al., 1990). They reported that articular chondrocytes isolated from rabbit knee joint and growth plate chondrocytes isolated from rib showed differential responses to increased gravitational force from 1.3 to 27 g. Thymidine incorporation was stimulated 1.6-fold in articular chondrocytes from 4-week-old rabbits, whereas centrifugal force of up to 27 g for 24 h had little effect on growth plate chondrocytes. Conversely, sulfate incorporation into GAGs was increased in chondrocytes from growth plate, more so than that from articular chondrocytes. The investigators basically verified the work of De Witt and co-workers (De Witt et al., 1984) but could not supply cyclic load with the centrifugation system.

F. Ligament Cells: Periodontal Ligament and Medial Collateral Ligament

Tipton and co-workers used the medial collateral ligament in the canine knee to investigate the effects of exercise and immobilization on ligament healing (Tipton et al., 1970). In the wound model, the left medial collateral ligament (MCL) was incised through the superficial and deep portions of the joint, then ligament ends were approximated with two or three 4-0 silk sutures. Immobilized joints were prepared with Steinman pins and the limb cast in plaster. Exercise was provided to dogs via a motor-driven treadmill with controlled speed, grade, and duration for up to 6 weeks. Three days per week were devoted to endurance sessions, and 3 days were used for speed activities. After 3 weeks, animals were exercising 1 h/day, with strenuous exercise achieving heart rates of 270 beats/min, V_{O_2} maximum approaching 90 l/min and rectal temperatures of 40.5–42.2° C. Results indicated that immobilization of intact or incised MCL weakened the ligament (increased failure in tensile strength tests) and decreased the collagen control and that exercise strengthened the MCL and increased the collagen content (Tipton et al., 1970). Muscle mass was also increased in the exercised groups.

Tipton and co-workers were one of the initial groups who used specific training regimens to define the nature of controlled repetitive loading (training) effects on bone, ligament, tendon, and muscle (Tipton et al., 1975). They reported that a 10-week endurance training schedule resulted in significant changes in heart and organ weights, heart rate, serum cholesterol, adipocyte diameter, and enzymatic activity. The result of endurance training in rats or dogs resulted in development of a heavier, thicker, wider ligament. On the other hand, an exercise schedule of sprint training consisting of 2-min duration runs up a 20% grade with rest periods of 3 s for a total of up to 2 h/day did not increase the junction strength, although the weight and weight/length ratio of ligaments were markedly increased. Immobilization of a ligament without exercise resulted in a decrease in the width of fiber bundles of knee ligaments in dogs or in hypophyosectomized rats. The junction strength of the bone–ligament–bone preparation achieved normal values after 15–18 weeks (Tipton et al., 1975).

Noyes and co-workers later substantiated the thought that immobilization reduced the performance of ligaments in wild primates to 32–39% of the control exercise group (Noyes et al., 1974). The anterior cruciate ligament in rats also increased in resistance to failure after 8 weeks of endurance exercise (Cabaud et al., 1980). Vailas, working with Tipton, extended their earlier findings with exercise and ligament healing to the rat model (Vailas et al., 1981). Total collagen and tensile strength were increased 2.35- and 2.06-fold, respectively, in the right leg MCL in rats subjected to continuous running for 65 min at 70% of their maximum aerobic capacity for 8 weeks compared to values for immobilized groups. DNA and collagen synthesis were 51% and 61% of the immobilized group levels, similar to that of the control group. The authors concluded that exercise returned MCL to more normal limits of cell division and collagen synthesis more quickly than did immobilization.

Woo and co-workers investigated biochemical and morphologic alterations in the rabbit MCL in response to 9–12 weeks of immobilization followed by 9–52 weeks of remobilization (Woo et al., 1987). Immobilization significantly decreased MCL strength after 9–12 weeks. Although remobilization increased MCL strength, there were one third more tibial avulsions than in the group that was exercised but had not been immobilized. The authors concluded that during remobilization, mechanical integrity of the uninjured immobilized MCL returns rapidly to normal strength levels, but that the integrity of the ligament–bone junction requires a much longer time (months).

Brunette demonstrated that epithelial cells from the rest of Malassez in the periodontal ligament of the pig increased DNA synthesis 1.9-fold after application of strain for 2 h (calculated 4.2% increase in surface area with flexible-bottomed plates stretched by applying an orthodontic screw to the substratum similar to the model due to Harell) (Brunette, 1984; Harell et al., 1977). Cells increased DNA synthesis (detected autoradiographically) after 30 min of applied strain. Brunette also reported that electron-microscopic observations indicated that the number of desmosomes, total length of cell membrane, volume fraction of microtubules, and volume fraction of cytoplasmic filaments in stretched cells increased 1.4-, 1.8-, 1.7- and 1.3-fold respectively. The author concluded that the in vitro experiments substantiated the in vivo observation that orthodontic movement of teeth stimulates cell division in the epithelial cells in the PDL under tension, but not compression.

Sato and co-workers isolated human PDL fibroblasts and subjected them for 24 h to compression by applying two 7-mm sections of 15-mm glass tubing, each weighing 1.14 g, or by intermittent tension (60 min of tension followed by relaxation for up to 12 h) on a flexible substratum (Sato et al., 1989). Mechanical strain on PDL induced appearance of a 60-kD protein that was not apparent in control cells. DNA synthesis was markedly inhibited 18–28% in cells treated with conditioned medium from stretched cells, whereas alkaline phosphatase activity was markedly enhanced. The authors concluded that mechanical force may inhibit cell growth while stimulating differentiation.

Sutker and co-workers subjected rat MCL cells to cyclic strain on flexible bottomed plates at 10 cpm at a maximum of 13% elongation for supercycles of 12-h cyclic tension and 12-h rest (Sutker et al., 1990). Cell division was increased 2.7-fold in exercised cells and relative collagen synthesis increased 1.4-fold. The authors concluded that passive tension in vitro stimulated cell division and matrix synthesis in similar fashion to the trend observed in animal models in vivo.

Saito and co-workers recently reported that intermittent mechanical stress increased DNA synthesis in PDL cells (Saito et al., 1991).

G. Skin Cells: Dermal Fibroblasts, Keratinocytes

The effects of tension on skin are of interest since skin is subjected to biaxial tension during growth, but especially in areas of hyperelasticity such as the axial regions, joints, and neck. The latter areas are more likely to develop hypertrophic scar after wounding, especially injuries resulting from burns. Moreover, skin over the abdomen or the breasts is particularly prone to stretch during pregnancy or in the case of obesity. Clinically, skin expansion is performed during breast-augmentation mammoplasty or when skin expanders are used to create skin in an area destined to receive a skin flap.

Francis and Marks reported in 1977 that epidermopoiesis could be stimulated by skin expansion (Francis and Marks, 1977). They injected silicone foam elastomer or implanted gel-filled testicular prostheses under the flank skin and positioned them toward the ribs in guinea pigs or in the testicular sack. The thymidine-labeling index, as judged by autoradiography and skin thickness, was increased in both sites, but the testicular site had less of an inflammatory reaction. The investigators concluded that an initial stretch stimulus increases the epidermal germinative cell pool size, but that the stimulus effect diminishes, leaving a normal rate of division thereafter. They reasoned that epidermal division in the skin of obese subjects does not differ from that of normal patients.

Squier next showed that dorsal skin of hairless mice loaded with springs, developing a force of 100 g, had increased epidermopoiesis after 4 days in vivo (Squier, 1980). The force exerted decreased exponentially to 20 g as a result of compliance in the skin. The mitotic index at the dermal–epidermal interface for stretched and nonstretched skin was increased 1.7-, 1.5-, and 2.6-fold, respectively on days 1, 2, and 4 during application of strain. There was a twofold increase in skin thickness in the stretched group after 4 days. The authors concluded that stretch stimulated division in the progenitor population since mitotic figures were present in the three deepest cell layers.

Görmar and co-workers utilized electromagnets to apply pressure via a pestle device to cultured, transformed keratinocytes (Görmar et al., 1990). The weights covered the entire culture surface and placed a strain of 0.015 N/cm^2 at a frequency of 0.83 Hz with a 1-h rest every 2 h, yielding approximately 100 cycles every 2 h. The investigators indicated that DNA synthesis was diminished, unlike protein synthesis, which increased with particular increases in cytokeratins at 44, 49.5, and 67 kD in mechanically stimulated cells. Moreover, they reported that mechanically stimulated cultures were more differentiated with keratinized cells that resisted rounding during trypsinization. The latter characteristic is a sign of the differentiated state.

Henderson-Brown, working in Banes's laboratory, found that both human dermal fibroblasts and scar fibroblasts from the same patient responded to applied strain by aligning perpendicularly to the strain field and polymerizing actin in a period of hours (Henderson et al., 1988). She utilized a Flexercell Strain Unit to apply cyclic strain to cells grown on collagen substrata subjected to 0.05 Hz (10 s on, 10 s off) at 13.8% maximum elongation (37.6 mm Hg), the pressure used in JOBST pressure garments used to control scar formation. She reported that although both cell types aligned normal to the strain field, the scar cells did so more rapidly and were morphologically flatter, shorter cells with short microvilli, parallel with the strain field, protruding from the cell plasma membrane. In contrast, the normal human dermal Fibroblasts (NHDFs) became much thinner and longer with extremely fine lateral processes that intimately intertwined with adjacent cells. Moreover, DNA synthesis in the scar fibroblasts was diminished by applied strain by day 6 in low-density cultures when NHDF were still in linear growth. The latter data indicated that applied strain may down-regulate cell division in scar fibroblasts and that compression applied to healing skin that sustained a burn wound may act to normalize the strain field (Henderson, 1988).

H. Lung Cells: Type II Epithelial Cells

The lungs are organs that maintain a cyclic inflation–deflation regimen to transfer oxygen to and carbon dioxide from tissues. Lung surfactant is necessary for normal function of respiration. Wirtz and Dobbs recently showed that a single stretch of 25% in cellular surface area of rat type II epithelial cells of lung triggered a 60-s intracellular calcium burst and concomitant rise in surfactant secretion that persisted for 30 min. Phosphatidylcholine is the major component of surfactant, and type II lung cells are the major storage entities of the intracellular surfactant granules. A 16% increase in cellular surface area induced a 1.7-fold increase in phosphatidylcholine secretion over a 1-h period in culture compared with the level from nonstretched controls. The effect was not inhibited by indomethacin, indicating that prostaglandins did not play a role in the action. The authors concluded that a single stretch event, such as a yawn or sigh in vivo, evoked surfactant secretion necessary for alveolar patency.

I. Examples of Other Mechanically Active Tissues

Every tissue in the body is subjected to tension, compression, shear stress, or some combination of all three types of deformation. Some examples of tissues that have not been well studied from the standpoint of effects of mechanical activity in vitro include the following: the cornea and cells lining the eyelid; cells from oral cavity tissues, the tongue, lips, gingival tissue, and the esophagus; tissues of the gut, epithelial, and smooth-muscle cells; cells lining the peritoneal wall; tissues in the urogenital tract, including the ureters and urethra, vagina, uterus, and bladder; and cells of the immune system, such as macrophages and lymphocytes.

Investigators have studied many of the tissue systems that are mechanically active in vivo and have cultured cells to use in experiments in vitro. Therefore, cells from striated skeletal, cardiac, and aortic smooth muscle, tendon, ligament, bone, cartilage, skin, lung, and endothelium have been used in some experiments, generally to show some phenomenological effect. However, few studies have addressed actual mechanisms by which mechanical deformation acts on tissues or cells in vitro.

III. PROPOSED MECHANISMS OF ACTION OF MECHANICAL STRAIN, AND A UNIFIED THEORY RELATING MECHANICAL STRAIN TO CELLULAR RESPONSES

A. General Theories Relating Mechanical Strain and Biologic, Biochemical, and Cellular Responses

Wolff was the first investigator to theorize that mechanical strain was responsible for remodeling forces in bone (Wolff, 1892). He did not propose a specific mechanism by which strain acted but provided examples from the then-current solutions for engineering construction to explain the perpendicular construction of trabecular bone. His treatise was written much like that of Darwin's *Origin of the Species* in its scope and implication, but, of course, was directed at the role that mechanical force must play during bone development, maturation, and remodeling.

Other investigators, notably Bassett and Tipton, as well as many others recognized the importance of mechanical forces in the organization and reorganization of bone and other tissues. However, few of these investigators actually attempted to reduce observations to a testable equation or proposed a mechanism by which strain might act.

The simple experiment of Hong involving the chemical response to mechanical scraping of cultured cells was the first to involve a biochemical mechanism to strain (Hong et al., 1976). Hong and co-workers found that prostaglandin release occurred during mechanical removal of MC5-5 cells, a strain of methylcholanthrene transformed mouse BALB/3T3 cells, from a plastic culture substratum (Hong et al., 1976). There is little cell storage of prostaglandins; therefore, their detection of PGE_2 in culture medium was indicative of new synthesis. The investigators found that thrombin, bradykinin, serum, and mechanical perturbation by scraping, stimulated prostaglandin synthesis. Synthesis could be inhibited by indomethacin, a prostaglandin synthase inhibitor. They remarked that PGE_2 synthesis was highly substrate-dependent, and that addition of arachidonic acid was stimulatory. Phospholipase A_2 plays a regulatory role in PGE_2 synthesis. The investigators argued that chemical mediators and mechanical disruption could act by disrupting the cell membrane, exposing phospholipids to phospholipase hydrolysis, releasing arachidonic acid as a substrate for prostaglandin synthase. Alternatively, phospholipase A_2 might be activated directly by dynamic change in the plasma membrane. Stimulation of PGE_2 production was additive with chemical mediators and mechanical strain; therefore, different mechanisms or sites might be activated.

Rodan, in 1975 and 1976, using the chick tibia system or cells isolated from the proliferative or hypertrophic zones of epiphyseal cartilage; and a year later, Harell and co-workers, employing the in vitro osteoblast stretching model, invoked the phospholipase A_2-membrane perturbation theory postulated by Hong to explain decreased glucose utilization and increased PGE_2, cAMP, and Ca^{2+} secretion respectively (Rodan et al., 1975b; Bourret and Rodan, 1976; Harell et al., 1977). The experiment by Bourret and Rodan with cultured epiphyseal cells indicated that 60,000 dyn/cm^2 hydrostatic pressure decreased cAMP production 20% and decreased adenyl cyclase activity by a calcium-dependent mechanism (Bourret and Rodan, 1976). Results of the simple membrane-scraping experiment, in conjunction with those of the stimulation of PGE_2 by chemical mediators, were the first reports invoking a biochemical mechanism to explain how mechanical strain might invoke a cellular response. It had particular relevance to the finding that osteoblasts were stimulated to make PGE_2 within minutes after static tension was applied, since this was the same finding and molecular mechanism invoked by Hong.

In 1985, Harold Frost published a treatise on the responses of bone to mechanical strain (Frost, 1987). In this tome, he reasoned that bone would remodel when challenged with a change in applied strain. He termed this thesis the "Mechanostat Theory." Basically, he drew an analogy between a thermostat's design to detect a change in temperature settings and either heat or cool depending on how ambient temperature changes. Bone, he argued, could react to applied strain like a thermostat. It could accrete mineral when challenged with increased mechanical deformation, yet could lose mineral when inactive. Much like Wolff, he drew upon a lifetime of observation and experiments involving parathyroid hormone and bone metabolism to formulate his hypothesis.

Julius Wolff was the first to write of the association of mechanical forces in development and remodeling of bone. Beginning in 1973, Carter and Hayes have been recent proponents of this theory and have applied the mathematics of materials science to strain effects on bone (Carter and

Hayes, 1977a; 1977b; Carter et al., 1981). They have applied the principles of engineering materials science, including equations for beam theory, bending moments and shear stress, to bone, as a material, to predict how bone fails under load (Carter and Hayes, 1977b).

In 1948, Evans published a report on stresscoat deformation on statistically loaded femurs and was the first investigator to directly measure strain on bone from strain gauges placed on living bone (Evans and Lissner, 1948, 1953). Lanyon also measured bone strain analytically and with his co-workers has determined that a *minimum effective strain* is required to maintain bone density (Pead et al., 1988). As little as one deformation event in vivo, within the physiologic range may stimulate bone cells to divide (Pead et al., 1988). Hence, these latter groups applied principles of mechanics and materials to analyze the responses of bone in quantitative ways.

Diamond, Eskin, and co-workers indicated that with endothelial cells, secretion of tissue plasminogen activator increased linearly from 5 to 25 dyn/cm^2 flow-induced shear stress (Diamond et al., 1989). Upchurch, Banes, and co-workers have shown that endothelial cell division may be regulated in threshold manner by strain (Upchurch et al., 1990). Strain regimens of 0.18, 0.24 or 0.27 maximum strain with 1 Hz cycling increased the rate of cell division in vitro in aortic endothelial cells. In thesis work, Wright and co-workers showed that when cardiac and striated skeletal muscle cells were subjected to two levels of cyclic strain, a 20% increase in the degree to which the cells were stretched (tension term) caused an increase in the rate of alignment of 3.2-fold (Wright, 1990). The latter data indicated that the tension term is most likely a log function response for the cells. In support of the latter thought, Sachs and Guharay demonstrated the existence of a stretch-activated membrane channel present in most cells, including bacteria (Guharay and Sachs, 1984, 1985; Sachs, 1987, 1988). They showed that increasing negative or positive pressure on a $3\text{-}\mu\text{m}$-diameter patch of membrane induced a current output that is a square function of the applied strain.

These and other data support the hypothesis that cells respond to applied strain in a reproducible manner.

Banes and co-workers have formulated a mathematical expression that relates the mechanical factors that impact on cells to a given cellular response (Banes et al., 1986, 1990). The expression relates the response of the cell R as a summation of a nonlinear function of the amplitude of the applied strain A; t_1, the duration of A; t_2, the time between alternate deformation events; t_3, the duration of a rest period after many deformations; C, the number of cycles; $*e_1$, the strain rate ascending to maximum A; $*e_2$, the strain rate descending from maximum A; T, shear stress if flow is present; s, a substrate chemistry term; and n, a term indicating that the process can be repeated. The expression is a collection of terms represented as

$$R = [A, (t_1, t_2, t_3), C, (^*e_1, ^*e_2, T), s]_n \qquad (2)$$

Although a simple collection of terms, the expression permits isolation of a given mechanical factor such as tension (A term), to discern its role in altering cell metabolism. The response R of the cell may be the rate of cell migration, ion transport, second-messenger production, RNA or DNA synthesis, specific protein production, all or part of which may be affected by a particular strain regimen delineated by the expression.

B. Chemical Mediators of Mechanical Strain

To date, no one has elucidated the mediator cascade involved in the cellular responses to applied mechanical strain, but evidence suggests that the pathways may be diverse. Hong, Rodan, and Harell, separately, initiated the field using isolated 3T3 cells, chondrocytes and osteoblasts outlining a biochemical response system involving PGE_2 alone, or PGE_2, cAMP, and Ca^{2+} linked to cell division as a final response, respectively; Frangos demonstrated IP_3 secretion from fluid shear stressed endothelial cells (Frangos et al., 1985); and Sumpio, Upchurch, and Banes demonstrated prostacyclin secretion stimulated by strain (Sumpio and Banes, 1988a; Upchurch et al., 1989). Strain may be transduced via a stretch-activated channel utilizing cytoplasmic filaments as signal amplifiers but may involve G proteins close to the channels to activate a second-messenger system that may have diverse effects depending on the type and biochemical state of the cell, strain history (environment or strain regimen), and the hierarchy of signals sent to a target cell. Strain signals superimposed on a cell that has been exposed to insulin or platelet-derived growth factor (PDGF) may have little detectable mitogenic effect in the face of a massive PDGF-induced response to divide. Alternatively, in cells that respond to strain by retarding division, a growth factor may supersede the initial response. Few of the latter scenarios have been tested in the laboratory yet but may be involved in mechanisms explaining how development proceeds, growth and remodeling occur or how wounds heal in bone, ligament, tendons, skin, bowel, lung, heart, the vascular system, and other mechanically active tissues. Moreover, the mechanism of cardiac muscle hypertrophy may be elucidated using applied strain models.

Possible mechanisms of action of mechanical deformation on cells have been summarized in Figure 1. A mechanical stimulus originating from the environment or from cell contraction itself (muscle contraction, for

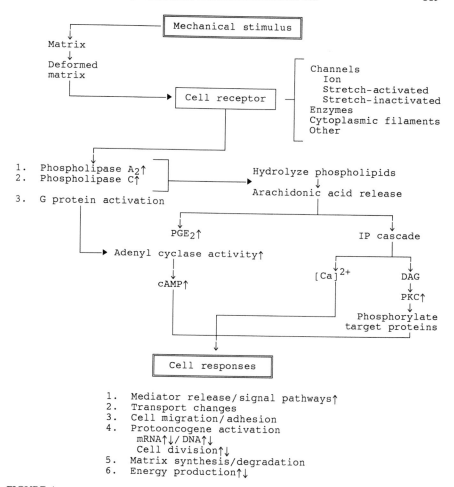

FIGURE 1   Diagram of possible mechanisms of action of mechanical deformation on cells.

instance) acts either on (1) matrix, which then impacts on a cell, or (2) directly on a cell. The cell strain receptor may be the proposed stretch-activated receptor discovered by Guharay and Sachs, or some other membrane protein, such as phosphorylase, as suggested initially by Hong and substantiated by Binderman and more recently Komuro (Guharay and Sachs, 1984, 1985; Harell et al., 1977; Binderman et al., 1988; Komuro et al., 1991).

A phospholipase may be activated (PLA_2 or PLC) and may be coupled with a G protein. A G protein could be directly coupled to adenyl cyclase to increase cAMP activity, or phospholipase action could release phospho-lipids, including arachidonic acid. Arachidonic acid could then enter the PGE_2 pathway as seems likely in osteoblasts (Binderman et al., 1988) or follow the inositol phosphate cascade (Harell et al., 1977; Somjen et al., 1980; Binderman et al., 1988) as appears likely for endothelial cells (Frangos et al., 1985) and cardiac myocytes (Komuro et al., 1991). If adenyl cyclase is activated, then [cAMP] is increased, whereas if the inositol phosphate pathway is activated, $[Ca]^{2+}$ directly, or diacylglycerol through protein kinase C may mediate a host of cell responses, including cell migration, adhesion, up- or down-regulation of division, protooncogene activation, matrix synthesis or degradation, or perturbation of energy-generating pathways. Alterations may occur in milliseconds to seconds: ion flux, stretch-activated channel activity, membrane depolarization, second-messenger activity, minutes to hours: second-messenger activity, cytoplas-mic filament polymerization, protooncogene induction, mRNA up- and down-regulation, protein synthesis, DNA synthesis, enzyme secretion, factor production, cell movement, division; or require days: matrix produc-tion–degradation, cell division, cell migration, production of strain-oriented matrix. Such responses may be dissimilar for cells from different tissues. Different cells may require different strain regimens to induce a response, and cells may respond utilizing different mediator pathways.

Data accumulating in this new field of cell biomechanics indicate that cells respond in vitro to applied strain. Cells that originate in the body in highly active tissues such as muscle, bone, tendon, ligament, lung, and aorta should be studied in vitro under conditions that most closely simulate living conditions. Supplying a mechanically active environment in vitro is another variable we must now all consider.

ACKNOWLEDGMENTS

Thanks to Cathy Ward for preparation of the manuscript and to Drs. G. W. Link, Jr. and Jerry Gilbert for their helpful discussions and assistance with preparation of the manuscript.

REFERENCES

Amiel, D., Foulk, R., Harwood, F., and Akeson, W. (1989). Quantitative assessment by competitive ELISA of fibronectin (Fn) in tendons and ligaments. Matrix 9, 421–427.
Anderson, J., and Eriksson, C. (1968). Electrical properties of wet collagen. Nature 218, 166–168.
Anderson, J., and Eriksson, C. (1970). Piezoelectric properties of dry and wet bone. Nature 227, 491–492.

Ashman, R., Cowin, S., Van Buskirk, W., and Rice, J. (1984). A continuous wave technique for the measurement of elastic properties of cortical bone. *J. Biomech.* **17**, 349–361.

Banes, A., Gilbert, J., Taylor, D. and Monbureau, O. (1985). A new vacuum-operated stress-providing instrument that applies static or variable duration cyclic tension or compression to cells *in vitro*. *J. Cell Sci.* **75**, 35–42.

Banes, A., Sumpio, B., Levin, L., and Buckley, M. (1986). Regulatory aspects of physical deformation on cultured cells: A general theory. *Proc. Am. Soc. Bone Min. Res.* **2**, Supplement 1, 418.

Banes, A., Dolon, K., Link, G., Gillespie, Y., Bevin, A., Peterson, H., Bynum, D., Watts, S., and Dahners, L. (1988a). A simplified method for isolation of tendon synovial cells and internal fibroblasts: Conformation of origin and biological properties. *J. Orthop. Res.* **6**, 83–94.

Banes, A., Link, G., Bevin, A., Peterson, H., Gillespie, Y., Bynum, D., Watts, S., and Dahners, L. (1988b). Tendon synovial cells secrete fibronectin *in vivo* and *in vitro*. *J. Orthop. Res.* **6**, 73–82.

Banes, A. J., Link, G. W., Gilbert, J. W., Tay, R. T. S., and Monbureau, O. (1990). Culturing cells in a mechanically active environment. *Am. Biotechnol. Lab.* **8**, 12–22.

Barnes, W. S. (1987). Respiration and lactate production in isolated frog skeletal muscle: Effects of passive stretch. *Comp. Biochem. Physiol.* **86A**, 229–232.

Barnes, W. S., and Worrell, G. A. (1985). Glycogen utilization in isolated frog muscle: An effect of passive mechanical stretch. *Comp. Biochem. Physiol.* **81A**, 243–246.

Bassett, C., and Becker, R. (1962). Generation of electrical potentials in bone in response to mechanical stress. *Science* **137**, 1063–1064.

Binderman, I., Zor, U., Kaye, A. M., Shimshoni, Z., Harell, A., and Somjen, D. (1988). The transduction of mechanical force into biochemical events in bone cells may involve activation of phospholipase A_2. *Calcif. Tis. Int.* **42**, 261–266.

Binkhorst, R. (1969). The effect of training on some isometric contraction characteristics of a fast muscle. *Pfluegers Arch.* **309**, 193.

Booth, F., and Gould, E. (1975). Effects of training and disuse on connective tissue. *Exercise Sports Sci. Rev.* **3**, 83–112.

Bourret, L. A., and Rodan, G. (1976). The role of calcium in the inhibition of cAMP accumulation in epiphyseal cartilage cells exposed to physiological pressure. *J. Cell. Physiol.* **88**, 353–362.

Brunette, D. M. (1984). Mechanical stretching increases the number of epithelial cells synthesizing DNA in culture. *J. Cell Sci.* **69**, 35–45.

Buckley, M. J., Banes, A. J., Levin, L. G., Sumpio, B. E., Sato, M., Jordan, B., Gilbert, J., Link, G. W., and Tay, R. T. S. (1988). Osteoblasts increase their rate of division and align in response to cyclic, mechanical tension *in vitro*. *Bone Miner.* **4**, 225–236.

Buckley, M. J., Banes, A. J., and Jordan, R. (1990). Effects of mechanical strain on osteoblasts *in vitro*. *J. Oral Maxofac. Surg.* **48**, 276–282.

Burr, D., Martin, B., Schaffler, M., and Radin, E. (1985). Bone remodeling in response to *in vivo* fatigue microdamage. *J. Biomech.* **18**, 189–200.

Burridge, K. (1986). Substrate adhesions in normal and transformed fibroblasts: Organization and regulation of cytoskeletal, membrane and extracellular matrix components at focal contacts. *Cancer Rev.* **4**, 18–78.

Cabaud, H. E., Chatty, A., Gildengorin, V., and Feltman, R. J. (1980). Exercise effects on the strength of the rat anterior cruciate ligament. *Am. J. Sports Med.* **8**, 79–86.

Carter, D. (1982). The relationship between *in vivo* strains and cortical bone remodeling. *CRC Crit. Rev. Biomed. Eng.* **8**, 1–28.

Carter, D., and Hayes, W. (1977a). Compact bone fatigue damage. A microscopic examination. *Clin. Orthop.* **127**, 265–274.

Carter, D., and Hayes, W. (1977b). Compact bone fatigue damage. I. Residual strength and stiffness. *J. Biomech.* **10**, 325–337.

Carter, D., Smith, D., Spengler, D., Daly, C., and Frankel, V. (1980). Measurement and analysis of *in vivo* bone strains on the canine radius and ulna. *J. Biomech.* **13**, 27–38.

Carter, D., Caler, W., Spengler, D. and Frankel, V. (1981). Fatigue behavior of adult cortical bone: the influence of mean strain and strain range. *Acta Orthop. Scand.* **52**, 481–490.

Caterson, B., and Lowther, D. A. (1978). Changes in the metabolism of the proteoclycans from sheep articular cartilage in response to mechanical stress. *Biochim. Biophys. Acta* **540**, 412–422.

Dartsch, P. C., and Hammerle, H. (1986). Orientation response of arterial smooth muscle cells to mechanical stimulation. *Eur. J. Cell Biol.* **41**, 339–346.

Davidovitch, Z., Montgomery, P. C., Eckerdal, O., and Gustafson, G. T. (1976). Cellular localization of cyclic AMP in periodontal tissues during experimental tooth movement in cats. *Calcif. Tiss. Res.* **19**, 317–329.

De Witt, M. T., Handley, C. J., Oakes, B. W., and Lowther, D. A. (1984). *In vitro* response of chondrocytes to mechanical loading. The effect of short term mechanical tension. *Conn. Tiss. Res.* **12**, 97–109.

Diamond, S. L., Eskin, S. G., and McIntire, L. V. (1989). Fluid flow stimulates tissue plasminogen activator secretion by cultured human endothelial cells. *Science* **243**, 1483–1485.

Dowell, R. T., Haithcoat, J. L., and Hasser, E. M. (1983). Metabolic enzyme response in the pressure-overloaded heart of weanling and adult rates. *Proc. Soc. Exp. Biol. Med.* **174**, 368–376.

Dustin, P. (1980). Microtubules. *Sci. Am.* **243**, 66–76.

Evans, F. G. (1953). Methods of studying the biomechanical significance of bone form. *Am. J. Phys. Anthropol.* **11**, 413–416.

Evans, F. G., and Lissner, H. (1948). "Stresscoat" deformation studies of the femur under static vertical loading. *Anat. Rec.* **100**, 159–190.

Fakuda, E., and Yasuda, I. (1957). On the piezoelectric effect of bone. *J. Phys. Soc. Jpn.* **12**, 1158–1169.

Feng, T. P. (1932). The effect of length on the resting metabolism of muscle. *J. Physiol.* **74**, 441–454.

Forwood, M., and Parker, A. (1986). Effects of exercise on bone morphology. *Acta Orthop. Scand.* **57**, 204–208.

Forwood, M., and Parker, A. (1987). Effects of exercise on bone growth: Mechanical and physical properties studied in the rat. *Clin. Biomech.* **2**, 185–190.

Francis, A. J., and Marks, R. (1977). Skin stretching and epidermopoiesis. *Br. J. Exp. Pathol.* **58**, 35.

Frangos, J. A., Eskin, S. G., McIntire, L. V., and Ives, C. L. (1985). Flow effects on prostacyclin production by cultured human endothelial cells. *Science* **227**, 1477–1479.

Frost, H. M. (1987). The mechanostat: A proposed pathogenic mechanism of osteoporoses and the bone mass effects of mechanical and nonmechanical agents. *Bone Miner.* **2**, 73–85.

Gilbert, J. A., Banes, A. J., Link, G. W., and Jones, G. L. (1989). Surface strain on living cells in a mechanically active *in vitro* environment. *ANSYS Conf. Proc.*, Dietrich, D. E. (ed.), pp. 13.2–13.7. Swanson Analysis Systems, Houston, PA.

Gillard, G. C., Reilly, H. C., Bell-Booth, P. G., and Flint, M. H. (1979). The influence of mechanical forces on the glycosaminoglycan content of the rabbit flexor digitorum profundus tendon. *Conn. Tiss. Res.* 7, 37–46.

Goldberg, A., and Goodman, H. (1969). Amino acid transport during work-induced growth of skeletal muscle. *Am. J. Physiol.* 216, 1111.

Gorfein, S. F., Winston, F. K., Thibault, L. E., and Macarak, E. J. (1989) Effects of biaxial deformation on pulmonary artery endothelial cells. *J. Cell. Physiol.* 139, 492–500.

Görmar, F. E., Bernd, A., Bereiter-Hahn, J., and Holzmann, H. (1990). A new model of epidermal differentiation: Induction by mechanical stimulation. *Arch. Dermatol. Res.* 282, 22–32.

Gross, D., and Williams, W. (1982). Streaming potentials and the electromechanical response of physiologically-moist bone. *J. Biomech.* 15, 277–295.

Guharay, F., and Sachs, F. (1984). Stretch activated single ion-channel currents in tissue-cultured embryonic chick skeletal muscle. *J. Physiol.* 352, 685–701.

Guharay, F., and Sachs, F. (1985). Mechanotransducer ion channels in chick skeletal muscle: The effects of extracellular pH. *J. Physiol.* 363, 119–134.

Guzelsu, N., and Walsh, W. (1990). Streaming potential of intact wet bone. *J. Biomech.* 23, 673–685.

Harell, A., Dekel, S., and Binderman, I. (1977). Biochemical effect of mechanical stress on cultured bone cells. *Calcif. Tiss. Res.* (Suppl.) 22, 202–209.

Hasegawa, S., Sato, S., Saito, S., Suzuki, Y., and Brunette, D. M. (1985). Mechanical stretching increases the number of cultured bone cells synthesizing DNA and alters their pattern of protein synthesis. *Calcif. Tiss. Int.* 37, 431–436.

Henderson, R., Banes, A., Solomon, G., Lawrence, W., and Peterson, H. (1988). Human scar fibroblasts react to applied tension *in vitro* by aligning and increasing polymerized actin content. *FASEB J.* 2, A574.

Henderson, R. (1988). Effect of strain on human dermal fibroblasts. Thesis, University of North Carolina at Chapel Hill.

Hong, S. L., Polsky-Cynkin, R., and Levine, L. (1976). Stimulation of prostaglandin biosynthesis by vasoactive substances in methylcholanthrene-transformed mouse BALB/3T3. *J. Biol. Chem.* 251, 776–780.

Inoue, H., Hiasa, Y. S., Nakamura, O., Sakuda, M., Iwamoto, M., Suzuki, F., and Kato, Y. (1990). Stimulation of proteoglycan and DNA syntheses in chondrocytes by centrifugation. *J. Dent. Res.* 69, 1560–1563.

Ives, C. L., Eskin, S. G., and McIntire, L. V. (1986). Mechanical effects on endothelial cell morphology: *In vitro* assessment. *In Vitro Cell. Devel. Biol.* 22, 500–507.

Kiiskinen, A., and Heikkinen, E. (1973a). Effects of physical training on development and strength of tendons and bones in growing mice. *Scand. J. Clin. Lab. Invest.* 29 (Suppl. 123), 20.

Kiiskinen, A., and Heikkinen, E. (1973b). Effect of prolonged physical training on the development of connective tissues in growing mice. *Proc. 2nd Int. Symp. Exercise Biochem., Abstract* 25.

Klein-Nulend, J., Veldhuijzen, J. P., and Burger, E. H. (1986). Increased calcification of growth plate cartilage as a result of compressive force *in vitro*. *Arthrit. Rheum.* 29, 1002–1009.

Komuro, I., Kaida, T., Shibazaki, Y., Kurabayashi, M., Katoh, Y., Hoh, E., Takaku, F., and Yazaki, Y. (1990). Stretching cardiac myocytes stimulates protooncogene expression. *J. Biol. Chem.* 265, 3595–3598.

Komuro, I., Katoh, Y., Kaida, T., Shibazaki, Y., Kurabayashi, E. H., Takaku, F., and Yazaki, Y. (1991). Mechanical loading stimulates cell hypertrophy and specific gene expression in cultured rat cardiac myocytes. *J. Biol. Chem.* 266, 1265–1268.

Lane, N., Bloch, D., Jones, H., Marshall, W., Wood, P., and Fries, J. (1986). Long-distance running, bone density, and osteoarthritis. *JAMA* 255, 1147–1151.

Lanyon, L. (1987). Functional strain in bone tissue as an objective and controlling stimulus for adaptive bone remodelling. *J. Biomech.* 20, 1083–1093.

Leung, D. Y. M., Glagov, S., and Mathews, M. B. (1976). Cyclic stretching stimulates synthesis of matrix components by arterial smooth muscle cells *in vitro*. *Science* 191, 475–477.

Leung, D. Y. M., Glagov, S., and Mathews, M. B. (1977). A new *in vitro* system for studying cell response to mechanical stimulation. *Exp. Cell. Res.* 109, 285–298.

Mann, D., Kent, R., and Cooper, I. (1989). Load regulation of the properties of adult feline cardiocytes: Growth induction by cellular deformation. *Circ. Res.* 66, 1079–1090.

Martin, R., and Burr, D. (1979). *Structure, Function, and Adaptation of Compact Bone.* Raven Press, New York.

Matheson, G., Clement, D., McKenzie, D., Taunton, J., Lloyd-Smith, D., and MacIntyre, J. (1987). Stress fractures in athletes. *Am. J. Sports Med.* 15, 46–58.

Meikle, M. C., Reynolds, J. J., Sellers, A., and Dingle, J. T. (1979). Rabbit cranial sutures *in vitro*: A new experimental model for studying the response of fibrous joints in mechanical stress. *Calcif. Tiss. Int.* 28, 137–144.

Meikle, M. C., Sellers, A., and Reynolds, J. J. (1980). Effect of tensile mechanical stress on the synthesis of metalloproteinases by rabbit coronal sutures *in vitro*. *Calcif. Tiss. Int.* 30, 77–82.

Merrilees, M. J., and Flint, M. H. (1980). Ultrastructural study of tension and pressure zones in a rabbit flexor tendon. *Am. J. Anat.* 157, 87–106.

Noyes, F. R., Torvik, P. J., Hyde, W. B., and DeLucas, J. L. (1974). Biomechanics of ligament failure. II. An analysis of immobilization, exercise, and reconditioning effects in primates. *J. Bone Joint Surg.* 56A, 1406–1418.

Oakes, B. W., and Bialkower, B. (1977). Biomechanical and ultrastructural studies on the elastic wing tendon from the domestic fowl. *J. Anat.* 123, 369–387.

Oakes, B. W., Handley, C. J., Lisner, F., and Lowther, D. A. (1977). An ultrastructural and biochemical study of high density primary cultures of embryonic chick chondrocytes. *J. Embryol. Exp. Morphol.* 38, 239–263.

Palmer, R. M., Reeds, P. J., Lobley, G. E., and Smith, R. H. (1981). The effect of intermittent changes in tension on protein and collagen synthesis in isolated rabbit muscles. *Biochem. J.* 198, 491–498.

Palmer, R. M., Reeds, P. J., Atkinson, T., and Smith, R. H. (1983). The influence of changes in tension on protein synthesis and prostaglandin release in isolated rabbit muscles. *Biochem. J.* 214, 1011–1014.

Pead, M. J., Skerry, T. M., and Lanyon, L. E. (1988). Direct transformation from quiescence to bone formation in the adult periosteum following a single brief period of bone loading. *J. Bone Miner. Res.* 3, 647–656.

Pienkowski, D., and Pollack, S. (1983). The origin of stress-generated potentials in fluid-saturated bone. *J. Orthop. Res.* 1, 30–41.

Pollack, G., and Sugi, H. (eds.) (1984). *Contractile Mechanisms in Muscle.* Plenum Press, New York.

Price-Jones, C. (1927). The effects of exercise on the growth of white rats. *Quart. J. Exp. Physiol.* 16, 61–67.

Reitsma, W. (1969). Skeletal muscle hypertrophy after heavy exercise in rats with surgically reduced muscle function. *Am. J. Phys. Med.* **48**, 237.

Rodan, G. A., Mensi, T., and Harvey, A. (1975a). A quantitative method for the application of compressive forces to bone in tissue culture. *Calcif. Tiss. Res.* **18**, 125–131.

Rodan, G., Bourret, L., Harvey, A., and Mensi, T. (1975b). Cyclic AMP and cyclic GMP: Mediators of the mechanical effects on bone remodeling. *Science* **189** 467–469.

Rubin, C., and Lanyon, L. (1985). Regulation of bone mass by mechanical strain magnitude. *Calcif. Tiss. Int.* **37**, 411–417.

Rubin, C., Pratt, G., Porter, A., Lanyon, L., and Poss, R. (1987). The use of ultrasound *in vivo* to determine acute change in the mechanical properties of bone following intense physical activity. *J. Biochem.* **20**, 723–727.

Sachs, F. (1987). Baroreceptor mechanisms at the cellular level. *Fed. Proc.* **46**, 12–16.

Sachs, F. (1988). Mechanical transduction in biological systems. *CRC Crit. Rev. Biomed. Eng.* **16**, 141–169.

Saito, S., Ngan, P., Rosol, I., Saito, M., Shimizu, H., Shinjo, N., Shanfeld, J., and Davidovitch, Z. (1991). Involvement of PGE synthesis in the effect of intermittent pressure and interleukin-1β on bone resorption. *J. Dent. Res.* **70**, 27–33.

Sandy, J. R., Meghji, S., Scutt, A. M., Harvey, W., Harris, M., and Meikle, M. C. (1989). Murine osteoblasts release bone-resorbing factors of high and low molecular weights: Stimulation by mechanical deformation. *Bone Miner.* **5**, 155–168.

Sato, S., Endo, N., Yamauchi, M., Takeuchi, M., Kawase, T., and Saito, S. (1989). Effects of compression and stretching forces on the activity of alkaline phosphatase in the bone and periodontium. In *Mechanobiological Research on the Masticatory System*, pp. 221–226, K. Kubota (ed.). VEB Verlage für Medizin und Biolgie, Berlin.

Saville, P., and Whyte, M. (1969). Muscle and bone hypertrophy. *Clin. Orthop.* **65**, 81–88.

Scott, J. E., and Hughes, E. W. (1985). Proteoglycan-collagen relationships in developing chick and bovine tendons. Influence of the physiological environment. *Conn. Tiss. Res.* **14**, 267–278.

Scott, G., and Korostoff, E. (1990). Oscillatory and step response electromechanical phenomena in human and bovine bone. *J. Biomech.* **23**, 127–143.

Shirinsky, V. P., Antonov, A. S., Birukov, K. G., Sobolevsky, A. V., Romanov, Y. A., Kabaeva, N. V., Antonova, G. N., and Smirnov, V. N. (1989). Mechano-chemical control of human endothelium orientation and size. *J. Cell Biol.* **109**, 331–339.

Skerry, T., Pead, M., Suswillo, R., Vedi, S., and Lanyon, L. (1988). Strain-related remodeling in bone tissue: Early stages of the cellular response to bone loading *in vivo*. *Trans. Orthop. Res. Soc.* **13**, 97.

Somjen, D., Binderman, I., Berger, E., and Harell, A. (1980). Bone remodeling induced by physical stress is prostaglandin E₂ mediated. *Biochim. Biophys. Acta* **627**, 91–100.

Sorrentino, V., Drozdoff, V., McKinney, M. D., Zeitz, L., and Fleissner, E. (1986). Potentiation of growth factor activity by exogenous c-*myc* expression. *Proc. Natl. Acad. Sci. USA* **83**, 8167–8171.

Sottiurai, V., Kollros, P., Glagov, S., Zarins, C., and Mathews, M. (1983). Morphologic alteration of cultured arterial smooth muscle cells by cyclic stretching. *J. Surg. Res.* **35**, 490–497.

Squier, C. A. (1980). The stretching of mouse skin *in vivo*: Effect on epidermal proliferation and thickness. *J. Invest. Dermatol.* **74**, 68–71.

Stopak, D., and Harris, A. (1982). Connective tissue morphogenesis by fibroblast traction. *Devel. Biol.* **90**, 383–398.

Summers, P. J., Ashmore, C. R., Lee, Y. B., and Ellis, S. (1985). Stretch-induced growth in chicken wing muscles: role of soluble growth-promoting factors. *J. Cell. Physiol.* 125, 288–294.

Sumpio, B. E., and Banes, A. J. (1988a). Prostacyclin synthetic activity in cultured aortic endothelial cells undergoing cyclic mechanical deformation. *Surgery* 104, 383–389.

Sumpio, B. E., and Banes, A. J. (1988b). Response of porcine aortic smooth muscle cells to cyclic tensional deformation in culture. *J. Surg. Res.* 44, 696–701.

Sumpio, B. E., Banes, A. J., Levin, L. G., and Johnson, G. (1987). Mechanical stress stimulates aortic endothelial cells to proliferate. *J. Vasc. Surg.* 6, 252–256.

Sumpio, B. E., Banes, A. J., Buckley, M., and Johnson, G. (1988a). Alterations in aortic endothelial cell morphology and cytoskeletal protein synthesis during cyclic tensional deformation. *J. Vasc. Surg.* 7, 130–138.

Sumpio, B. E., Banes, A. J., Link, G. W., and Johnson, G. (1988b). Enhanced collagen production by smooth muscle cells during repetitive mechanical stretching. *Arch. Surg.* 123, 1213–1266.

Sumpio, B. E., Banes, A. J., Link, W., and Iba, T. (1990). Modulation of endothelial cell phenotype by cyclic stretch: Inhibition of collagen production. *J. Surg. Res.* 48, 415–420.

Sutker, B. D., Lester, G. E., Banes, A. J., and Dahners, L. E. (1990). *J. Bone Joint Surg.* 14, 35–36.

Swain, J. L., Stewart, T. A., and Leder, P. (1987). Parental legacy determines methylation and expression of an autosomal transgene: A molecular mechanism for parental imprinting. *Cell* 50, 719–727.

Terracio, L., Miller, B., and Borg, T. K. (1988). Effects of cyclic mechanical stimulation of the cellular components of the heart: In vitro. *In Vitro Cell. Develop. Biol.* 24, 53–57.

Tidball, J. G. (1986). Energy stored and dissipated in skeletal muscle basement membranes during sinusoidal oscillations. *Biophys. J.* 50, 1127–1138.

Tipton, C. M., James, S. L., Mergner, W., and Tcheng, T.-K. (1970). Influence of exercise on strength of medial collateral knee ligaments of dogs. *Am. J. Physiol.* 218, 894–902.

Tipton, C., Matthes, R., and Maynard, J. (1972). Influence of chronic exercise on rat bones. *Med. Sci. Sports Exercise* 4, 55.

Tipton, C. M., Matthes, R. D., Maynard, J. A., and Carey, R. A. (1975). The influence of physical activity on ligaments and tendons. *Med. and Sci. in Sports* 7, 165–175.

Upchurch, G. R., Banes, A. J., Wagner, W. H., Ramadan, F., Link, G. W., Henderson, R. H., and Johnson, G. (1989). Differences in secretion of prostacyclin by venous and arterial endothelial cells grown *in vitro* in a static versus a mechanically active environment. *J. Vasc. Surg.* 10, 292–298.

Upchurch, G., Ramadan, F., Solomon, G., S. Klemmer, G. Link, Johnson, G., and Banes, A. (1990). Cyclic strain alters endothelial cell DNA synthesis and division in a threshold manner. *FASEB J* 4, A911, 3743.

Vailas, A. C., Tipton, C. M., Matthes, R. D., and Gart, M. (1981). Physical activity and its influence on the repair process of medial collateral ligaments. *Conn. Tiss. Res.* 9, 25–31.

Vailas, A., Zernicke, R., Matsuda, J., Curwin, S., and Durivage, J. (1986). Adaptation of rat knee meniscus to prolong exercise. *J. Appl. Physiol.* 60, 1031–1034.

Vandenburgh, H. (1987). Motion into mass: How does tension stimulate muscle growth? *Med. and Sci. in Sports and Exercise* 19, 5142–5149.

Vandenburgh, H. (1988). A computerized mechanical cell stimulator for tissue culture: Effects of skeletal muscle organogenesis. *In Vitro Cell. Develop. Biol.* 24, 609–618.

Vandenburgh, H., and Kaufman, S. (1979). *In vitro* model for stretch-induced hypertrophy of skeletal muscle. *Science* 203, 265–268.

Vogel, K., and Evanko, S. P. (1988). Proteoglycans of fetal bovine tendon. *Trans. Orthop. Res. Soc.* 13, 182.

Wirtz, H., and Dobbs, L. (1990). Calcium mobilization and exocytosis after on mechanical stretch of lung epithelial cells. *Science* 250, 1266–1269.

Wolff, J. (1892). *The Law of Bone Remodelling* (transl. Maquet, P., and Furlong, R.), pp. 1–126. Springer-Verlag, New York.

Woo, S. L.-Y., Gomez, M. S., Sites, T. J., Newton, P. O., Orlando, C. A., and Akeson, W. H. (1987). The biomechanical and morphological changes in the medial collateral ligament of the rabbit after immobilization and remobilization. *J. Bone Joint Surg.* 69A, 1200–1211.

Wright, E., MacMurray, R., and Banes, A. (1988). Alignment rates of skeletal myocytes subjected to cyclic stretch *in vitro*. *J. Cell Biol.* 107, 453a.

Wright, E. E. (1990). Responses of isolated avian striated skeletal and cardiac myocytes subjected to cyclic tension *in vitro*: A possible model for training cultured myocytes. A thesis research proposal, University of North Carolina at Chapel Hill.

Yeh, C. K., and Rodan, G. A. (1984). Tensile forces enhance prostaglandin E synthesis in osteoblastic cells grown on collagen ribbons. *Calcif. Tiss. Int.* 36, S67–S71.

Young, E., and Sparks, H. (1980). Prostaglandins and exercise hyperemia of dog skeletal muscles. *Am. J. Physiol.* 238, H191–H195.

CHAPTER 4

Hemodynamic Forces in Relation to Mechanosensitive Ion Channels in Endothelial Cells

■ ■ ■ ■ ■

Peter F. Davies and Randal O. Dull

I. INTRODUCTION

The detection and transduction of mechanical forces is an important set of mechanisms for the physiological control of locomotion and balance, hearing, respiration, gastrointestinal motility, urinary bladder function, tactile sensation, and regulation of the cardiovascular system. With respect to the last, cardiac contractility (and hence cardiac output) is dependent on myocardial stretch, while arterial blood pressure is intricately controlled by the arterial wall reactions to changes in pressure and flow.

While numerous mechanisms have evolved to ensure the smooth integration of mechanical information, several common sensing mechanisms have been identified that are found in lower life forms, as well as higher mammals. Such mechanisms require virtually instantaneous activation in order to provide rapid accommodation to a changed mechanical environment. Vascular cells and cell membranes exhibit certain mechanical responses that appear identical to those identified in simpler life forms. In arteries there exists a local acute response to changes of blood flow. As first reported by Schretzenmayr (1933), there is a rapid relaxation of the vessel when flow is increased. Such effects were shown to be endothelium-dependent in the 1980s (Pohl et al., 1986) and recent work indicates that hyperpolarization of the endothelial cells is required to facilitate the vasorelaxation response (Olesen, personal communication). In this chapter, we will confine the bulk of our discussion to mechanical sensing by the arterial wall in the context of signal transduction of flow forces. The question then is: How does the arterial wall "sense" and transduce information regarding blood pressure and flow into a signal capable of creating appropriate wall tone adjustments?

II. HEMODYNAMIC FORCES

It is useful to first consider the nature of forces created by a flowing liquid that can act on the arterial wall. Movement of blood through an artery can be resolved into two principal vectors (Fig. 1): one is perpendicular to the wall and is referred to as *blood pressure*; the second vector acts parallel to the wall and creates a tangential dragging force over the surface of the arterial lumen. This frictional force, which usually presents in laminar flow profile, is called *shear stress*. Thus, two kinds of forces act on the vessel wall; the endothelium is subjected to most of the shear stress forces along the vessel lumen, while smooth-muscle cells, together with the arterial extracellular matrix materials elastin, collagen, and proteoglycans, absorb most of the transmural pressure. There is some evidence (see Section III) that there exist two separate sensing mechanisms in blood vessels, one responding to changes in wall shear stress, and the other sensitive to pressure–stretch. The mechanisms involved in these disparate

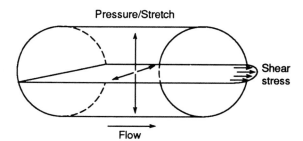

Pressure/Stretch

Shear stress

Flow

FIGURE 1   Schematic of hemodynamic force vectors normalized to pressure–stretch at right angles to the arterial wall, and shear stress, the tangential frictional force acting primarily at the endothelial lining of the vessel in the direction of flow.

responses appear to involve direct effects of the force on the vascular cell membrane, as well as indirect effects mediated by the flow characteristics in the bulk medium adjacent to the cell membrane. The primary candidates for direct effects of physical forces on vascular cell responsiveness are ion channels activated by either stretch or shear stress.

III. MECHANICAL SENSING ION CHANNELS

Originally described in specialized skeletal muscle cells, the existence of mechanotransducing ion channels has been demonstrated in a variety of cells. For example, stretch-activated ion channels have been found in skeletal muscle (Guhary and Sachs, 1985), ventricular cells (Craeluis et al., 1988), renal tubular epithelium (Sackin, 1987), choroid plexus epithelium (Christensen, 1987), and vascular endothelial cells (Lansman et al., 1987). A second type of mechanotransducing ion channel has been described in vascular endothelial cells (Olesen et al., 1988) and skeletal muscle cells (Burton and Hutter, 1990); these appear, however, to be regulated more by shear stress than by membrane stretch.

A. Stretch-Activated Ion Channels

When a micropipette is applied to the cell surface and attached to the cell membrane by suction to form a seal, distension of the membrane patch captured in the pipette can be controlled by negative pressure. The degree of stretch has been found to be related to the electrical activity of the membrane, specifically the opening of transmembrane ion channels.

All stretch-activated ion channels (SAC) characterized in mammals are cation-specific, although a few discriminate for a particular ionic species, usually potassium (Sachs, 1988). The conductances of several of these channels are listed in Table 1.

B. Stretch-Activated Ion Channels in Vascular Endothelial Cells

In 1987, Lansman and colleagues described mechanosensitive ion channels in membrane patches obtained from arterial endothelial cells. Using the cell-attached patch-clamp configuration, single-channel data were collected during suction pulses of 1–20 mm Hg pressure. The stretch-sensitive channel was cation-specific ($P_{Ca^{2+}}/P_{Na^+} = 6.0$) with a conductance of approximately 40 pS. Open time histograms revealed both fast and slow components. Lansman et al. (1987) speculated that the role of blood flow

TABLE 1

Stretch-Activated Ion Channels

Cell type	Conductance (pS)	Ionic selectivity	Reference
Endothelial cells (vascular)	40	$P_{Ca^{2+}}/P_{N^{2+}}= 60$	Lansman et al., 1987
Ventricular myocytes	120	Nonselective	Craeluis et al., 1988
Renal epithelium	25	$P_{K^+}/P_{Na^+} = 25 : 1$	Sackin, 1987
Renal medullary thick ascending limb	146	Ca^{2+}-activated K^+	Tanaguchi and Guggino, 1989
Pectoralis muscle	70	$P_K/P_{Na} = 2.0$	Guhary and Sachs, 1984

in modulating arterial tone may occur via Ca^{2+} influx through such channels.

C. Shear-Stress-Activated K^+ Ion Channels in Endothelial Cells

As noted above, the endothelial cells lining blood vessels are uniquely exposed to shear stresses associated with flowing blood, the magnitude of which varies considerably throughout the arterial circulation. In these cells, shear-stress-activated ion channels have recently been identified by Olesen et al. (1988) using the patch-clamp whole-cell recording method. Cells were grown in flow tubes such that the pressure differential was quite small over a range of shear stress values; shear stresses were favored over pressure–stretch forces. In this flow arrangement, stretch-activated cation-selective channels (see Section III.B) were not activated. Instead, there developed a membrane current that varied as a function of shear stress (Fig. 2) with a half-maximal activation near 1 dyn/cm² and approached a plateau near 20 dyn/cm². On the basis of ion selectivity and reversal potential, the current appears to result from the activity of an inward rectifying K^+ channel (Fig. 3) and is designated $I_{K,S}$. It activates rapidly in response to flow and does not rapidly desensitize; it inactivates when flow is stopped.

This channel, which is blocked by Ba^{2+} and Cs^{2+}, causes hyperpolarization of the endothelial cell. It is unclear at present whether G-binding proteins are involved in its activity. Atrial myocytes and vascular smooth-muscle cells did not express shear stress-activated currents. It may be recalled from Section I that endothelial hyperpolarization is associated with

a

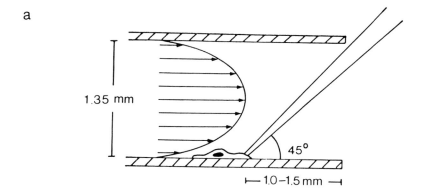

b

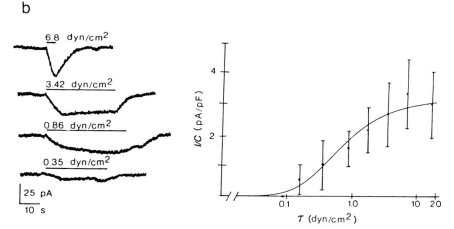

FIGURE 2   Shear-stress-activated K^+ current in bovine aortic endothelial cells. (a) Single cell cultured in a laminar flow tube. Patch-clamp whole-cell recording technique. (b) Relationship between shear stress (τ) and whole-cell current $I_{K,S}$. Holding voltage -30 mV. Tracing shows the time course of whole-cell currents from one cell induced by four different flow rates. The mean whole-cell currents normalized by cell capacitance for seven experiments are plotted as a function of τ. [From Olesen et al. (1988), *Nature* **331**, 168–170, with permission.]

vasorelaxation; thus $I_{K,S}$ is the fastest response to flow identified to date that is consistent with endothelium-mediated regulation of vascular tone. Indirect confirmation of endothelial hyperpolarization in response to shear stress has been reported by Nakache and Gaub (1988) using membrane-potential-sensitive fluorescent dyes. The mechanism by which a physical force, shear stress, elicits ion-channel activation is unclear. If it operates

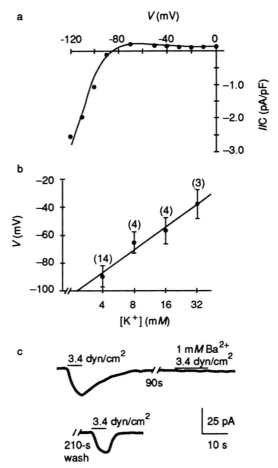

FIGURE 3   Shear-stress-activated current in bovine aortic endothelial cells. (a) Current–voltage curve (extracellular $[K^+] = 4$ mM) showing inward rectification with a reversal potential of -90 mV and slope conductance of 91 pS/pF. (b) K^+ selectivity of shear-stress-induced current. Reversal potentials are plotted as a function of extracellular K^+ concentration. Slope = 55 mV per 10-fold increase in $[K^+]_0$. (c) Blocking of $I_{K.S}$ by Ba^{2+}. [From Olesen et al. (1988), *Nature* 331, 168–170, with permission.]

indirectly via membrane stretch, why are endothelial stretch-activated channels (tending to depolarization) not seen? Conversely, why is $I_{K.S}$ in addition to stretch-sensitive channels not detected when membrane patches are stretched? Two possibilities are that (1) shear stress-activated channels are secondary to a flow-sensitive structure (mechanosensor) on the cell surface that escapes lateral deformation when suction is applied and (2) that the polarity of the endothelium is such that $I_{K.S}$ is activated on the

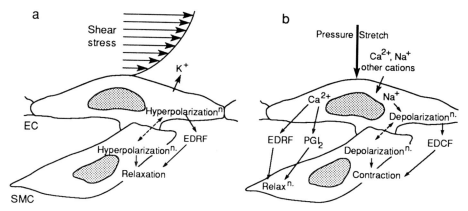

FIGURE 4   Schematic of possible ion-channel-mediated endothelial cell (EC) regulation of vascular smooth-muscle cell (SMC) tone as a result of flow forces. (a) Shear stress activation of K^+ channels induces hyperpolarization of EC. Hyperpolarization appears necessary to stimulate the synthesis and release of endothelium-derived relaxing factors (EDRFs) that stimulate SMC relaxation via guanylate cyclase. In some vessels direct electrical coupling via open-gap junctions may facilitate repolarization (and hence relaxation) of SMC. (b) Hemodynamic pressure–stretch forces activate stretch-sensitive, cation-specific channels. Ca^{2+} and Na^+ influx and endothelial depolarization may transduce conflicting signals to the SMC via release of relaxing factors EDRF and prostacyclin (PGI_2) and endothelium-derived contracting factors (e.g., endothelin) or by direct electrical coupling. [From Davies (1989), *News in Physiol. Sci.* 4, 22–25, with permission.]

basal side of the cell only, the flow forces undergoing apical to basal transfer via the cytoskeleton.

The physiological significance of separate channels sensitive to stretch and shear stress in endothelium may be related to the high variability of hemodynamic forces in the arterial circulation. Near branches, curves, and bifurcations there exist complex regions of disturbed laminar flow where shear stresses and pressures vary over short distances and throughout the cardiac cycle (Ku et al., 1985; Davies, 1988). At the boundaries of regions where pressure variations are prevalent, but shear stress is low, stretch activation may be more important. Where pressure variation is smaller, but where shear stresses range up to 50 dyn/cm^2 or higher, shear activation may predominate. Corresponding regional hyperpolarization–depolarization responses [Davies (1989) and Fig. 4] may have some physiological basis in regulating local tone at such sites.

D. Shear-Stress-Inactivated K^+ Ion Channels

In a recent report, Burton and Hutter (1990) have literally turned the preceding discussion inside-out and upside-down! In what started out as a routine perfusion protocol to deliver agonists onto inside-out patches containing inwardly rectifying K^+ channels from mammalian skeletal muscle, they noted in controls (no agonists) that the open probability of K^+ channels decreased when superfusion was started; that is, inactivation of inward rectifying K^+ channels by flow over the *inner* surface of the membrane patch at negative holding potentials. The response rapidly reversed when flow was stopped and was restored when flow was restarted. The lowest flow rates in their system (0.5 cm/s) elicited the response. When the membrane was stretched by suction, the channels remained open (unaffected), leading the authors to conclude that shear forces and not pressure–stretch were responsible for the channel gating behavior. Furthermore, concentration gradient effects were controlled for and eliminated as a potential cause, and neither Ca^{2+}-activated K^+ channels nor ATP-sensitive K^+ channels were sensitive to flow conditions. Although less physiologically relevant, a flow effect persisted and indeed reversed when the patches were tested at positive membrane potentials, with open probability increasing in the presence of flow. It is noteworthy that Burton and Hutter detected these effects by perfusing the inner side of the membrane patch (apparently they did not test the outer membrane) and that Olesen et al. (1988) were unable to detect activation of $I_{K.S}$ in outside-out patches of endothelial membrane. They raise the intriguing suggestion that flow-induced endothelial $I_{K.S}$ may operate via the generation of shear stresses at the *inner* surface of the membrane, perhaps via submembranous cytoskeletal components.

E. Factors Influencing Stretch-Activated Channels

The mechanisms by which mechanical forces control ion channel gating are at present unknown. The open probability of SACs increases exponentially with the square of differential pressure across the membrane (Guhary and Sachs, 1985). Intuitively it might be expected that changes in membrane tension are transferred from the lipid bilayer to the channel protein. While the amount of energy needed to open a channel is considerable, an expansion of only 2% of the membrane area would be required to produce such a force. It has been calculated, however, that the force needed to open a single SAC would require a large area of membrane (> 800 A), a requirement inconsistent with membrane distension alone. This calculation suggests that the channels are attached to cytoskeletal elements in order to generate the tension required for channel activation, typically up to 3 dyn/cm (Sachs, 1988). Such a scenario is consistent with the anion transporter of the red blood cell that is linked to spectrin. Interestingly, however, excised inside-out patches and cell-attached patches show identical stretch sensitivity, an observation that suggests that gating does not require an *intact* cytoskeleton throughout the cell (Guhary and Sachs, 1985). Submembranous cytoskeletal interactions within the membrane patch, however, can probably provide a local tensile force equivalent to that operating when the cell is intact. The density and distribution of channels (approximately 3 channels per square micrometer of membrane surface area) favors a model of stretch-activated channels distributed at the nodes of a network of submembranous cytoskeletal elements (Sachs, 1988).

Membrane tension can be directed to the mechanosensitive ion channels through elastic and/or nonelastic elements. The delayed response of SAC following the application of suction or pressure suggests that *elastic* elements are involved in directing membrane tension to the channel (Sachs, 1988). Treatment of cells with the actin-disrupting agent, cytochalasin, markedly reduces the delay time and shifts the open-probability curve toward lower pressures. Conversely, phalloidin, a toxin that stabilizes the actin polymer, had no effect on channel pressure–sensitivity. Such results strongly implicate actin as an important cytoskeletal element involved in tension distribution.

Further evidence for the role of actin in the distribution of membrane tension comes from studies of cellular biomechanical properties (Sato et al., 1987). For example, in the vascular endothelium, actin is the primary element responsible for the viscoelastic properties of the cell. Treatment of endothelial cells with cytochalasin reduces the time constant for membrane relaxation (following release of tension loading), while pretreatment with colchicine, a microfilament toxin, has no effect on the membrane time constant. These results support the belief that actin is a prime candidate that directs membrane tension to stretch-activated channels.

1. Voltage-Dependent Gating

While stretch-activated channels are regulated primarily by membrane tension, they also demonstrate a dependence on transmembrane voltage. For example, depolarization (making the membrane potential more positive) increases stretch-activated channel open probability in chick skeletal muscle (Guhary and Sachs, 1985), oocytes (Methfessel et al., 1986) and *Escherichia coli* (Martinac et al., 1987).

Kinetic analyses demonstrate that stretch-activated channels can be modeled with the following transitions between the open and closed states (Guhary and Sachs, 1985):

$$C_1 \underset{K_{2,1}}{\overset{K_{1,2}}{\rightleftarrows}} C_2 \underset{K_{3,2}}{\overset{K_{2,3}}{\rightleftarrows}} C_3 \underset{K_{4,3}}{\overset{K_{3,4}}{\rightleftarrows}} O_4$$

where C_1–C_3 represent closed states and O_4 represents the single open state. The tension-dependent rate constant is $K_{1,2}$. Interestingly, $K_{1,2}$ is also the voltage-dependent rate constant (Sachs, 1988). Thus the pressure-dependent rate constant and the voltage-dependent rate constant can be added to yield a single constant demonstrating that tension does not alter the channel's electric dipole.

2. Effect of pH on Channel Gating

Extracellular pH also affects the tension and voltage-dependent rate constant ($K_{1,2}$) for channel state transition. Increasing extracellular pH from 7.4 to 10.0 increases the voltage sensitivity from 45 to 20 mV/$\log_e$ (probability of channel opening); at pH = 7.4, 3.12 mm Hg were required for an e-fold change in open probability, while at pH = 9.4 only 0.83 mm Hg were required for an identical change in open probability. This change in pH does not affect the voltage and pressure-independent component of $K_{1,2}$. These results were interpreted to suggest that changing pH from 7.4 to 10.0 (1) titrates a site with a pK_a = 9.1, probably a single lysine, (2) does not alter the local surface charge, and (3) does not alter channel conductance or reversal potential; thus the titration site(s) are not close to the mouth of the channel.

F. Stretch-Sensitive Ca^{2+}-Activated K^+ Channels

Another example of mechanosensitive K^+ channels is found in the renal medullary thick ascending limb (MTAL). Excised membrane patches stretched by negative pressure showed increased open probability of K^+ channels (Tanaguchi and Guggino, 1989). Removal of Ca^{2+} from both bath and pipette solutions resulted in a loss of channel activity. When MTAL cells were exposed to hypotonic bath solution during cell-attached patch clamping, K^+ channels were likewise activated; removal of extracellular Ca^{2+} abolished channel activity. These channels were therefore identified as being Ca^{2+}-activated K^+ channels. Because of their location and response to osmotic swelling, these mechanosensitive channels appear to be involved in volume regulation.

G. Stretch-Inactivated Ion Channels

A recent unexpected finding is the existence in some cells of K^+-selective stretch-activated channels side-by-side with K^+-selective stretch-inactivated channels (Morris and Sigurdson, 1989). The latter presumably exist to counteract the effects of the SAC and thus provide a finer regulation of K, presumably at intermediate membrane tensions. At low

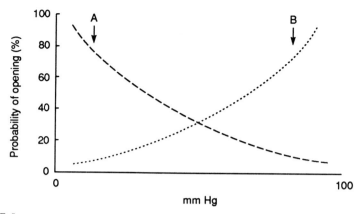

FIGURE 5   Stretch-activated and stretch-inactivated ion channels in the same membrane patch. Probability of opening of snail neuron stretch-activated (B) and stretch-inactivated (A) K$^+$-selective ion channels as a function of pressure. [Condensed from Morris and Sigurdson (1989), *Science* **243**, 807–809].

tensions, stretch-inactivated ion channels (SIC) are fully open, whereas SACs are closed. As tension increases, SICs close as SACs open (Fig. 5). It is unclear at present whether SICs are as widely distributed throughout the evolutionary tree as SACs; they have not yet been reported in higher animals.

H. Single Channels in Relation to Whole-Cell Currents

In a recent paper, Morris and Horn (1991) applied physical forces to isolated neuronal growth cones using a variety of methods to create mechanical stimulation. They reasoned that although tension in the membrane of stretched cells is neither measurable nor homogeneous, when mechanical stimulation is taken to an extreme (including membrane rupture), there must be high levels of tension generated at the membrane, which should elicit massive changes in stretch-activated current arising from the abundant mechanosensitive K$^+$ channels present in these cells. When subjected to vigorous spritzing, however, the maximum currents recorded were always below 5 pA, two orders of magnitude lower than expected from single-channel data. When cells were distended by intra-cellular pressure until they ruptured, there were no significant changes of membrane current until just before the lytic pressure. Other mechanical stimuli also produced little overall effect on "macroscopic" currents even though the sensitivity of mechanosensitive channels to stretch in the same cells was repeatedly demonstrated by patch clamp. While these results at first seem inconsistent with the concept of supersensitive mechanosensitive membrane channels, they are a predictable consequence of the association of SACs and SICs with the cortical cytoskeleton (Sachs, 1989). It is likely that tension is distributed throughout the cell, particularly among the cytoskeletal components, rather than concentrated primarily at the plasma membrane itself. Alternatively, tensions induced by the patch pipette are very local and intense, and may disassociate the channels from the cortical cytoskeleton (Milton and Caldwell, 1990). In this respect, Morris and Horn rightly point out the implications of these discrepancies in interpreting the mechanical responses of the entire cell with respect to its normal physiology.

IV. CONCLUSION

Force-sensing mechanisms in cells are found throughout the evolutionary tree; they convert a mechanical stimulus into electrical and biochemical responses. Some, such as those involved in hearing and touch, are highly specialized; others, such as visceral sensors and those involved in local regulation of blood flow, are not connected directly to the nervous system. Flow regulation of vascular tone via the endothelium and smooth-muscle cells involves mechanical signal transduction via ion channels. The data, however, are recent and limited; we do not know if accessory membrane proteins are required as either effectors (sensing the primary force as "agonist") or cofactors for channel activation (Dull and Davies, 1991), nor even if the effects of flow-activated channels are antagonized in the same cell by flow-*in*activated channels, as occurs in the snail. Practical limitations of flow experiments have delayed progress. The studies are cross-disciplinary, requiring cooperation between fluid dynamicists and biologists, and the application of precise flow forces to vascular cells is complex, thereby limiting routine screening of many flow variables. Alternative strategies of those used for the isolation and cloning of membrane receptors are required when the "agonist" is a physical force; fortunately, such strategies are evolving. There is great interest in flow and force-related responses of cells, particularly with respect to the regulation of blood vessel tone where the initial electrical changes in endothelial cells appear to be associated with the stimulus for a series of biochemical transduction pathways that regulate smooth-muscle relaxation (Pohl et al., 1986; Olesen et al., 1988; Davies, 1989). With focus on the ion channels responsible for signal-response coupling in the endothelium, a better understanding of this important regulatory mechanism will surely follow.

ACKNOWLEDGMENTS

The work presented in this chapter was supported by grants to Dr. Davies from the NHLBI (HL 36049 and HL 36028). We thank Charlene Murphy for preparation of the manuscript.

REFERENCES

Burton, F. L., and Hutter, O. F. (1990). Sensitivity to flow of intrinsic gating in inwardly rectifying potassium channel from mammalian skeletal muscle. *J. Physiol.* **424**, 253–261.

Christensen, O. (1987). Mediation of cell volume regulation by calcium influx through stretch-activated channels. *Nature* **330**, 66–68.

Craeluis, W., Chen, V., and El-Sharif, N. (1988). Stretch-activated ion channels in ventricular myocytes. *Biosci. Rep.* **8**, 407–414.

Davies, P. F. (1988). Endothelial cells, hemodynamic forces, and the localization of atherosclerosis. In *Endothelial Cells*, Vol. II, pp. 123–138, Ryan, U. S. (ed.). CRC Press, Boca Raton, FL.

Davies, P. F. (1989). How do vascular endothelial cells respond to flow? *News in Physiol. Sci.* **4**, 22–25.

Dull, R. O., and Davies, P. F. (1991). Flow modulation of agonist (ATP)-response (Ca^{2+}) coupling in vascular endothelial cells. *Am. J. Physiol.* **261**, 149–156.

Guhary F., and Sachs, F. (1984). Stretch-activated single ion channel currents in tissue cultured embryonic check skeletal muscle. *J. Physiol.* **352**, 685–701.

Guhary, F., and Sachs, F. (1985). Mechanotransducer ion channels in chick skeletal muscle: The effect of extracellular pH. *J. Physiol.* **363**, 119–134.

Ku, D. N., Giddens, D. P., Zarins, C. K., and Glagov, S. (1985). Pulsatile flow and atherosclerosis in the human carotid bifurcation. Positive correlation between plaque location and low and oscillating shear stress. *Arteriosclerosis* **5**, 293–301.

Lansman, J. B., Hallam, T. J., and Rink, T. J. (1987). Single stretch-activated ion channels in vascular endothelial cells as mechanotransducers? *Nature* **325**, 811–813.

Martinac, B., Buechner, M., Delcour, A. H., Adler, J., and Kung, C. (1987). Pressure-sensitive ion channels in *Escherichia coli*. *Proc. Natl. Acad. Sci. USA* **84**, 2279–2301.

Methfessel, C., Witzemann, V., Takahashi, T., Mishina, M., Numa, S., Sakmann, B. (1986). Patch clamp measurements on xenopus laevis oocytes: currents through endogenous channels and implanted acetycholine receptor and sodium channels. *Pflug Arch.* **407**, 577–588.

Milton, R. L., and Caldwell, H. J. (1990). How do patch clamp seals form? A lipid bleb model. *Pflug Arch.* **416**, 758–762.

Morris, C., and Sigurdson, W. S. (1989). Stretch-inactivated ion channels coexist with stretch-activated ion channels. *Science* **243**, 807–809.

Morris, C. E., and Horn, R. (1991). Failure to elicit neuronal macroscopic mechanosensitive currents anticipated by single-channel studies. *Science* **251**, 1246–1249.

Nakache, M., and Gaub, H. E. (1988). Hydrodynamic hyperpolarization of endothelial cells. *Proc. Natl. Acad. Sci. USA* **85**, 1841–1844.

Olesen, S. P., Clapham, D. E., and Davies, P. F. (1988). Hemodynamic shear stress activates a K^+ current in vascular endothelial cells. *Nature* **331**, 168–170.

Pohl, V., Holtz, J., Busse, R., and Barrenge, E. (1986). Crucial role of endothelium in the vasodilator response to increased flow in vitro. *Hypertension* **8**, 37–47.

Sachs, F. (1988). Mechanical transduction in biological systems. *CRC Crit. Rev. Biomed. Eng.* **16**, 141–169.

Sachs, F. (1989). Ion channels as mechanical transducers. In *Cell Shape: Determinants, Regulation and Regulatory Role*, pp. 63–92, Stein, W. D., and Bronnes, F. (eds.). Academic Press, New York.

Sackin, H. (1987). Stretch-activated potassium channels in renal proximal tubule. *Am. J. Physiol.* **253**, F1253–F1262.

Sato, M., Levesque, M. J., and Nerem, R. M. (1987). Micropipette aspiration of cultured bovine aortic endothelial cells exposed to shear stress. *Arteriosclerosis* **7**, 276–286.

Schretzenmayr, A. (1933). Uber kreislaufregulatorische Vorgange an den grossen Arterien bei der Muskelarbeit. *Pflug. Arch.* **232**, 743–757.

Taniguchi, J., and Guggino, W. B. (1989). Membrane stretch: A physiological stimulator of Ca^{2+}-activated K^+ channels in thick ascending limb. *Am. J. Physiol.* **257**, F347–F352.

CHAPTER 5

■ ■ ■ ■ ■

Effects of Flow on Anchorage-Dependent Mammalian Cells—Secreted Products

■ ■ ■ ■ ■

François Berthiaume and John A. Frangos

I. INTRODUCTION

All mammalian cells are subjected to fluid shear stress in vivo. Endothelial cells are directly exposed to the circulating blood in the vasculature, while cells in other tissues, although they do not experience flows of similar magnitude, are subjected to the shear stress arising from the flow of interstitial fluid. In the case of bones, an externally applied force can cause flow of the extracapillary fluid filling the trabecular meshwork. Osteoblasts, which line the surfaces in trabecular bone, are then exposed to these load-induced shear stresses.

Fluid flow effects on anchorage-dependent mammalian cells have been studied primarily on endothelial cells, and such investigations may help to clarify the mechanisms regarding endothelial dysfunction, hypertension, and atherogenesis. We give here an extensive summary of all the work published so far on the effect of fluid shear stress on secreted products and second messengers in endothelial cells. Flow studies on other cell types are fewer. We report the work that has been published on osteoblasts and also the isolated studies on smooth-muscle cells, epithelial cells, and fibroblasts.

139

The last section gives an overview of the possible mechanisms by which fluid shear stress may be transduced into a biochemical signal inside the cell, eventually leading to a physiological response. This particular section will be more speculative since little data are available on the early mechanisms involved in the transduction of the shear stress signal across the cell membrane, but should give a survey of the current concepts proposed to describe this phenomenon.

II. ENDOTHELIAL CELLS

Endothelial cells line the luminal side of all blood vessels and are directly exposed to blood flow in vivo. They produce a variety of substances involved in the control of vascular tone, platelet aggregation, and cell proliferation. Most flow studies have been carried out using human umbilical-vein endothelial cells (HUVECs), bovine aortic endothelial cells (BAECs), and pig aortic endothelial cells (PAECs). The first flow-induced physiological response of endothelial cells identified was the stimulation of histamine-forming capacity. Later studies showed that fluid flow modulates the release of several vasoactive compounds such as prostacyclin, endothelium-derived relaxing factor, and endothelin and increases transendothelial permeability. The release of several other compounds, such as tissue-type plasminogen activator, platelet-derived growth factor, fibronectin, interleukin 6, and proteoglycans, is affected, too. We also report the recent findings on the effects of fluid flow on the levels of intracellular second messengers, which should lead to a better understanding of the early steps in the mechanism of flow-induced stimulation of endothelial cells. Other investigations include work on adhesion of suspended cells to endothelial cell monolayers and on metabolic activities such as endocytosis and endothelial cell turnover under flow conditions.

A. Intracellular Second Messengers

Endothelial cells have hormone receptors linked to guanine-binding proteins (G proteins) which modulate the function of enzymes that produce intracellular messengers such as adenosine $3':5'$-cyclic monophosphate (cAMP), inositol 1,4,5-triphosphate (IP$_3$), and 1,2-diacylglycerol [reviewed by Flavahan and Vanhoutte (1990)]. These intracellular agonists trigger a cascade of reactions inside the cell eventually leading to the physiological response. One of the most important signal transduction pathways involves the activation of phospholipase C, which cleaves membrane-bound phosphatidylinositol 4,5-diphosphate into IP$_3$ and diacylglycerol. This mechanism has been shown to mediate several agonist-induced endothelial cell responses (Newby and Henderson, 1990).

1. Inositol 1,4,5-Triphosphate

Inositol 1,4,5-triphosphate (IP_3) causes the release of calcium into the cytosol from intracellular stores and diacylglycerol activates protein kinase C in the presence of elevated levels of calcium in the cytoplasm (reviewed by Berridge and Irving, 1989). IP_3 levels in HUVEC monolayers subjected to a shear stress of 22 dyn/cm^2 were increased 2.1-fold at 30 s and remained elevated for up to 6 min (Nollert et al., 1990). At 15 min, the levels were nearly back to the values for unstimulated cells. A small but not statistically significant increase was also seen using a shear of 2 dyn/cm^2. Similar cells were also stimulated with histamine, and the response was of the same intensity although IP_3 levels appeared to return to basal values faster than in the flow-stimulated cells. We also performed a similar study with HUVECs and found increased levels of IP_3 on stimulation by flow, with a half-maximal response below 1.4 dyn/cm^2 (Bhagyalakshmi and Frangos, 1989b). The response was oscillatory with a small peak at 30 s and a maximum at 10 min. These observations clearly indicate that flow activates phospholipase C in HUVECs, but contrast with findings in agonist-stimulated endothelial cells, where the IP_3 levels usually peak within 1 min after the addition of agonist and then decrease rapidly (e.g., Halldórsson et al., 1988, Derian and Moskowitz, 1986, Brock et al., 1988).

The IP_3 response is consistent with the findings of Ando et al. (1988), who observed that flow-induced increases in intracellular calcium in BAECs are mainly due to the release of calcium from intracellular stores, although the latter data are still controversial (see Section II.A.3). These results and the IP_3 data do not seem to be entirely consistent, however, since intracellular calcium levels go back to resting values within 5 min while IP_3 levels remain elevated several minutes after the onset on shear. This may be the result of the desensitization of intracellular calcium mobilization or due to interspecies differences. The implications of the increase in IP_3 levels by shear is very important, considering the central role played by this second messenger in the activation of the calcium-dependent pathways in the endothelial cell. As will be seen in the rest of this chapter, several flow-induced responses in endothelial cells can be ultimately traced back to the activation of the phospholipase C pathway.

2. Adenosine 3':5'-Cyclic Monophosphate

The adenosine 3':5'-cyclic monophosphate (cAMP) compound is generated by the action of adenylate cyclase on adenosine triphosphate (ATP), and the activity of adenylate cyclase is modulated by stimulatory and inhibitory guanine-binding proteins (G_s and G_i). Exogeneous ATP and adenosine diphosphate (ADP) increase cAMP levels in HUVECs (Pirotton et al., 1987b). In HUVECs subjected to a shear stress of 4.3 dyn/cm^2 for 15 min, the levels of cAMP increased from 0.80 ± 0.06 pmol/mg protein in the stationary controls to 3.2 ± 1.2 pmol/mg protein in the sheared cells (Reich et al., 1990). Pharmacological concentrations of prostacyclin induce increases in cAMP levels (Hopkins and Gorman, 1981) in HUVECs, and elevated cAMP attenuates ATP-mediated increases in intracellular calcium and thereby reduces the production of prostaglandin I_2 (PGI$_2$) (Lückhoff et al., 1990b), which suggests that cAMP may serve as a feedback loop to control PGI$_2$ production. Direct G protein activation by aluminum fluoride leads to prostacyclin synthesis in HUVECs (Magnússon et al., 1988), but recent findings show that aluminum fluoride does not increase cAMP levels in PAECs (Graier et al., 1990), which is consistent with other data demonstrating that agonist-induced release of PGI$_2$ does not have an autocrine action on endothelial cells (Schröder et al., 1990, Makarski, 1981). These results suggest that cAMP does not generally act as a feedback loop for PGI$_2$ production. The mechanism by which flow induces an increase in cAMP levels in HUVECs is still unknown.

The physiological implications of the flow-induced cAMP increase are unclear at this time since little has been uncovered on the role of cAMP in endothelial responses, but it may be involved in the control of transendothelial permeability (Stelzner et al., 1989; Oliver, 1990).

3. Calcium

Resting cytosolic calcium levels in endothelial cells are normally less than 100 nM, whereas the concentration in the culture medium is about 2 mM. Extracellular calcium continuously leaks into the cell due to the overall permeability of the cell plasma membrane and the negative transmembrane potential, but the calcium gradient across the membrane is maintained by a calcium pump. In BAECs subjected to shear stresses ranging between 3.7 and 10.3 dyn/cm^2, a sharp increase in intracellar calcium was observed at the onset of flow, followed by a decrease close to the preshear levels within 3 min as measured by the fluorescent dye Fura 2 (Ando et al., 1988). The steady-state concentration reached in Medium 199, which contains calcium ions, was slightly above the preshear values, whereas in nominally calcium-free media, the levels appeared to return to the initial levels. The peak values were similar in both cases, going up to 1 μM (according to their calibration curve), although the peak values for similar experiments seemed to vary significantly. Step increases in shear stress resulted in smaller transient increases in intracellular calcium levels.

The contribution of shear stress as a physical force to the cytosolic calcium response has recently been questioned. Dull and Davies (1991) and Mo et al. (1991) could not repeat some of the experiments of Ando et al. (1988), as they saw no flow-induced changes in intracellular calcium in ATP-free media. They show that during a typical experiment in ATP-containing medium, such as Medium 199, as soon as the flow is stopped or ATP is washed out of the system, intracellular calcium levels instantaneously drop back to basal levels. This suggests that when the flow is stopped, ATP is rapidly removed near the cell surface by ATPases, thereby terminating the stimulation. With medium supplemented with a slowly hydrolyzable analog of ATP, a similar response was seen in both the presence and the absence of flow. These results tend to indicate that flow by itself does not increase intracellular calcium levels in BAECs. On the other hand, flow-induced increases in prostacyclin production and endothelium-derived relaxing factor release have been observed in BAECs in ATP-free media (Grabowski et al., 1985, Buga et al., 1991). The presence of calcium in the extracellular medium is essential for a sustained release of endothelium-derived relaxing factor in agonist-stimulated endothelial cells (White and Martin 1989, Lückhoff et al., 1988), and flow-induced prostacyclin production by HUVECs is inhibited when extracellular calcium is removed (Bhagyalakshmi and Frangos, 1989a), which tends to support the hypothesis that there is an influx of calcium into the cell during shear, at least in some intracellular compartment. There is evidence that there are several calcium pools in endothelial cells (Ishihata and Endoh, 1991). Also using Fura 2, other authors (Shen et al., 1991, Geiger et al., 1991) have seen increased intracellular calcium levels induced by flow alone. Depending on the incubation conditions for loading the cells with Fura 2, the localization of the dye may be different, leading to inconsistent results. There is evidence that sometimes Fura 2 preferably associates with mitochondria in BAECs (Steinberg et al., 1987). The synthesis of shear-induced prostacyclin production in HUVECs is inhibited by Quin 2/AM (Bhagyalakshmi and Frangos, 1989a), an intracellular calcium chelator that enters the cell by the same mechanism as Fura 2. It suggests that either Quin 2/AM is able to enter the compartment where calcium levels would be presumably increased or it acts as a nonspecific inhibitor. At this time, it is therefore unclear why flow-induced intracellular calcium changes are sometimes undetectable with Fura 2 or if intracellular calcium levels are affected by fluid flow alone at all.

Some evidence in the literature suggests that calcium released from intracellular stores may trigger an influx of potassium ions from calcium-activated channels, which then hyperpolarizes the membrane, thereby facilitating the entry of calcium through ion channels in BAECs (Lückhoff and Busse, 1990a, 1990c). A hyperpolarizing potassium current activated by flow has been identified in BAECs (Olesen et al., 1988), but is not inhibited by tetrabutylammonium, an inhibitor of calcium-activated potassium channels (Alevriadou et al., 1991). Direct activation of G proteins by aluminum fluoride in PAECs causes a calcium influx without stimulating the phospholipase C pathway, suggesting that calcium channels may also be linked to G proteins (Graier et al., 1990). It is possible that flow could activate these G proteins or their parent hormone receptors. Fluid flow might also increase the passive permeability of the plasma membrane to calcium ions as observed in red blood cells (Larsen et al., 1981) or activate stretch-activated channels (Lansman et al., 1987). If extracellular calcium does enter the cell, whether this occurs before or after the potassium flux at the onset of fluid shear stress is not known.

B. Mediators of Blood Vessel Tone and Permeability

1. Prostacyclin

Endothelial cells produce prostacyclin (PGI_2), a prostanoid derived from arachidonic acid with a half-life of 3 min at physiological pH (Moncada et al., 1976). This vasodilator inhibits platelet aggregation and is thought to play a major role in keeping a nonthrombogenic surface on the vascular endothelium.

The rate of PGI_2 synthesis is increased by fluid flow in HUVECs (Frangos et al., 1985, 1988) and BAECs (Grabowski et al., 1985). This is consistent with an in vivo study that revealed increased PGI_2 production for increasing stenosis in canine femoral and carotid arteries (Ovarfordt et al., 1985), a situation where a higher wall shear rate was produced. In the case of BAECs in serum-free medium, the rate of PGI_2 production was transiently increased up to 3.9-fold after a step increase of shear stress, going back to nearly zero 10–20 min after the step change (Fig. 1). The total amount of PGI_2 released was independent of shear stress, and the lack of exogenous arachidonic acid seemed to be responsible for the decline in the rate of PGI_2 production. Results with HUVECs in serum-containing medium show a sustained production of PGI_2 at a rate above that of the static controls (Fig. 2). The steady state rate was increased up to 11-fold for shear stresses up to 24 dyn/cm^2 relative to stationary controls. Cell monolayers subjected to pulsatile flow of 1-Hz frequency with a mean shear stress of 10 dyn/cm^2 maintained a steady production rate of $168 \pm 64 \, pg \, PGI_2/min/10^6$ cells, a 2.5-fold increase compared to 64 ± 16 pg $PGI_2/min/10^6$ cells for cells exposed to a steady shear stress of 10 dyn/cm^2.

Free arachidonic acid is generally considered to be a rate-limiting step in PGI_2 synthesis (Irvine, 1982). Phosphatidylcholine, phosphatidylinosi-

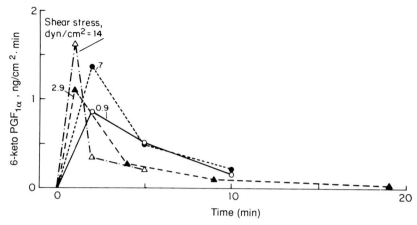

FIGURE 1   Production of 6-keto-prostaglandin $F_{1\alpha}$ by BAECs at different levels of shear stress. BAECs were subjected in serum-free medium to the levels of shear stress shown and the rate of release of 6-keto-prostaglandin $F_{1\alpha}$, the stable breakdown product of prostacyclin, was determined at different times. [Reprinted from Grabowski et al. (1985) by permission of Mosby-Year Book, Inc. Copyright © 1985.]

tol, and phosphatidylethanolamine contain respectively 24, 43, and 28 mol% of the total arachidonic acid pool in human umbilical vein endothelial cells or HUVECs (Takamura et al., 1990).

The most direct way to release arachidonic acid is by the single action of calcium-sensitive phospholipase A_2 on membrane phospholipids. This enzyme has generally been considered as the major producer of arachidonic acid in hormone-stimulated endothelial cells since agonist-stimulated PGI_2 production is dependent on elevated intracellular calcium levels (Hallam et al., 1988; Jaffe et al., 1987; Resink et al., 1987; Whorton et al., 1984); however, these studies have not clearly demonstrated whether the observed phospholipase A_2-like activity is in fact due to phospholipase A_2 or due to the combined action of other phospholipases and lipases. Stronger evidence for phospholipase A_2 activation comes from the detection of inositol, ethanolamine, and choline lysophospholipids in HUVECs and porcine aortic endothelial cells after stimulation by various calcium-mobilizing agonists (Hong and Deykin, 1982; et al., 1985). In these experiments, the major sources of arachidonic acid appeared to be phosphatidylinositol and phosphatidylethanolamine.

Endogeneous arachidonic acid can be released from diacylglycerol by the action of di- and monoacylglycerol lipases. Diacylglycerol is usually generated by the hydrolysis of phosphatidylinositol by phosphatidylinosi-

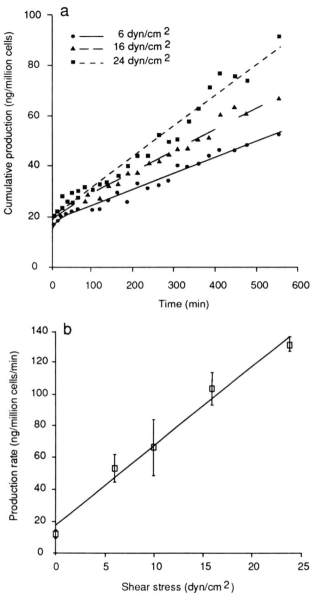

FIGURE 2   Long-term production of 6-keto-prostaglandin $F_{1\alpha}$ by HUVECs at different levels of shear stress. HUVECs were subjected to flow in Medium 199 supplemented with 20% fetal bovine serum. (a) At the onset of shear, the initial burst in the release of 6-keto-prostaglandin $F_{1\alpha}$ was followed by a sustained rate of production for almost 10 h. (b) The steady-state rate of synthesis was proportional to the level of shear stress used. Error bars represent the standard error of the mean (SEM). [Reprinted from Frangos et al. (1988) by permission of John Wiley and Sons, Inc. Copyright © 1988.]

tol-specific phospholipase C as a result of receptor-mediated events (Jaffe et al., 1987; Lambert et al., 1986; Pirotton et al., 1987b; Resink et al., 1987). In endothelial cells, the diacylglycerol lipase step is calcium-dependent, but not the activation of phospholipase C (Martin and Wysolmerski, 1987). In addition, a phospholipase C active on phosphatidylcholine has been identified in endothelial cells (Martin et al., 1987), but its role in agonist-stimulated cells is still unclear (Martin and Michaelis, 1988). Alternatively, diacylglycerol may be produced by the consecutive actions of phospholipase D and phosphatidate phosphate on phosphatidylcholine with the production of the intermediate phosphatidic acid (Martin, 1988; Martin et al., 1990).

The proportion of exogenous labeled arachidonic acid incorporated into the diacylglycerol and phosphatidylinositol pools in HUVECs is increased by flow (Nollert et al., 1989), which suggests that flow stimulates the turnover of both lipids. A significant reduction in the flow-induced PGI_2 synthesis by HUVECs is observed in the presence of the diacylglycerol lipase inhibitor RG-80267 (Bhagyalakshmi and Frangos, 1989a). These findings suggest that the combined actions of phospholipase C and diacylglycerol lipase may provide a significant part of the free arachidonic acid for PGI_2 synthesis.

Shear-induced PGI_2 production in HUVECs is inhibited by Quin 2/AM, an intracellular calcium chelator, and in the presence of ethylene glycol bis(β-aminoethyl ether)N,N, N', N'-tetraacetic acid (EGTA), which chelates the calcium present in the culture medium, suggesting that both intra- and extracellular calcium are essential (Bhagyalakshmi and Frangos, 1989a). The elevation of intracellular calcium levels in agonist-stimulated endothelial cells is believed to be sufficient for PGI_2 synthesis (Brotherton and Hoak, 1982, Hallam et al., 1988). Fluid flow may cause a rise in intracellular calcium levels in BAECs, and the ions are apparently mostly released from intracellular stores (see Section II,A,3). The latter is generally mediated by IP_3, suggesting that phospholipase C (PLC) activation would occur before the increase in intracellular calcium levels in the pathway leading to PGI_2 production by flow. In fact, the PLC pathway seems to be the main regulator of arachidonic release and PGI_2 synthesis in HUVECs treated with fluoride, a general activator of G proteins (Magnússon et al., 1989). Similar observations were obtained with histamine-, thrombin- and bradykinin-stimulated HUVECs (Dudley and Spector, 1986). In addition, one may consider the possibility of activation by flow of receptor-operated calcium channels, which have been found in PAECs (Graier et al., 1990).

This suggests that increased intracellular calcium levels due to IP_3 and/or to the opening of calcium channels would occur first, followed by the stimulation of calcium-sensitive phospholipases and other lipases. It is also possible that phospholipase-2 (PLA_2) may be stimulated independently from the PLC pathway, as this enzyme has been found to be directly linked to G proteins in other cell types (Burch and Axelrod, 1987; Fuse and Tai, 1987). It has been suggested that G proteins are involved in PLC activation as well as in an other step in PGI_2 production in BAECs (Pirotton et al., 1987a). In bovine endothelial cell membranes, leukotriene D_4 activates PLA_2, but not PLC (Clark et al., 1986a) and induces PGI_2 synthesis in bovine endothelial cells (Clark et al., 1986b). In addition, both PLC and PLA_2 activities were stimulated by streptokinase in bovine pulmonary artery endothelial cells, resulting in a marked release of arachidonic acid (Kawaguchi et al., 1990).

An increase in diacylglycerol turnover by shear stress suggests that, as a secondary response, protein kinase C (PKC) may also be activated during shear. The fluoride-induced PGI_2 synthesis in rat aortic endothelial cells (Jeremy and Dandona, 1988) and HUVECs (Magnússon et al., 1989) was suppressed by inhibition of PKC, and stimulated by activation of PKC in HUVECs (Magnússon et al., 1989). Phorbol esters, which activate PKC, have been shown to enhance PGI_2 release in response to increased Ca^{2+} levels in BAECs (Demolle and Boeynaems, 1988). Similar results have been obtained with HUVECs, where it was also found that stimulation of PLA_2 was responsible for the enhanced release of PGI_2, while PGH synthase and PGI_2 synthase activities were not increased (Zavoico et al., 1990). Conversely, specific inhibitors of PKC resulted in the abolition of the enhancing effects of phorbol esters. It appears that activation of PKC lowers the threshold Ca^{2+} levels required for PGI_2 synthesis in HUVECs, possibly at the level of arachidonic release from the phospholipids by PLA_2 (Carter et al., 1989). Arachidonic acid metabolism is enhanced by phorbol esters in calcium-ionophore-stimualted HUVECs (Halldórsson et al., 1988). On the other hand, phorbol esters reduce IP_3 formation and calcium mobilization induced by thrombin and histamine, which may serve as a down-regulation mechanism (Brock and Capasso, 1988).

HUVECs stimulated by ATP, bradykinin, and thrombin release PGI_2 in a transient manner, probably as a result of desensitization at the receptor level, which is specific for each agonist (Toothhill et al., 1988). In fact, ATP receptors can be uncoupled from PLC in BAECs by a stable analog of GTP, GTPγS, which permanently binds on the G protein that activates PLC (Brock et al., 1988). The initial burst in PGI_2 release by HUVECs at the onset of shear may be analogous to hormonal stimulation of endothelial cells. Close inspection of the data of Frangos et al. (1988) with sheared HUVECs reveals that the accumulated PGI_2 in the perfusing medium reached a plateau at approximately 1 h, and PGI_2 production started again at 3 h. There are no published reports of the effect of hormonal agonists on PGI_2 production over several hours. G protein activation by aluminum fluoride leads to increased levels of IP_3 and a fairly sustained production of PGI_2 over more than 50 min in HUVECs (Magnússon et al., 1989). The presence of a sustained component of PGI_2 production in both flow- and fluoride-stimulated HUVECs suggests that shear stress may be able to bypass the receptor-G protein coupling, which would be a major difference between the signal transduction mechanism for shear stress and hormone stimulation of endothelial cells. This possibility has not been explored experimentally. Alternatively, the initial events occurring during the initial burst in flow-induced PGI_2 production may trigger other events which lead to the sustained phase of PGI_2 production after a delay of a few hours.

The increases in intracellular calcium and inositol triphosphate are usually transient and can be correlated to the initial burst in PGI_2 production, but the mechanism mediating the sustained release of PGI_2 remains unclear. The delay between the end of the first phase and the beginning of the sustained component of the flow-induced PGI_2 synthesis is suggestive of the involvement of protein synthesis. It seems that the enhanced PLA_2 activity induced by leukotriene D_4 in bovine endothelial cells is dependent on protein synthesis (Clark et al., 1986a, 1986b). Enhanced PLA_2 activity has also been observed in PAECs incubated with pertussis toxin, and correlated with a decrease in lipocortin I synthesis, an inhibitor of PLA_2 (Fujimoto et al., 1990). The sustained release of PGI_2 may be regulated at the level of PLA_2, PGH, or PGI_2 synthase activities.

While a lot of knowledge has been gained in the shear-induced mechanism of PGI_2 production, the very first step involved in the stimulation has not been clearly identified yet. In addition, the mechanism that controls the steady rate of PGI_2 production, which exhibits a dose response covering a wide range of shear stresses (0–24 dyn/cm^2), is unknown. Our previous discussion, including the effects of shear stress on phospholipid metabolism and intracellular calcium, is summarized in Figure 3. The most plausible mechanism would first involve the activation of PLC with concomitant phosphatidylinositol (PI) turnover. The activation may be at the level of hormone receptors, G proteins or shear could activate the enzyme directly. The second messenger IP_3 then presumably triggers the release of calcium from intracellular stores, and the resulting increased cytosolic calcium levels would act synergistically with diacylglycerol, the other phospholipase C metabolite, to activate protein kinase C. Both the elevated intracellular calcium levels and the active protein kinase

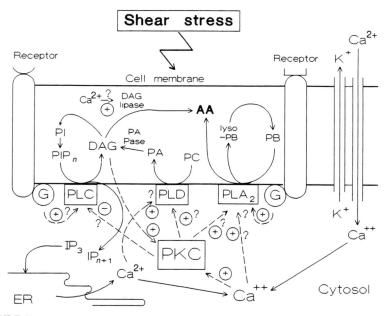

FIGURE 3   Pathways for the liberation of arachidonic acid in sheared endothelial cells. See text for explanations. Solid arrows denote chemical transformations or transport between different intracellular compartments. Broken arrows denote allosteric effects. Question marks signify that the information was inferred from results with agonist-stimulated cells. Symbols used are as follows: phosphatidylinositol (PI), phosphatidylinositol n-phosphate (PIP$_n$, n = 0–2), phospholipase C (PLC), inositol (n + 1)-phosphate (IP$_{n+1}$), diacylglycerol (DAG), protein (G), phospholipid (PB), phospholipase A$_2$ (PLA$_2$), arachidonic acid (AA), phosphatidylcholine (PC), phosphatidic acid (PA), phospholipase D (PD), phosphatidate phosphatase (PA Pase), protein kinase C (PKC), and endoplasmic reticulum (ER).

C would stimulate PLA$_2$ and/or phospholipase D. Another possibility is that PLA$_2$ be independently activated by a G protein or via the entry of calcium through receptor-operated channels. Arachidonic acid could be liberated from phosphatidylinositol by the actions of phospholipase C and diacylglycerol lipase, the latter of which is another calcium-sensitive enzyme. PLA$_2$ would directly release arachidonic acid from phosphatidylinositol, phosphatidylethanolamine, or phosphatidylcholine. Phospholipase D, followed by the action of a phosphatidate phosphatase and diacylglycerol lipase, would produce arachidonic acid mostly from phosphatidylcholine. Diacylglycerol may also inhibit lysophosphatide acyltransferase as in the case of agonist-stimulated platelets (Goppelt-Strübe et al., 1987), thereby reducing the rate of reincorporation of arachidonic acid into the phospholipids and increasing the availability of free arachidonic acid for PGI_2 synthesis.

2. Endothelin

Endothelin is a potent vasoconstrictor produced by endothelial cells from the proteolytic processing of preproendothelin, a product of messenger RNA translation (Yanagisawa et al., 1988) that stimulates smooth-muscle cell growth (Komuro et al., 1988). Endothelin-1, -2, and -3 have been identified. Bovine carotid artery endothelial cells have receptors specific for endothelin that activate phospholipase C (Emori et al., 1990), and endothelin may have an effect on human endothelial cell growth (Takagi et al., 1990).

Investigations on the effect of flow on endothelin synthesis by endothelial cells have given contradictory results. In a first study by Yoshizumi et al. (1989), there was a transient increase in endothelin messenger RNA in PAECs subjected to 5 dyn/cm² after 1 to 4 h, followed by a decrease to initial levels after 12–24 h. Increased secretion of endothelin in the culture medium was also detected. Pulsatile flow with a 15-min cycle did not yield a different response from steady flow. Addition of thrombin further enhanced endothelin mRNA levels, suggesting that the pathways from thrombin and flow stimulation of endothelin mRNA levels may be different, and that fluid flow alone did not cause full expression of endothelin mRNA. Endothelin mRNA levels of aligned cells after 48 h of exposure to 8 dyn/cm² were similar to those cells kept under stationary conditions. Milner et al. (1990a) studied the effect of flow by using rabbit aortic endothelial cells placed on 3-μm filters that were perfused across at different flow rates. The release of endothelin was nearly instantaneous at the onset of flow and sustained during the 3 min of stimulation. A second step increase of flow rate 10 min after the first stimulation resulted in a similar release of endothelin. A study on the effect of transforming growth factor β on endothelin mRNA levels in PAECs shows that the stimulation occurs only 1 h after addition of the agonist (Kurihara et al., 1989), suggesting that changes in endothelin mRNA levels occur over a time scale of an hour. This suggests that the source of the rapid flow-induced release of endothelin may have been from intracellular pools of preformed peptide or its precursor resulting from new mRNA synthesis. No arginine vasopressin was released during the stimulation, suggesting that the release of endothelin was not the result of nonspecific effects such as cell damage. The perfusion system used by Milner et al. (1990a) did not allow an accurate estimate of the shear stresses to which the cells were exposed. In addition, no justification on the choice of flow rates was given.

More recent evidence is in contradiction with this data, since Nollert et al. (1991) have found that endothelin gene expression and release were almost completely suppressed in HUVECs subjected to flow for up to 27 h. In this study, the rates of secretion were 0.549 ± 0.046 ng/10^6 cells/h and 0.239 ± 0.099 ng/10^6 cells/h for static controls and cells sheared at 25 dyn/cm^2, respectively. It appears that nitrovasodilators inhibit endothelin production in HUVECs (Saijonmaa et al., 1990); therefore, it is possible that the flow-induced release of endothelium-derived relaxing factor (see Section II.B.4) could mediate this response. Endothelin production by endothelial cells is markedly reduced by substances secreted by cells from the vascular media, such as smooth-muscle cells and fibroblasts (Stewart et al., 1990); therefore, in vitro investigations with pure cultures of endothelial cells may reveal only a very partial picture of the role of flow on endothelial endothelin secretion in vivo.

3. Endothelium-Derived Relaxing Factors

Endothelium-derived relaxing factors (EDRFs) are usually described as short-half-life vasodilatory compounds released by endothelial cells stimulated by a variety of agonists. BAECs produce EDRFs when stimulated by calcium ionophore, bradykinin, acetylcholine, and substance P (Cocks et al., 1985). EDRFs are thought to be derived from the enzymatic oxidation of L-arginine followed by cleavage of the molecule (nitric oxide NO). In both cultures endothelial cells and in situ, the enzymes responsible for NO production appear to be mostly membrane-associated and calcium-dependent (Förstermann et al., 1991; and Mitchell et al., 1991). The NO produced can diffuse out and cause smooth-muscle cell relaxation or can be converted into S-nitrosothiols, which are more stable. EDRFs are usually identified by their ability to dilate norepinephrine precontracted, deendothelialized vascular rings. The biological effect of EDRFs is suppressed by inhibitors of guanylate cyclase such as methylene blue. It is also inhibited by hemoglobin, which is a scavenger of EDRFs, and is potentiated by agents that reduce superoxide radicals, such as superoxide dismutase. A review on NO and EDRFs has recently been published (Ignarro, 1990).

The flow-mediated production of EDRFs by endothelial cells from vascular rings is well documented (see Chapter 7 in this book), but only more recently some studies have shown that cultured endothelial cells produce EDRFs when subjected to fluid flow (Cooke et al., 1990a, 1990b; Stamler et al., 1989). In these investigations, BAECs were grown on microcarrier beads and placed in a beaker where a vascular ring precontracted with norepinephrine was present. A magnetic stirrer was used to produce the flow. The onset of stirring corresponded to a reversible relaxation of the bioassay ring. This flow-induced relaxation was inhibited by methylene blue and hemoglobin, and no relaxation was observed at the onset of flow in the presence of one of these reagents or the competitive inhibitor of NO synthase, N-monomethyl-L-arginine. The response was potentiated by N-acetyl-L-cysteine, which promotes the formation of the more stable S-nitrosocysteine, a compound with vasorelaxant properties thought to be closer to EDRFs than NO (Myers et al., 1990). These results suggest that a nitrovasodilator with short half-life was produced by the endothelial cells when stimulated by flow. Prostacyclin was not involved since the cyclooxygenase inhibitors indomethacin and aspirin had no effect on the relaxations observed. A more recent study by Buga et al. (1991), using a packed bed of BAECs grown on microcarrier beads and perfused at different flow rates, also shows a flow-dependent release of EDRFs. The shear stresses generated in this particular system are difficult to evaluate because of the complex nature of the flow between the beads; therefore, the shear stress dose-response could not be investigated.

The mechanism of EDRF production by cultured endothelial cells in response to flow has not been studied yet, but could involve intracellular calcium, as is the case in agonist-stimulated endothelial cells. EDRFs stimulate soluble guanylate cyclase in endothelial cells (Martin et al., 1988; Schmidt et al., 1989; Boulanger et al., 1990). Kuchan and Frangos (1991) recently reported that shear stress increases the levels of intracellular cyclic guanosine monophosphate (cGMP) in HUVECs in the presence of the phosphodiesterase inhibitor, 3-isobutyl-1-methylxanthine, indicating that guanylate cyclase was activated by shear stress. This could lead to cGMP-dependent protein kinase activation and other responses, and possibly cause feedback inhibition of inositol triphosphate formation (Lang and Lewis, 1991). It does not appear, however, that EDRF release is feedback regulated by cGMP since increased cGMP levels do not influence agonist-induced release of EDRFs in BAECs (Kuhn et al., 1991). Nitrovasodilators inhibit endothelin production in HUVECs (Saijonmaa et al., 1990); therefore, it is possible that the flow-induced release of EDRFs could play a role in the reduction of endothelin synthesis by flow in HUVECs (see Section II,B,2).

4. Adenosine Trisphosphate, Acetylcholine, and Substance P

It is known that endothelial cells release adenine nucleotides when they are exposed to a variety of stimuli (Pearson and Gordon, 1979) and have the capability of synthesizing neurotransmitters which also act as vasodilators, such as acetylcholine and substance P (Milner et al., 1989). In addition, there is evidence that fluid flow stimulates the release of substance P from the endothelium of rat arteries in vivo (Ralevic et al., 1990).

Milner et al. (1990b) has performed studies with HUVECs grown on microcarrier beads placed in 1-mL disposable syringes. The packed columns were perfused with buffer at a rate of 0.5 mL/min for approximately 30 min, after which there was a step change in the flow rate to 1.5 mL/min. Adenosine triphosphate was continuously released at the lower flow rate, but a burst of ATP occurred at the onset of the higher flow rate. The release rate of ATP returned to nearly basal levels within a minute. A second stimulation, 10 min after the flow rate was reduced to 0.5 mL/min, led to a smaller peak of ATP in the effluent. The release of substance P and acetylcholine was also measured, and increases in the release rate of these neurotransmitters were found in 25–33% of all microcarrier columns tested. The rates were highly variable from column to column. It is not specified if the preparations that released substance P also released acetylcholine. No lactate dehydrogenase was released from the packed beds, suggesting that cell integrity was not compromised in the process.

A study by Bodin et al. (1991) with rabbit aortic endothelial cells placed on 3-μm-pore filters that were perfused at different flow rates led to a similar observation of flow-induced release of ATP. There was a small reduction in the cell-associated lactate dehydrogenase activity after subjecting the cells to flow.

5. Platelet-Derived Growth Factor

Platelet-derived growth factor (PDGF) is a potent stimulator of smooth-muscle cell growth and a vasoconstrictor produced by endothelial cells (Ross et al., 1986; Berk et al., 1986). It consists of dimers made of so-called A and B chains, each expressed by a different gene (Betsholtz et al., 1986; Swan et al., 1982). A-A, A-B, and B-B dimers have been isolated.

In a study by Hsieh et al. (1991), HUVECs subjected to fluid flow exhibited a transient increase in PDGF mRNA A and B chains, with peak levels observed for both 1.5–2 h after the flow was started, then returning to the stationary control values after 4 h. The peak levels were approximately 10 and 3–4 times the control levels for PDGF mRNA A and B chains, respectively. The response at 2 h was also dependent on the intensity of the shear stress. For the A chain, the maximum response was observed at 6 dyn/cm^2, whereas the response for the B chain exhibited a maximum at 6 dyn/cm^2 and a minimum at 31 dyn/cm^2 (Fig. 4). mRNA levels of glyceraldehyde 3-phosphate dehydrogenase, a constitutively expressed protein, were measured as well and used as internal standards as in the work of Diamond et al. (1990). No secreted PDGF could be measured

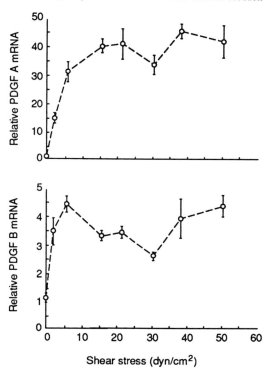

FIGURE 4   Peak levels of platelet-derived growth factor mRNA in sheared HUVECs. HUVECs were subjected to various levels of shear stress for 2 h, after which a Northern blot analysis on PDGF A and PDGF B was performed. The mRNA levels of GAPDH, a constitutively expressed protein, were used to normalize the data. Data shown represent the normalized optical density of the hybridization bands ± SEM. The results show a complex shear-dependent behavior of PDGF A and B gene expression. [Reprinted from Hsieh et al. (1991) by permission of the American Physiological Society. Copyright © 1991.]

in the medium, however, presumably because of the very small amounts released (Hsieh, unpublished data), but the increased mRNA levels suggest that the actual production of PDGF could be enhanced by fluid flow. The mechanism for the stimulation of PDGF gene expression by fluid flow has not been studied yet, but is likely to be dependent on the activation of protein kinase C. Thrombin (Harlan et al., 1986) and endotoxin (Albeda et al., 1989) stimulate the production of PDGF by HUVECs and human pulmonary endothelial cells, respectively. These agonists induce phospholipid turnover, and the resulting increased levels of diacylglycerol may activate protein kinase C. Moreover, the stimulation of this kinase by phorbol ester leads to an enhanced release of PDGF in microvascular endothelial cells (Starksen et al., 1987). As mentioned earlier, there is a strong possibility that protein kinase C is activated by shear stress.

The significance of the increase in PDGF gene expression by shear stress is twofold. Physiologically, the endothelium mediates the adaptation of blood vessel size to long-term changes in blood flow rate, with the result that an optimal wall shear stress is maintained (Kamiya and Togawa, 1980; Langille and O'Donnell, 1986). PDGF may be involved in this process since it is a mitogen for smooth-muscle cells. Pathologically, for the same reason, PDGF may also play a role in atherogenesis since overgrowth of smooth-muscle cells is an important factor in the formation of atherosclerotic plaques.

It is generally believed that endothelial cells do not possess PDGF receptors; however, it has been recently found that human microvascular endothelial cells express receptors for PDGF (Beitz et al., 1991). It is not known if such cells respond to flow similarly to HUVECs, but if it is the case, flow-induced PDGF release may possibly have an autocrine action in microvessels. This would suggest that in capillaries, which do not have smooth-muscle cells, endothelial cells may perform some smooth-muscle as well as endothelial functions.

C. Mediators of Inflammatory Responses and Endothelial Permeability

1. Interleukin 6

Cytokines are proteins produced by a variety of cells and are involved in inflammatory and immune responses. Interleukin 6 (IL-6) is an inflammatory mediator secreted by endothelial cells and is inducible by a variety of inflammatory substances, such as endotoxin and interleukin 1 (Jirik et al., 1989). No effect of IL-6 on endothelial cells has been found yet (for review, see Pober and Cotran, 1990).

Wright et al. (1991) found that shear stresses in the range 1–50 dyn/cm^2 reduce IL-6 secretion and mRNA levels in human aortic endothelial cells. Inhibition of IL-6 secretion was approximately threefold at 26 dyn/cm^2 after 24 h. The decrease in mRNA levels was only about half of that, suggesting that other regulatory pathways may be involved. Interestingly, IL-6 by itself was also shear-sensitive, as its decomposition rate in the flow system was proportional to shear stress. Since atherosclerotic plaques preferentially localize in regions of low wall shear stress, the higher secretion of IL-6 in areas of low shear may contribute to the atherogenetic process.

2. Histamine-Forming Capacity and Permeability to Albumin

The endothelium selectively controls the transport of macromolecules across the vascular wall, and the reduced effectiveness of this barrier resulting from endothelial damage may be involved in the pathogenesis of atherosclerosis.

The first flow studies on endothelial permeability were based on the hypothesis that histamine released by endothelial cells could be a major factor mediating vascular permeability. Primary BAECs subjected to shear stresses between 2.8 and 6.2 dyn/cm^2 for 1.5 h showed at 2.3–3.7-fold increase in histidine decarboxylase activity when compared to static controls (Rosen et al., 1974). Experiments in perfused rabbit aorta showed that increasing the shear stress from 8 to 27 dyn/cm^2 led to a fivefold enhancement in the activity of this enzyme, but a further increase in shear stress to 800 dyn/cm^2, causing the removal of the endothelium, resulted in a 50% decrease of histidine decarboxylase activity, thereby showing that most of the histamine-forming capacity was located in the endothelium (Hollis and Ferrone, 1974). There was a strong correlation between histamine-forming capacity and permeability to fluorescently labeled bovine serum albumin in rabbit aorta subjected to pulsatile flow with mean shear stresses in the range of 7–21 dyn/cm^2 (DeForrest and Hollis, 1980) and 22–109 dyn/cm^2 (DeForrest and Hollis, 1978). However, Caro et al. (1985) later found that the histamine receptor antagonist mepyramine does not significantly inhibit the effect of shear stress on transmural flux in excised arteries, which rules out the hypothesis that flow-induced changes in transendothelial permeability are mediated by histamine. Increased intracellular calcium and activation of protein kinase C increase transendothelial permeability (Lynch et al., 1990; Oliver, 1990), while cAMP causes a decrease (Stelzner et al., 1989; Oliver, 1990). Since fluid flow has been shown to induce increases in cAMP levels and that protein kinase C is likely to be activated during flow, it seems that two antagonistic pathways seem to be induced at the same time. Given the overall effect of shear on endothelial permeability, it suggests that protein kinase C activation prevails, but that cAMP could act as a moderator.

Detailed studies on transendothelial permeability in vitro have been done using BAEC monolayers grown on polycarbonate filters. The cells were subjected to flow on their luminal side while the amount of albumin that crossed the monolayer to the abluminal region could be determined (Jo et al., 1991). The hydrostatic pressure difference across the monolayer was maintained equal to zero to prevent any convective flux. The transendothelial permeability to (fluorescein isothicyanate) FITC-labeled albumin was increased 10-fold at 10 dyn/cm^2 and approximately fivefold

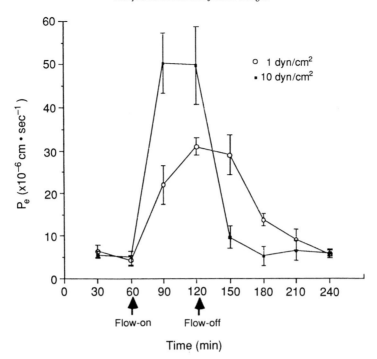

FIGURE 5   Transendothelial permeability of BAECs subjected to flow. BAECs were grown on polycarbonate filters and subjected to flow for up to 4 h in minimal essential medium with 1% bovine serum albumin. The permeability of the endothelial monolayer was determined by measuring the diffusive flux of fluorescently labeled albumin across the plane of the cells. Arrows indicate the times at the onset and the termination of shear. Transendothelial permeability was both shear- and time-dependent ($n = 6$ and 9 for data at 1 and 10 dyn/cm^2, respectively). Error bars represent the SEM. [Reprinted from Jo et al. (1991) by permission of the American Physiological Society. Copyright © 1991.]

at 1 dyn/cm^2 (Fig. 5). The return to basal levels after the flow was stopped was slower in the low-shear experiments. Given the magnitude of the increase of transendothelial flux, the most plausible explanation seems to be the widening of intercellular junctions during exposure to shear stress. The precise mechanism by which intercellular gaps could form during flow is not clear, but alterations in the structure of the microfilament network correlate with increased vascular permeability in endothelial cell monolayers exposed to ethchlorvynol (Wysolmerski and Lagunoff, 1985) and thrombin (Garcia et al., 1986), and may therefore also be involved in flow-induced changes in transendothelial permeability. It is well known that fluid flow induces microfilament reorganization in HUVECs within 2 h (Franke et al., 1984).

D. Cell–Substrate Adhesion and Fibrinolysis

1. Fibronectin

Fibronectin is a protein that enters in the composition of the extracellular matrix, which allows endothelial cells to adhere and spread on the substrate. Endothelial cells are attached to their matrix by a chain of proteins that link intracellular cytoskeletal actin to fibronectin receptors located on the basal side of the cells (Pytela et al., 1985). Glass surfaces precoated with fibronectin promote endothelial cell adhesion significantly more than surfaces treated with albumin or serum (Ivarsson et al., 1989). Fibronectin is secreted by endothelial cells preferentially on their basal side (Unemori et al., 1990) and may be important in modulating capillary endothelial cell shape and growth (Ingber, 1990).

Intracellular fibronectin content, the amount of fibronectin released in the medium and that incorporated in the extracellular matrix were all found to be reduced in HUVECs subjected to flow compared to stationary controls (Gupte and Frangos, 1990). The fibronectin levels were lowest at 12 h and increased after 48 h in the flow system. Static controls incubated in the presence of the protein synthesis inhibitor cycloheximide exhibited reduced intracellular levels of fibronectin and an inhibition of its release. The extent of inhibition was very similar to the effect flow on the cells, suggesting that flow may have completely inhibited fibronectin synthesis. Protein kinase C may be involved in the mechanism, since activation of this enzyme by phorbol ester inhibits the release of fibronectin from HUVECs in the culture medium (Reinders et al., 1985).

The significance of the inhibition of fibronectin synthesis by fluid flow is unclear. In fact, we have observed in our laboratory that HUVECs subjected to flow are more strongly attached to their substrate. It is possible that flow-induced spatial redistribution of the adhesion receptors and reorganization of the microfilament network (Franke et al., 1984) in confluent cell monolayers are more important factors in the regulation of cell–substrate adhesion than the absolute amount of fibronectin synthesized.

2. Tissue Plasminogen Activator

The fibrinolytic system of endothelial cells has been reviewed by Podor et al. (1988). Tissue plasminogen activator (tPA) converts plasminogen into plasmin in the presence of blood clots to break them up. The synthesis of tPA in human endothelial cells is stimulated by butyrate (Kooistra et al., 1987), histamine, and thrombin (Levin and Santell, 1988; Hanss and Collen, 1987), and by basic fibroblast growth factor in BAECs (Sato and Rifkin, 1988). tPA binds on the surface of HUVECs by interacting with plasminogen activator inhibitor-1 (PAI-1) to form complexes released in the culture medium (Russell et al., 1990) and can induce the synthesis of PAI-1 in HUVECs (Fuji et al., 1990).

HUVECs exposed to shear stresses up to 25 dyn/cm^2 secreted tPA at a nearly constant rate after a lag period of 6 h (Diamond et al., 1989). The rates of production were 0.83, 2.1, and 3.1 times the basal rate of 0.168 $\pm$ 0.053 ng/10^6 cells/h for cells subjected to 4, 15, and 25 dyn/cm^2, respectively. The authors also reported that pulsatile flow at a frequency of 1 Hz did not further enhance the production rate. The production of plasminogen activator inhibitor type 1 (PAI-1) was unaffected by fluid flow (basal rate was 53 $\pm$ 37 ng/10^6 cells/h), although a statistically non-significant flow-induced decrease was reported in a later publication (Diamond et al., 1990), with an overall result that high shear stress would enhance the capacity of the endothelium to activate plasminogen and contribute to its nonthrombogenicity. This is consistent with the observed increases in fibrinolytic potential in blood after exercise (Johnson et al., 1986). Similar observations have been made on human saphenous-vein endothelial cells subjected to cyclic stretch (Iba et al., 1991). Enhanced rates of tPA production correlated with increased levels of mRNA when extracted from the cells after 24 h of exposure to flow (Diamond et al., 1990). mRNA levels of glyceraldehyde 3-phosphate dehydrogenase were unchanged by shear stress and were used as an internal standard to normalize the data. mRNA levels of basic fibroblast growth factor were found not be be affected by fluid flow.

The mechanism of the flow-induced enhancement of tPA production is still unclear. Prostacyclin and cAMP have been found not to affect tPA synthesis (Diamond et al., 1989), but there are indications that protein kinase C may be involved. The activation of this enzyme by phorbol ester enhances the expression of tPA mRNA in HUVECs (Levin and Santell, 1988), and more recent findings show that the production of tPA is stimulated by diacylglycerol, phorbol ester, and thrombin, while only the last two agonists lead to an increased production of PAI-1 (Grulich-Henn and Müller-Berghaus, 1990). The authors suggest that diacylglycerol may induce a transient activation of protein kinase C, whereas phorbol ester and thrombin would lead to a more sustained stimulation, explaining the difference in the responses for PAI-1. Alternatively, different isozymes of protein kinase C could be activated by the different stimuli, leading to various responses. Data for the stimulation by diacylglycerol seems to be most consistent with the results of the flow studies, and diacylglycerol levels are likely to be transiently increased by flow since diacylglycerol is produced concommitently with inositol triphosphate by phospholipase C (see Section II.A.1). This suggests that protein kinase C activation resulting from the increased diacylglycerol levels may mediate the flow-induced increase in tPA synthesis. This higher tPA production rate could be important for wound healing of blood vessels, which requires cell division and migration to restore the integrity of the endothelium. tPA causes the formation of plasmin and has been shown to facilitate endothelial cell migration (Inyang and Tobelem, 1990; Mawatari et al., 1991). Low levels of shear stress (≤ 1.7 dyn/cm^2) have been shown to stimulate migration and proliferation of BAECs into a denuded area (Ando et al., 1987). The mechanism may involve basement membrane degradation and protease activation (Moscatelli and Rifkin, 1988). Shear stress may therefore promote the restoration of a nonthrombogenic surface by stimulating cell migration via tPA synthesis as well as cell division in nonconfluent regions as suggested by studies on DNA synthesis in BAECs (Ando et al., 1990).

E. Adhesion of Suspended Cells to Endothelial Monolayers

1. Neutrophils

Neutrophils are normally present in the blood and are involved in inflammatory responses and host defense. They cross the endothelial cell barrier in order to reach inflammatory sites, a process that follows their adhesion to the endothelium and sometimes their rolling along the luminal surface of the blood vessel.

The adherence of bovine neutrophils on bovine endothelial cell monolayers has been studied by Worthen et al. (1987). The neutrophils were deposited on bovine pulmonary endothelial cell monolayers in a cone-and-plate viscometer, where shear stresses of up to 2 dyn/cm^2 were used. There was a shear-dependent reduction in the percentage of adherent neutrophils to the endothelial cells, a trend that was also observed on serum-coated plastic without endothelial cells. Similar results were obtained with and without the cyclooxygenase inhibitor meclofenamate, indicating that cyclooxygenase products were not responsible for the effects seen. Bovine zymosan-activated plasma enhanced adherence at near-zero shear stress, but no difference in adherence was found with cells in normal platelet-poor plasma at shear stresses of 0.5 dyn/cm^2 and above. The use of low- and high-viscosity media showed that neutrophil adherence was correlated with the shear stress levels, but not the shear rates

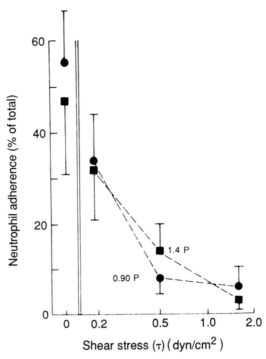

FIGURE 6   Adhesion of neutrophils to BAEC monolayers at several shear stresses. Adherence was reduced as the wall shear stress was increased. Using media of two different viscosities, the adherence–flow curves could be superposed if plotted as a function of the wall shear stress. [Reprinted from Worthen et al. (1987) by permission of the American Physiological Society. Copyright © 1987.]

(Fig. 6). The authors suggest that shear stress, mainly acting as a dragging force, was the cause of the shear-dependent decrease in adhesion. No conclusions could be drawn yet on the respective contributions of cell shape and cell–cell adhesion in the phenomena observed.

A similar study was performed with HUVECs and human polymorphonuclear neutrophils (PMNLs) using a parallel plate flow chamber (Lawrence et al., 1987). The results obtained were qualitatively similar to those of Worthen et al. (1987). At shear stresses below 3 dyn/cm^2, interleukin 1 enhanced the adherence of PMNLs on endothelial monolayers, but above 3 dyn/cm^2, there was no significant difference in the adhesion of PMNLs on activated and normal endothelial cells. PMNLs can roll along the endothelium of postcapillary venules in vivo. Rolling velocities of PMNLs could be measured in the parallel-plate flow system, and these values (< 1 μm/s) were at least two orders of magnitude less than the velocity of unattached PMNLs, thereby indicating that strong adhesive forces were present. The mean rolling velocity was doubled when the wall shear stress was increased from 1 to 2 dyn/cm^2.

The work on PMNLs and HUVECs was extended to study the roles played by glycoprotein CD11/18 on PMNLs and their respective ligand on the endothelial cell, intercellular adherence molecule-1 (ICAM-1) (Lawrence et al., 1990). Adherence of PMNLs to endothelial cell monolayers is increased by interleukin 1 and lipolysaccharide, which enhance the expression of ICAM-1 at the surface of endothelial cells (Bevilacqua et al., 1985), and f-Met-Leu-Phe (FMLP), which upregulates CD18 expression in PMNLs (Anderson et al., 1986). Using HUVECs pretreated with interleukin 1, the antibody against CD18 (a subunit of CD11/18) significantly inhibited adhesion under static conditions, whereas antibodies against CD18 and ICAM-1 did not inhibit adhesion of neutrophils at a wall shear stress of 2 dyn/cm^2. This suggests that the interaction of CD11/18 with ICAM-1 is not likely to be involved in the initial attachment of neutrophils to endothelial cells in regions where the wall shear stress is around 2 dyn/cm^2. This interaction may be important, however, in transendothelial migration of neutrophils since both antibodies inhibited the migration of PMNLs across the endothelial monolayer under flow conditions. FMLP significantly enhanced the adherence of PMNLs on endothelial monolayers that were not activated by interleukin 1, but the enhancing effect was gradually reduced as the wall shear rate was increased, and was completely inhibited at a wall shear stress of 1 dyn/cm^2 and above. Adherence of FMLP-treated and unstimulated PMNLs was inhibited by antibodies to CD18 at 0.5 dyn/cm^2, demonstrating the important role of the CD11/18–ICAM-1 interaction at low levels of shear stress. Adhesion of neutrophils could be observed at shear stresses as high as 3 dyn/cm^2, suggesting that other mechanisms bind PMNLs to HUVECs. Rolling velocities were reduced four- to fivefold by FMLP on interleukin 1-treated HUVECs, but were not different from that on nonactivated HUVECs, and therefore seemed to be independent of the interaction of CD11/18 with ICAM-1. FMLP tends to cause the spreading of PMNLs, therefore the inhibitory effect of FMLP on the rolling velocity was attributed to its effect on cell shape.

2. Erythrocytes

Interactions between erythrocytes and endothelial cells have also been investigated. Previous studies have suggested that the frequent occurrence of obstruction of blood vessels in the microvasculature in sickle cell anemia may be caused by the unusually high adherence of sickle cells to the endothelium.

Barabino et al. (1987) subjected HUVEC monolayers precoated with red blood cells for 10 min to a shear stress of 1 dyn/cm^2 in a parallel-plate flow chamber, after which the number of cells that remained attached on the monolayer surface was determined. Homozygous sickle cells were approximately 6–7 times as adherent as normal red blood cells. Cells carrying only one sickle cell gene behaved similarly to normal cells. Presheared red blood cells in a shear field of 1500 dyn/cm^2 for 2 min were more adhesive than normal erythrocytes, but not as much as sickle cells. The adherence after 20 min of washout was determined at different shear rates. The difference between the two phenotypes was largest at low shear rates, especially below 100 s^{-1}, which corresponds to a shear stress of 1 dyn/cm^2, suggesting that vasoocclusion would occur most likely in vessels where low wall shear stresses exist, such as venules. In addition, analysis of the velocity distribution of normal and sickle cells moving on the endothelial monolayer showed that sickle cells travel at approximately half the average speed of the normal cells. Hemoglobin molecules of sickle cells tend to polymerize and precipitate at low-oxygen partial pressures, causing the red blood cells to take a sickle shape and become stiff; therefore, the higher transit time of sickle cells increases the risk of gel formation and clogging in capillaries and venules. The mechanism that renders sickle cells more adherent may be related to their younger age, which was reflected in the high reticulocyte content of the sickle cell suspensions, and also to their susceptibility to damage at physiological shear stress levels.

F. Proteoglycans

Protoheparan sulfates are compounds secreted by endothelial cells that associate with the cell surface, enter in the composition of the extracellular matrix, and are also released in the culture medium (Keller et al., 1987a, 1987b). Three types have been identified: protoheparan sulfate I (HSI), bound to the plasma membrane and slowly released into the culture medium by a shedding mechanism, HSII, associated with the extracellular matrix and also released into the medium and HSIII, entirely secreted into the medium.

The production pattern of these compounds by endothelial cells in culture is very different from that of the endothelium of perfused excised aorta (Keller et al., 1987b). Grimm et al. (1988) hypothesized that this may come from the difference in the flow conditions for cultured cells and cells

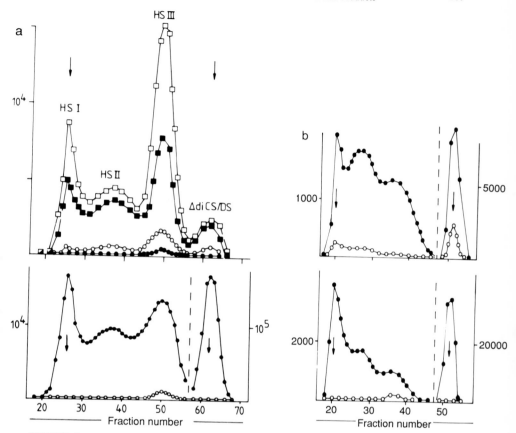

FIGURE 7   Effect of flow on endothelial protoheparan sulfates. BAECs were sheared in sulfate deficient DMEM + 15% fetal bovine serum for 60 min, after which ^{35}S was added, and the cells were sheared for a further 60 min. (a) and (b) represent protoheparan sulfates released in the medium and associated with the cells, respectively (ΔdiCS/DS = unsaturated disaccharides). These were separated on a Sepharose CL-6B column following digestion by chondroitinase ABC. (a) Upper graph: medium from BAECs exposed to shear rates of 5 s^{-1} ($\square$), 10 s^{-1} ($\blacksquare$), 50 s^{-1} ($\circ$), 100 s^{-1} ($\bullet$). Lower graph: medium from BAECs maintained under static conditions ($\bullet$) and perfusate from a bovine aorta ($\circ$). (b) Upper graph: elution pattern of protoheparan sulfates in BAECs exposed to a shear rate of 100 s^{-1} ($\bullet$) and maintained under static conditions ($\circ$). Lower graph: protoheparan sulfates released in the medium from the same cells. The ordinates represent the radioactivity in counts per min of ^{35}S in each fraction. Arrows indicate the void and total volumes of the column. [Reprinted from Grimm et al. (1988) by permission of F. K. Schattauer Verlagsgesellschaft mbH. Copyright © 1988.]

in vivo. BAECs were incubated in sulfate-depleted medium for 60 min, and then in [^{35}S]-sulfate-containing medium for a further 60 min. The incubations were performed under flow conditions using steady shear stresses ranging from 0.05 dyn/cm^2 up to 1 dyn/cm^2. The amounts of ^{35}S in the different protoheparan sulfates were determined in the medium and the cells together with their extracellular matrix. Flow completely inhibited the release of HSI and HSII in the medium and sharply reduced the secretion of HSII (see Fig. 7a). The production of protoheparan sulfates that remained associated with the cells was strongly inhibited (see Fig. 7b). Unsaturated chondroitin sulfate disaccharides formed the major group of proteoglycans present in the sheared cells, followed by HSI and HSII, and very little HSII. These findings show that low levels of shear stress cause a general inhibition of protoheparan sulfate production in BAECs, and also induce a polarized production pattern that resembles that of endothelial cells in aortic preparations.

The physiological relevance of these results is difficult to assess since the shear stress levels used in this study were very low. Grimm et al. (1988) suggested that the protoheparan sulfates released by static endothelial cells may bind antithrombin III and interfere with the coagulation cascade. The dramatic increase in the release of protoheparan sulfates by endothelial cells when the shear stress approaches zero may therefore serve as an antithrombogenic mechanism.

G. Endocytosis

1. Uptake of Low-Density Lipoproteins

Low-density lipoproteins (LDLs) are lipid–protein complexes composed of various types of lipids and apoproteins, and transport the lipids throughout the body. In order to reach hepatocytes and other cells where the LDLs are metabolized, they must necessarily go through the endothelium. Thus endothelial cells are potentially important regulators of the transport of LDLs. Sprague et al. (1987) studied the effect of fluid flow on the uptake of LDLs as well as on the expression of LDL receptors by BAECs. The cells were subjected to shear stress for 24 h, after which radiolabeled LDL was added. They found that cells exposed to 30 dyn/cm^2 of shear for a further 24 h internalized 69% more LDLs than cells subjected to a shear of less than 1 dyn/cm^2. In some experiments, radiolabeled LDLs were added concurrently with the onset of shear. In that case, high shear increased the uptake of LDLs by 34% after 24 h. The enhanced internalization appeared to be related to an increase in the receptor-mediated endocytosis of LDLs rather than an increase in nonspecific uptake by fluid-phase endocytosis.

The authors also verified that the number of LDL receptors at the surface of the cells was significantly increased in cells subjected to 30 dyn/cm^2 for 24 h compared to cells maintained under static conditions. Although the cells subjected to high shear exhibited notable changes in morphology, there was no significant cell turnover during the experiments described above. The mechanism of enhanced endothelial LDL uptake by shear stress is not known. Its physiological significance is also unclear, but the elevated transport of LDLs to the arterial wall resulting from higher shear may play a role in cholesterol homeostasis and affect the accumulation of lipids in the vascular intima.

2. Pinocytosis

The effects of shear stress on fluid-phase endocytosis, also called pinocytosis, have been studied in BAECs by measuring the uptake of horseradish peroxidase (Davies et al., 1984). The average pinocytosis rate for the first hour was increased approximately twofold when the shear stress was increased from 0 to 8 dyn/cm^2, and threefold further for an increase from 8 to 15 dyn/cm^2 (Fig. 8). After the onset of a shear stress of 8 dyn/cm^2, the pinocytosis rate was highest during the first 2 h and returned to control values following a second order oscillatory response (Fig. 8b), suggesting that complex feedback mechanisms occur inside the cell, eventually leading to its adaptation to the existing flow. Stopping the flow after a 48-h exposure could also increase the pinocytosis rate, demonstrating that changes in shear stress are more important in stimulating pinocytosis than the absolute value of shear stress used. The effect of pulsatile flow was investigated, and it was found that repeated step changes at 5-min intervals had no effect, whereas a 15-min cycle would induce a significantly higher pinocytosis rate, which was still very much sustained after 4 h of exposure. Oscillating shear stress between 3 and 13 dyn/cm^2 with a frequency of 1 Hz did not stimulate pinocytosis. This suggests that the time constant for the stimulation of endocytosis is probably between 5 and 15 min. It also shows that some endothelial responses are somehow adapted to recognize only relatively slow hemodynamic changes that may reflect changes in the environment that may necessitate adaptation, rather than to rapid variations of shear, which normally occur anyway because the pulsatile nature of blood flow.

Mathematical models suggest that shear stress may cause instabilities in the membrane leading to the formation of vesicles (Gallez, 1984; Steinchen et al., 1981); however, it is also believed that pinocytosis is an active phenomenon that requires energy and that is related to the function of the cytoskeleton. Since shear stress triggers the formation of several secondary messengers and leads to the reorganization of the cytoskeleton, it is most

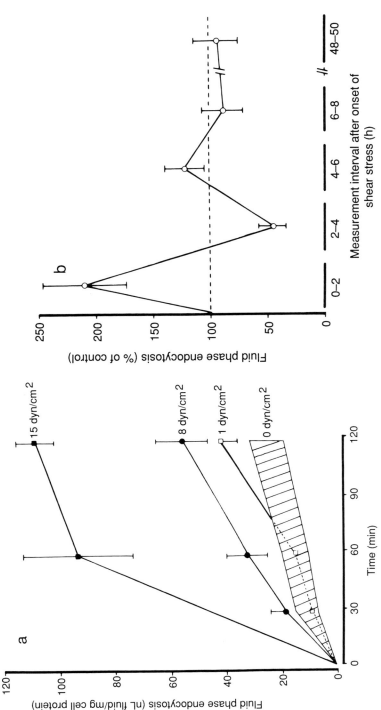

FIGURE 8 Effect of steady flow on pinocytosis in BAECs. (a) Five minutes after the imposition of the indicated shear stress, horseradish peroxidase (HRP) was injected in the perfusing medium (time 0). The uptake of HRP was determined at different times ($n = 4$). (b) Endothelial cells were exposed to a shear of 8 dyn/cm^2 for the indicated times, after which the uptake of HRP over a period of 2 h was determined. Static cultures were used as controls. Error bars represent the standard deviation. [Reprinted from Davies et al. (1984) by permission of the American Society of Clinical Investigation. Copyright © 1984.]

probable that flow modulates pinocytosis via an active physiological mechanism, the details of which still remain unknown.

H. Turbulent-Flow Studies

All the studies mentioned so far in this chapter were done under laminar flow conditions. Observations performed in vivo demonstrate the existence of zones where the flow is highly irregular and possibly turbulent. These regions are particularly prone to atherosclerotic lesions (Roach, 1977), therefore turbulent flow may be detrimental to endothelial cell viability.

BAECs were exposed to laminar and turbulent flow of low intensity (1–1.5 dyn/cm^2) and high intensity (14–15 dyn/cm^2) for times of up to 24 h (Davies et al., 1986). The response under laminar flow was similar to what has been reported by other authors, where cell alignment was observed at the higher shear stress. Turbulent flow, however, did not induce any cell orientation and caused significant cell loss after 24 h. DNA synthesis was simultaneously monitored by [^{3}H]-thymidine incorporation. Laminar flow did not affect the rate of DNA synthesis in confluent monolayers, whereas turbulent flow induced a significant increase in [^{3}H]-thymidine, even at shear stresses as low as 1.5 dyn/cm^2. The increased DNA synthesis suggests that an important fraction of the cells were in the mitotic phase, which the authors confirmed by cytofluorometry of the cells after labeling with the nuclear dye propidium iodide. Dividing cells are presumably less strongly attached to the substratum, which may explain the apparition of gaps observed in the monolayer after long exposure to turbulent flow.

I. Summary

The general effect of shear stress on the secretion of products by endothelial cells is shown in Figure 9. The early events in shear-induced stimulation of endothelial cells were previously summarized in Figure 3. Briefly, shear activates phospholipases with the concomitant release of 1,4,5-inositol trisphosphate, diacylglycerol and arachidonic acid, the latter of which being the precursor of prostaglandins. Whether or not intracellular calcium is increased by flow is still controversial, but it appears that the presence of calcium is important in the flow-induced response of prostacyclin production. Increased intracellular calcium would lead to the activation of the EDRF synthase pathway. EDRF may activate soluble guanylate

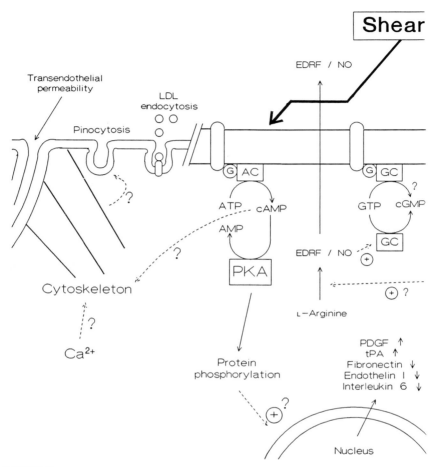

FIGURE 9   Summary of the effect of flow on the physiological response of endothelial cells. See text for explanations. Symbols used are as follows: PI, phosphatidylinositol; DAG, 1,2-diacylglycerol; P, phospholipid; AA, arachidonic acid; PGI_2, prostacyclin; G, G protein; PLC, phospholipase C; PLA_2, phospholipase A_2; PKC, protein kinase C; IP_3, inositol 1,4,5-triphosphate; GC, guanylate cyclase; ATP and GTP, adenosine and guanosine 5'-tri-

cyclase, leading to increased cGMP levels. There is a recent report that shear activates guanylate cyclase in HUVECs (Kuchan and Frangos, 1991). It is not known whether particulate guanylate cyclase is activated by shear. Protein kinase C is likely to be activated by calcium and the free diacylglyc-erol, and could modulate the function of phospholipase C and phospho-

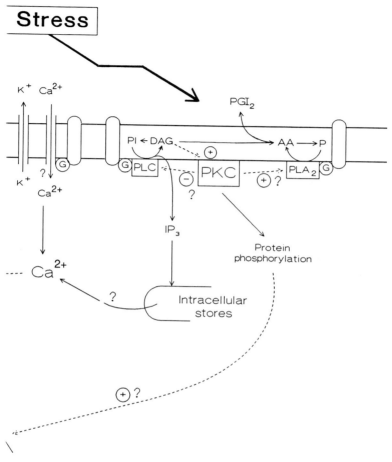

phosphate, respectively; cAMP and cGMP, adenosine and guanosine 3′:5′-cyclic-mono-phosphate, respectively; EDRF, endothelium-derived relaxing factor; AMP, adenosine 5′-monophosphate; AC, adenylate cyclase; PKA, protein kinase A; PDGF, platelet-derived growth factor; tPA, tissue-type plasminogen activator; LDL, low-density lipoprotein.

lipase A_2. In addition, protein kinase C is probably involved in triggering nuclear events. Another candidate that may be involved in gene expression is cAMP, which is increased by an unknown mechanism, possibly by direct activation of adenylate cyclase by shear. Cytosolic calcium and cAMP may be involved in the control of transendothelial permeability and possibly pinocytosis via a cytoskeleton-mediated mechanism. The rate of receptor-mediated LDL endocytosis is also increased by shear, but the mechanism is unknown at this time.

III. OSTEOBLASTS

Osteoblasts are the bone-producing cells that   line the surfaces in trabecular bone. When bones are mechanically loaded, flow of the extra-capillary fluid filling the trabecular meshwork is induced. It has been hypothesized that this flow stimulates bone-producing cells such as os-teoblasts and thereby could induce bone remodeling.

To study the effect of fluid flow on the metabolism of osteoblasts, the cells were cultured and exposed to flow in vitro. Osteoblasts were sub-jected to shear rates in the range of $10–3500$ s^{-1}, corresponding to shear stresses of $0.1–35$ dyn/cm^2, in the presence of the phosphodiesterase inhibitor 3-isobutyl-1-methylxanthine (Reich et al., 1990). The levels of adenosine $3':5'$-cyclic monophosphate (cAMP) after 15 min of exposure to flow were significantly increased compared to stationary controls, but there was no statistically significant difference between the cAMP levels of cells subjected to the different shear rates used. The observed increase in cAMP after 15 min of exposure to flow reached up to 16 times the control values. cAMP levels in cells subjected to a shear rate of 430 s^{-1} increased very rapidly as a function of time, followed by a further increase that was not statistically significant. The stimulation of cAMP levels by flow was inhibited by the cyclooxygenase inhibitor ibuprofen, indicating that prostaglandins mediate the cAMP response caused by shear stress in murine osteoblasts. Stretch-induced cAMP accumulation in osteoblasts has been also shown to be mediated by the cyclooxygenase-derived product prostaglandin E$_2$ (Binderman et al., 1984).

Prostaglandin E$_2$ synthesis was later shown to be stimulated by flow in shear-dependent manner (Reich and Frangos, 1991). The rate of release was increased 9- and 20-fold at 6 and 24 dyn/cm^2, respectively, when compared to static controls, which released 274 ± 15 pg/mg protein per hour. The flow-induced effect was inhibited by the cyclooxygenase in-hibitor ibuprofen.

The effect of flow on inositol 1,4,5-triphosphate (IP$_3$) levels was also investigated. Low shear stress (1 dyn/cm^2) did not affect IP$_3$ production, but high shear stress (24 dyn/cm^2) caused a transient increase in IP$_3$ levels, up to 17 ± 4 ng/mg protein. The stimulation of IP$_3$ production by flow in osteoblasts was significantly inhibited by ibuprofen and indomethacin, another cyclooxygenase inhibitor, which suggests that prostaglandin E$_2$

mediates the IP_3 as well as the cAMP responses. Addition of exogenous PGE_2 to stationary cultures of osteoblasts resulted in an increased production of IP_3, confirming the key role of PGE_2 as a mediator of phospholipase C activation.

These observations confirm that osteoblasts are sensitive to flow and support the hypothesis that extracapillary fluid flow induced by mechanical loading stimulates bone metabolism.

IV. OTHER CELL TYPES

Only four flow studies on other cell types have been reported. Stathopoulos and Hellums (1985) subjected human embryonic epithelial kidney cells to shear stresses in the range 2.6–54 dyn/cm^2 for up to 24 h in a parallel-plate flow chamber. Cell viability was 80% or more for cells subjected to shear stresses of 13 dyn/cm^2 or less, and decreased to 25% for cells exposed to 26 dyn/cm^2 or more after 24 h. The cells were clearly aligned to the flow after 24 h at 13 dyn/cm^2. The release of urokinase was measured after a 24-h exposure to flow, and was increased up to 2.5-fold compared to static controls.

In another investigation, baby hamster kidney fibroblasts were subjected to a shear stress of 4.3 dyn/cm^2 for 15 min (Reich et al., 1990). The levels of cAMP were increased fourfold relative to static controls.

Finally, the effects of flow on rat aortic smooth-muscle cells were studied by Garay et al. (1989) and Bodin et al. (1991). Upon washing the cells in Ringer's medium, Garay et al. (1989) observed intracellular calcium increases and a transient elevation in intracellular sodium levels. Intracellular potassium levels remained unchanged. The potential-dependent calcium channel blocker nitrendipine suppressed the effect of flow. To explain these results, the authors suggested that flow triggers the opening of voltage-dependent calcium channels, followed by internalization of extracellular medium by pinocytosis. The same treatment given to rat fibroblasts did not result in any response. Recently, Bodin et al. (1991) reported that rabbit aortic smooth-muscle cells do not release ATP when exposed to flow, whereas endothelial cells isolated from the same vessels do.

V. MECHANISM OF SHEAR STRESS ACTIVATION

There must be a mechanism by which physical forces are transmitted to the cell membrane from the moving fluid, causing some type of mechanical perturbation, which is then transduced into an intracellular chemical signal. Direct mechanical perturbation of cells can indeed lead to a physiological response: dimpling and poking at endothelial cells with a pipette results in transient increases in intracellular calcium levels while the membrane integrity is unaffected (Goligorsky, 1988). The actual mechanotransducers could be stretch-activated channels, which have been previously described in endothelial cells (Lansman et al., 1987), although their role in the response of mammalian cells to physical forces is controversial (Morris and Horn, 1991). A flow-activated potassium channel has also been characterized in endothelial cells (Olesen et al., 1988), but it is not yet clear whether it is directly activated by flow or is a secondary response to the triggering of other possible flow-induced biochemical events. While one may not exclude the possible role of specific mechanotransducers, it is also possible that the structures normally present in all cells may be sufficient to transduce the flow signal. In fact, any protein normally sensitive to conformational changes, such as G proteins, could be sensitive to mechanical perturbations and be potential candidates as flow sensors. In that case, the transduction mechanism would be rather nonspecific.

The mechanical forces may be sensed directly by a mechanotransducing protein to generate a signal, or alternatively cause an effect on the membrane, which would then mediate the activation of mechanotransducers. It is known that membrane properties affect membrane protein function (Viret et al., 1990, Carruthers and Melchior, 1986); therefore, it is conceivable that flow-induced changes in membrane shape and conformation could activate certain proteins and trigger enzymatic reactions. Recent evidence suggests that membrane bending caused by molecules specific for the inner and outer leaflets of the plasma membrane can generate forces sufficient to open ion channels (Martinac et al., 1991). When cell monolayers are subjected to flow, the kinetic energy is primarily dissipated in the moving fluid; however, as the cell membrane is directly in contact with it, it is plausible that some of the kinetic energy may be dissipated or stored in the cell membrane itself. The plasma membrane has a low bending modulus, which allows for the small diameter of curvature of vesicles (100–200 nm); therefore, it may be very susceptible to shape changes induced by external forces. Mechanical fluctuations in the plasma membrane of erythrocytes caused by thermal noise have been observed (Parpart and Hoffman, 1956). These cells also exhibit an increase in the passive permeability of the plasma membrane to calcium ions when subjected to flow (Larsen et al., 1981), which clearly suggests that flow can alter some basic properties of biological membranes. Berthiaume and Frangos (1991) have also found that endothelial cells subjected to flow have an increased permeability to the amphipath Merocyanine 540. Membrane fluidity itself appears to be an important modulator of endothelial cell function, but it is not known yet if it is affected by flow. Endothelium-dependent relaxation of vascular rings by unsaturated free fatty acids has been reported, the mechanism proposed being an effect on membrane fluidity (Cherry et al., 1983).

Other variations on the theme of this membrane perturbation hypothesis have been postulated. Most shearing forces are probably transmitted to the substrate where the cells are attached; therefore, the overall tangential force on a cell exposed to shear is concentrated on the adhesion plaques located on the basal side of the cell. This suggests that each adhesion site bears a significant amount of stress. It is possible that membrane domains in the vicinity of adhesion plaques are highly disturbed. While the activation of potassium channels could be observed in whole endothelial cells subjected to flow by Olesen et al. (1988), these flow-sensitive channels could not be seen in membrane patches isolated from the luminal side of the cells, which led the authors to hypothesize that they may be located on their basal side. Since adhesion proteins are linked to cytoskeletal actin filaments via transmembrane bridges, it has also been postulated that the internal tensions generated in the cytoskeleton may trigger physiological responses and may even directly send the shear signal to the nucleus (see Chapter 2). A more detailed discussion of this tensegrity mechanism is found in Chapter 2 in this book.

The glycocalyx bears a net negative charge (Vargas et al., 1989) which, in an electrolytic fluid, is covered by another layer of ions of opposite polarity tightly maintained by electrostatic forces. Additional ions loosely bind on the previous ionic layer, which result in a net opposite charge in the fluid in the vicinity of the surface of the cell monolayer. Fluid flow will cause these ions to travel along with the liquid, thereby creating an electric current, since a net charge is moving. The voltage difference associated with this current is, for a tube, $V = \epsilon \zeta L S / (\kappa_0 l)$, where ϵ and κ_0 are, respectively, the dielectric constant and the specific conductance of the liquid, l and L are the circumference and the length of the tube, S is the wall shear rate, and ζ is the zeta potential, defined as the potential difference between the hydrodynamic slip plane and the bulk of the fluid (Eriksson, 1974). It has been postulated that this "streaming potential" may stimulate cells in a flow field. One can recognize that the streaming potential varies as the flow rate is changed since it is directly proportional to the wall shear rate. The wall shear stress is the product of the wall shear rate by the viscosity of the perfusing medium; therefore, it is possible to study the individual effects of shear stress and streaming potential by using media of different viscosity, as in the study of Reich et al. (1990). Murine osteoblasts were subjected to shear rates of 215 s^{-1} with medium of 5 cP viscosity and of 1081 s^{-1} with medium of 1-cP viscosity, both producing the same shear stress. They observed that the response was not significantly

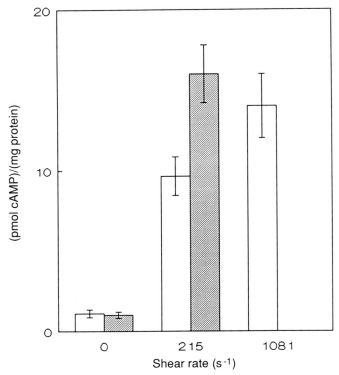

FIGURE 10   Effect of shear stress and shear rate on the flow-induced response of osteoblasts. Osteoblasts were subjected to flow in medium with a viscosity of 1 cP (□) and in
dextran-supplemented medium, with a viscosity of 5 cP (▨). The magnitude of the cAMP
flow-induced response increased as the shear stress was increased by changing the viscosity of
the medium. There was no significant difference between the responses when the wall shear
rate was increased while keeping the same shear stress by decreasing the viscosity of the
medium ($n = 5$). Error bars represent the SEM. [Reprinted from Reich et al. (1990) by
permission by Wiley-Liss, Inc. Copyright © 1990.]

different, as shown in Figure 10; however, values obtained at 215 s^{-1} with
medium of 1 cP were significantly lower than when dextran was added to
raise the viscosity to 5 cP, indicating that a reduction in shear stress at
constant shear rate resulted in a reduced stimulation. These results indicate
that shear stress modulates the flow-induced response in osteoblasts, but do
not support the hypothesis that streaming potentials are involved in
flow-induced stimulation.

Another effect of flow is to change the rate of delivery of an agonist
from the bulk of the fluid to the plasma membrane, where it may bind to
hormone receptors. In the case where the agonist is simultaneously broken

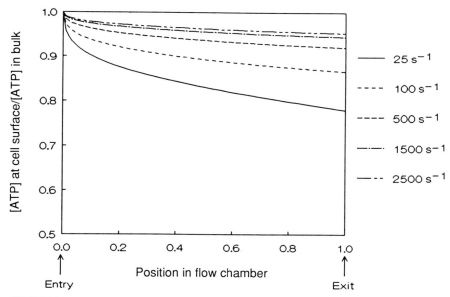

FIGURE 11   Effect of shear rate on the concentration of ATP at the cell surface in a parallel-plate flow channel. Theoretical calculations were performed using a simple partial-differential equation to calculate the concentrations of ATP throughout the flow field in the flow chamber. The kinetic parameters used for the endothelial ectoATPase were $V_{max} = 22$ nmol/10^6 cells/min, $K_m = 249$ μM; the cell density was 2×10^6 cells/25 cm^2; and the diffusivity of ATP was 6×10^{-6} cm^2/s. The flow channel length and thickness were 6 cm and 0.025 cm, respectively.

down by ectoproteases, its concentration at the surface of the cell will depend on its rates of transport and degradation. An increase in flow rate would increase the rate of delivery and effectively lead to a higher agonist concentration at the cell membrane, thereby stimulating the cell. Since endothelial cells possess a high ectoATPase activity and ATP receptors with high affinity, it has been postulated that convective transport of ATP from the bulk of the medium to the cell surface by flow can significantly increase its concentration in the vicinity of ATP receptors (Dull and Davies, 1991; and Mo et al., 1991). We performed theoretical calculations to predict the surface concentration of ATP on sheared cells in a parallel-plate flow chamber, the details of which are given in the appendix at the end of this chapter. Other authors have used slightly different parameters (Nollert et al., 1991), but the final results are fairly similar to ours, shown in Figure 11, which gives the concentration of ATP over the cell surface from the entrance ($Z = 0$) to the exit ($Z = 1$) of the flow chamber, assuming fully developed parabolic flow. Within the physiological range of shear stresses, the surface concentration of ATP is 75% of that in the bulk or more. Thus, if cells are maintained in ATP-free medium and then placed in a flow system with ATP containing medium, it is quite possible that ATP will stimulate the cells in addition to the flow. This may happen if the cells have been maintained in the same culture medium for a long time (> 12 h), since ATPases will have degraded all exogenous ATP present. It is therefore recommended to use old culture medium or ATP-free medium to carry out experiments. However, ATP is probably not involved in responses to changes in shear stress unless the affinity of the ATP receptor and/or the activity of ectoATPase is also modulated by flow. The bulk concentration of ATP in Medium 199 ($1.8~\mu M$) is at the lower end of the dose–response curve for ATP on endothelial cells as measured by PGI_2 production and IP_3 formation (Needham et al., 1987; Pirotton et al., 1987b), and the response to ATP is strongly down-regulated in HUVECs maintained in an ATP-containing medium (Toothill et al., 1988). This is consistent with our findings on the shear-induced IP_3 response in HUVECs, which was found to be the same in DMEM with 1% BSA (serum- and ATP-free medium) as well as in normal complete medium (with ATP) (Bhagyalakshmi and Frangos, 1989b). However, Dull and Davies (1991) and Mo et al. (1991) have found that calcium increases caused by flow in BAECs were present only in ATP-containing medium, suggesting that ATP may then be involved in several flow-induced responses. In the work of Dull and Davies (1991), flow inhibited calcium oscillations induced by a slowly hydrolyzable analog of ATP, an indication that flow may also directly affect the stimulus–response coupling. In addition, BAECs release EDRF and prostacyclin in ATP-free medium (Buga et al., 1991; Grabowski et al., 1985), both of which are usually dependent on calcium for their synthesis. These results suggest that flow by itself is a stimulus, and that ATP effects cannot explain the existence of a flow-dependent response in the physiological range of shear stresses. Considering that, in vivo, the average shear stress is above zero anywhere in the vasculature, the role of ATP in the normal flow-dependent responses of the blood vessels is uncertain. Moreover, the concentration of ATP in the blood is probably very low unless an injury causes the release of intracellular ATP from damaged cells. We have mentioned earlier that endothelial cells release ATP at the onset of step changes in flow. Some authors have suggested that this flow-induced release of ATP could be followed by an autocrine or paracrine effect on the same cells, as ATP is a well-known endothelium-dependent vasodilator. Bodin et al. (1991) measured the concentration of ATP in the outflow of his system, which peaked at $0.6~nM/10^6$ cells at 3 ml/min. For 2×10^6 cells/25 cm^2, this would correspond to a flux of 2.4×10^{-6} n-mol/cm^2/S from a sheared confluent monolayer. For a shear stress of 1 dyn, one can calculate a mean ATP level in the effluent of less than $0.1\mu M$ in a standard parallel plate flow chamber (channel dimensions: 6 cm $\times$ 2.5 cm $\times$ 0.025 cm). Our previous calculations suggested that the surface and bulk concentrations of ATP are nearly the same in the presence of flow; therefore, the surface concentration of ATP resulting from the flow-induced release should also be of the order of 0.1 μM. Since the K_m for the ATP receptor is approximately 10 μM, it appears that the release of ATP by endothelial cells would be too small to cause a response.

VI. CONCLUSIONS

There is no doubt that fluid flow alters the metabolism of mammalian cells of different types. This suggests that besides the role of flow in enhancing the transport of nutrients and waste products to and from the cells in a living organism, flow by itself is another stimulus that exists in the cell environment. In the case of endothelial cells, which are exposed to blood flow and are supposed to regulate it, the role of fluid shear stress appears to be obvious. In the case of cells solely exposed to interstitial flow, the physiological significance of shear is unknown at this time.

We have seen that endothelial cells chronically exposed to flow exhibit a higher production rate for certain products and a lower rate for other ones when compared to similar cells maintained under static conditions. As seen in Chapter 4, the morphology of endothelial cells is also sensitive to flow. This clearly demonstrates that the phenotype of endothelial cells is influenced by the presence of shear stress, and this may be also true for other cell types. It is well known that data obtained with cultured cells in vitro must be extrapolated to in vivo situations with great care since cultured cells are generally in an environment that is very different from their natural conditions. In general, the phenotype of cultured cells can be affected by age or passage number and culture conditions. The latter includes the type of substrate cells are grown on, the culture medium, and one might also include the physical forces present. Cells that are normally subjected to flow in vivo are usually cultured under static conditions, which introduces an additional difference. Cells maintained in the presence of shear stress may provide an in vitro model that would be closer to the real case in vivo.

The findings on cultured endothelial cells suggest that the function of the endothelium in vivo is modulated by the local levels of shear stress. It is well known that the location of the lesions on the endothelium, leading eventually to atherosclerotic plaque formation, is related to the flow characteristics in the arterial network. In vitro studies on the effect of flow on endothelial cell biology could help to understand the pathology of vascular disease.

More studies are also needed to understand the transduction mechanism by which the flow signal is transduced across the cell membrane. So far, no detailed mechanism has been proposed and experimentally verified. The elucidation of this mechanochemical transduction mechanism is an interesting scientific question by itself, and it is probably one of the most fundamental processes in biology.

APPENDIX

If we consider a solute A in an incompressible liquid moving between two parallel plates of finite length L and separated by a gap of width h. The concentration of A anywhere between the plates can be determined by solving the problem

$$v_z \frac{\partial C_A}{\partial z} = D \frac{\partial^2 C_A}{\partial x^2}$$

where v_z is the velocity of the fluid between the plates, a function of position; C_A is the concentration of A in the fluid, also a function of position; x is the vertical distance from the bottom plate (at the top plate, $x = h$); z is the distance measured in the direction of the flow field and taken from the point of entrance of the fluid into the space between the plates (at the exit, $z = L$); and D is the diffusivity of A in the fluid.

The entering fluid contains a uniform concentration of $A = C_{A0}$, and there is no flux of A through the top plate, while at the bottom plate, the flux of A is equal to a rate of disappearance of A per unit surface due to ectonucleotidase activity. The boundary conditions are then

$$z = 0, \quad C_A = C_{A0} \qquad \text{for all } x \text{ at the entrance of the flow chamber}$$

$$x = 0, \quad D \frac{\partial C_A}{\partial x} = kC_A \qquad \text{at the bottom plate}$$

$$x = h, \quad \frac{\partial C_A}{\partial x} = 0 \qquad \text{for all } z \text{ at the top plate}$$

where k is the first order rate constant for the consumption of A per unit surface.

The flow was assumed to be fully developed and at steady state; therefore, for two parallel plates $v_z = 4v_{max} x(h - x)/h^2$ and the wall shear rate is $4v_{max}/h^2$, where v_{max} is the velocity of the fluid in the center of the gap between the plates (Frangos et al., 1988). After implementing the dimensionless variables $C = C_A/C_{A0}$, $Z = z/L$, $X = x/h$, $\alpha = DL/(4h^2 v_{max})$ and $\gamma = kh/D$, the equation was solved numerically with an implicit method using a grid size $(\Delta X, \Delta Z) = (0.005, 0.01)$. The solution C_A can be expressed as a function of the three variables X, αZ, and γ.

The parameters used for the ATPase were $V_{max} = 22$ nmol/10^6 cells/min and $K_m = 249 \ \mu M$ at 37°C (Gordon et al., 1986). For $C_A \ll 249 \ \mu M$, we have a first-order reaction with specific rate given by $k = V_{max}/K_m$. Assuming that we have 2×10^6 cells/25 cm^2, $k = 1.18 \times 10^{-4}$ cm^3/cm^2/s. The diffusivity of ATP in the medium was estimated as follows. According to the Stokes–Einstein equation, $D = \kappa T/(6\pi R_A \mu_B)$, where κ is the Boltzmann constant, T is the absolute temperature, μ is the viscosity, and R_A is the radius of the diffusing particle (Bird et al., 1960). The subscripts A and B refer to solute and solvent, respectively. The diffusivity of ATP in phosphate buffer is 4.16×10^{-6} cm^2/s at 20°C (Nevo and Rikmenspoel, 1970). For water at 20°C and 370°C, $\mu_B = 0.01002$ and 0.006915 g/cm/s, respectively (Weast and Astle, 1980). From these data and the Stokes–Einstein equation, the diffusivity of ATP can be estimated to 6.38×10^{-6} cm^2/s at 37°C. For the purpose of the calculations, a more conservative estimate of the diffusivity for ATP was used, 6×10^{-6} cm^2/s, to take into account the slightly higher viscosity of the medium relative to water due to the proteins in solution. Using the above values for D and k, together with $L = 6$ cm and $h = 0.0250$ cm. For our particular flow chamber, we have $\gamma = 0.491$ and α can be expressed as a function of the wall shear rate S in s^{-1}: $\alpha = 2.3/S$. The gap width available for the flow between the plates is effectively the distance between the plates minus the thickness of the cell monolayer. As the latter of which is approximately 1 μm, it can be considered negligible.

Calculations were performed for a range of S between 25 and 2500 s^{-1}, which corresponds to shear stresses in the range of 0.25 to 25 dyn/cm^2, assuming a medium viscosity of 0.01 g/cm/s.

REFERENCES

Albeda, S. M., Elias, J. A., Levine, E. M., and Kern, J. A. (1989). Endotoxin stimulates platelet-derived growth factor production from cultured human pulmonary endothelial cells. Am. J. Physiol. 257, L65–L70.

Alevriadou, B. R., Mo, M., Rickman, D. S., Eskin, S. G., McIntire, L. V., and Schilling, W. P. (1991). Effect of shear stress on ^{86}Rb$^+$ efflux and cytosolic Ca^{2+} of calf pulmonary artery endothelial cells (CPAEs). FASEB J. 5, A697 (abstract).

Anderson, D. C., Miller, L. J., Schmalstieg, F. C., Rothlein, R., and Springer, T. A. (1986). Contributions of the Mac-1 glycoprotein family to adherence-dependent functions: Structure-function assessments employing subunit-specific monoclonal antibodies. *J. Immunol.* **137**, 15–27.

Ando, J., Komatsuda, T., Ishikawa, C., and Kamiya, A. (1990). Fluid shear stress enhanced DNA synthesis in cultured endothelial cells during repair of mechanical denudation. *Biorheology* **27**, 675–684.

Ando, J., Komatsuda, T., and Kamiya, A. (1988). Cytoplasmic calcium response to fluid shear stress in cultured vascular endothelial cells. *In Vitro Cell. Devel. Biol.* **24**, 871–877.

Ando, J., Nomura, H., and Kamiya, A. (1987). The effect of fluid shear stress on the migration and proliferation of cultured endothelial cells. *Microvasc. Res.* **33**, 62–70.

Barabino, G. A., McIntire, L. V., Eskin, S. G., Sears, D. A., and Udden, M. (1987). Endothelial cell interactions with sickle cell, sickle trait, mechanically injured, and normal erythrocytes under controlled flow. *Blood* **70**, 152–157.

Beitz, J. G., Kim, I.-S., Calabresi, P., and Frackelton, R., Jr. (1991). Human microvascular endothelial cells express receptors for platelet-derived growth factor. *Proc. Natl. Acad. Sci. USA* **88**, 2021–2025.

Berk, B. C., Alexander, R. W., Brock, T. A., Gimbrone, M. A., Jr., and Webb, R. C. (1986). Vasoconstriction: A new acivity for platelet-derived growth factor. *Science* **232**, 87–90.

Berridge, M. J., and Irvine, R. F. (1989). Inositol phosphates and cell signaling. *Nature* **341**, 197–205.

Berthiaume, F., and Frangos, J. F. (1991). Fluid flow increases membrane permeability to Merocyanine 540 in human endothelial cells. *FASEB J.* **5**, A1473 (abstract).

Betsholtz, C., Johnsson, A., Heldin, C.-H., Westermark, B., Lind, P., Urdea, M. S., Eddy, R., Shows, T. B., Philpott, K., Mellor, A. L., Knott, T. J., and Scott, J. (1986). cDNA sequence and chromosomal localization of human platelet-derived growth factor A-chain and its expression in tumor cell lines. *Nature* **320**, 695–699.

Bevilacqua, M. P., Pober, J. S., Wheeler, M. E., Cotran, R. S., and Gimbrone, M. A., Jr. (1985). Interleukin 1 (IL-1) activation of vascular endothelium: Effects on procoagulant activity and leukocyte adhesion. *Am. J. Pathol.* **121**, 393–403.

Bhagyalakshmi, A., and Frangos, J. A. (1989a). Mechanism of shear-induced prostacyclin production in endothelial cells. *Biochem. Biophys. Res. Commun.* **158**, 31–37.

Bhagyalakshmi, A., and Frangos, J. A. (1989b). Membrane phospholipid metabolism in sheared endothelial cells. In *Proc. 2nd Int. Symp. Biofluid Mechanics and Biorheology*, p. 249, Liepsch, D. (ed.). Munich, Germany, June 1989.

Binderman, I., Shimshoni, Z., and Somjen, D. (1984). Biochemical pathways involved in the translation of physical stimulus into biological message. *Calcif. Tiss. Int.* **36**, S82–S85.

Bird, R. B., Stewart, W. E., and Lightfoot, E. N. (1960). *Transport Phenomena*, pp. 513–516, Wiley, New York.

Bodin, P., Bailey, D., and Burnstock, G. (1991). Increased flow-induced ATP release from isolated vascular endothelial cells but not smooth muscle cells. *Br. J. Pharmacol.* **103**, 1203–1205.

Boulanger, C., Schini, V. B., Moncada, S., and Vanhoutte, P. M. (1990). Stimulation of cyclic GMP production in cultured endothelial cells of the pig by bradykinin, adenosine diphosphate, calcium ionophore A23187 and nitric oxide. *Br. J. Pharmacol.* **101**, 152–156.

Brock, T. A., and Capasso, E. A. (1988). Thrombin and histamine activate phospholipase C in human endothelial cells via phorbol ester-sensitive pathway. *J. Cell. Physiol.* **136**, 54–62.

Brock, T. A., Dennis, P. A., Griedling, K. K., Diehl, T. S., and Davies, P. F. (1988). GTPγS loading of endothelial cells stimulates phospholipase C and uncouples ATP receptors. *Am. J. Physiol.* **255**, C667–C673.

Brotherton, A. F. A., and Hoak, J. C. (1982). Role of Ca^{2+} and cyclic AMP in the regulation of the production of prostacyclin by the vascular endothelium. *Proc. Natl. Acad. Sci. USA* **79**, 495–499.

Buga, G. M., Gold, M. E., Fukuto, J. M., and Ignarro, L. J. (1991). Shear stress-induced release of nitric oxide from endothelial cells grown on beads. *Hypertension* **17**, 187–193.

Burch, R. M., and Axelrod, J. (1987). Dissociation of bradykinin-induced prostaglandin formation from phosphatidylinositol turnover in Swiss 3T3 fibroblasts: Evidence for G protein regulation of phospholipase A_2. *Proc. Natl. Acad. Sci. USA* **84**, 6374–6378.

Caro, C. G., Lever, M. J., and Tarbell, J. M. (1985). Effect of luminal flow rate on transmural fluid flux in the perfused rabbit common carotid artery. *J. Physiol. Lond.* **365**, 92.

Carruthers, A., and Melchior, D. L. (1986). How bilayer lipids affect membrane protein activity. *Trends Biochem. Sci.* **11**, 331–335.

Carter, T. D., Hallam, T. J., and Pearson, J. D. (1989). Protein kinase C activation alters the sensitivity of agonist-stimulated endothelial-cell prostacyclin production to intracellular Ca^{2+}. *Biochem. J.* **262**, 431–437.

Cherry, P. D., Furchgott, R. F., and Zawadzki, J. V. (1983). The endothelium-dependent relaxation of vascular smooth muscle by unsaturated fatty acids. *Fed. Proc.* **42**, 619 (abstract).

Clark, M. A., Littlejohn, D., Conway, T. M., Mong, S., Steiner, S., and Crooke, S. T. (1986a). Leukotriene D_4 treatment of bovine aortic endothelial cells and murine smooth muscle cells in culture results in an increase in phospholipase A_2 activity. *J. Biol. Chem.* **261**, 10713–10718.

Clark, M. A., Littlejohn, D., Mong, S., and Crooke, S. T. (1986b). Effect of leukotrienes, bradykinin and calcium ionophore (A 23187) on bovine endothelial cells: Release of prostacyclin. *Prostaglandins* **31**, 157–166.

Cocks, T. M., Angus, J. A., Campbell, J. H., and Campbell, G. R. (1985). Release and properties of endothelium-derived relaxing factor (EDRF) from endothelial cells in culture. *J. Cell. Physiol.* **123**, 310–320.

Cooke, J. P., Stamler, J., Andon, N., Davies, P. F., McKinley, G., and Loscalzo, J. (1990a). Flow stimulates endothelial cells to release a nitrovasodilator that is potentiated by reduced thiol. *Am. J. Physiol.* **259**, H804–H812.

Cooke, J. P., Stamler, J., Andon, N. A., Davies, P. F., Mendelsohn, M. E., and Loscalzo, J. (1990b). Flow-mediated endothelium-dependent effects on platelet and vascular reactivity. In *Endothelium-Derived Relaxing Factors*, pp. 244–253, Rubanyi, G. M., and Vanhoutte, P. M. (eds.). Karger, Basel.

Davies, P. F., Remuzzi, A., Gordon, E. J., Dewey, C. F., Jr., and Gimbrone, M. A., Jr. (1986). Turbulent fluid shear stress induces vascular endothelial cell turnover in vitro. *Proc. Natl. Acad. Sci. USA* **83**, 2114–2117.

Davies, P. F., Dewey, C. F., Jr., Bussolari, S. R., Gordon, E. J., and Gimbrone, M. A., Jr. (1984). Influence of hemodynamic forces on vascular endothelial function. In vitro studies of shear stress and pinocytosis in bovine aortic cells. *J. Clin. Invest.* **73**, 1121–1129.

DeForrest, J. M., and Hollis, T. M. (1978). Shear stress and aortic histamine synthesis. *Am. J. Physiol.* **234**, H701–H705.

DeForrest, J. M., and Hollis, T. M. (1980). Relationship between low intensity shear stress, aortic histamine formation, and aortic albumin uptake. *Exp. Mol. Pathol.* **32**, 217–225.

Demolle, D., and Boeynaems, J. M. (1988). Role of protein kinase C in the control of vascular prostacyclin: Study of phorbol ester effects in bovine aortic endothelium and smooth muscle. *Prostaglandins* 35, 243–257.

Derian, C. K., and Moskowitz, M. A. (1986). Polyphosphoinositide hydrolysis in endothelial cells and carotid artery segments. Bradykinin-2 receptor stimulation is calcium-independent. *J. Biol. Chem.* 261, 3831–3837.

Diamond, S. L., Sharefkin, J. B., Dieffenbach, C., Frasier-Scott, K., McIntire, L. V., and Eskin, S. G. (1990). Tissue plasminogen activator messenger RNA levels increase in cultured human endothelial cells exposed to laminar shear stress. *J. Cell. Physiol.* 143, 364–371.

Diamond, S. L., Eskin, S. G., and McIntire, L. V. (1989). Fluid flow stimulates tissue plasminogen activator secretion by cultured human endothelial cells. *Science* 243, 1483–1485.

Dudley, D. T., and Spector, A. A. (1986). Stimulated phosphatidylinositol turnover provides arachidonic acid for prostacyclin synthesis in human endothelial cells. *Circulation* 74, (Suppl. II), 231.

Dull, R. O., and Davies, P. F. (1991). Flow modulation of agonist (ATP)-response (Ca^{++}) coupling in vascular endothelial cells. *Am. J. Physiol.* 261, H149–H154.

Emori, T., Hirata, Y., and Marumo, F. (1990). Specific receptors for endothelin-3 in cultured bovine endothelial cells and its cellular mechanism of action. *FEBS Lett.* 263, 261–264.

Eriksson, C. (1974). Streaming potentials and other water-dependent effects in mineralized tissues. *Ann. N.Y. Acad. Sci.* 283, 321–338.

Flavahan, N. A., and Vanhoutte, P. M. (1990). G-Proteins and endothelial responses. *Blood Vessels* 27, 218–229.

Förstermann, U., Pollock, J. S., Schmidt, H. H. H. W., Heller, M., and Murad, F. (1991). Calmodulin-dependent endothelium-derived relaxing factor/nitric oxide synthase activity is present in the particulate and cytosolic fractions of bovine aortic endothelial cells. *Proc. Natl. Acad. Sci. USA* 88, 1788–1792.

Frangos, J. A., McIntire, L. V., and Eskin, S. G. (1988). Shear stress induced stimulation of mammalian cell metabolism. *Biotechnol. Bioeng.* 32, 1053–1060.

Frangos, J. A., Eskin, S. G., McIntire, L. V., and Ives, C. L. (1985). Flow effects on prostacyclin production in cultured human endothelial cells. *Science* 227, 1477–1479.

Franke, R. P., Grafe, M., and Schnittler, H. (1984). Induction of human vascular endothelial stress fibres by fluid shear stress. *Nature* 307, 648–649.

Fuji, S., Lucore, C. L., Hopkins, W. E., Billadello, J. J., and Sobel, B. E. (1990). Induction of synthesis of plasminogen activator inhibitor type-1 by tissue-type plasminogen activator in human hepatic and endothelial cells. *Thromb. Haemostas.* 64, 412–419.

Fujimoto, M., Sakata, T., Tsuruta, Y., Iwagami, S., Teraoka, H., Mihara, S.-I., Fukiishi, Y., and Ide, M. (1990). Enhancement of bradykinin-induced prostacyclin synthesis in porcine aortic endothelial cells by pertussis toxin. Possible implication of lipocortin I. *Biochem. Pharmacol.* 40, 2661–2670.

Fuse, I., and Tai, H.-H. (1987). Stimulations of arachidonate release and inositol-1,4,5-trisphosphate formation are mediated by distinct G-proteins in human platelets. *Biochem. Biophys. Res. Commun.* 146, 659–665.

Gallez, D. (1984). Cell membranes after malignant transformation. Part I: Dynamic stability at low surface tension. *J. Theor. Biol.* 111, 323–340.

Garay, R., Rota, R., and Rosati, C. (1989). Non-laminar flow as initiating factor of atherosclerosis: Evidence for a role of membrane ion transport. In *Membrane Technology*, pp. 137–141, Verna, R. (ed.). Raven Press, New York.

Garcia, J. G. N., Siflinger-Birnboim, A., Bizios, R., Del Vecchio, P. J., Fenton, J. W., II, and Malik, A. B. (1986). Thrombin-induced increase in albumin permeability across the endothelium. *J. Cell. Physiol.* **128**, 96–104.

Geiger, R. V., Berk, B. C., Alexander, R. W., and Nerem, R. M. (1992). Flow induced calcium transients in single endothelial cells: spatial and temporal analysis. *Am. J. Phys.* **262**, C1411–C1417.

Goligorsky, M. S. (1988). Mechanical stimulation induces Ca^{2+i} transients and membrane depolarization in cultured endothelial cells. *FEBS Lett.* **240**, 59–64.

Goppelt-Strübe, M., Pfannkuche, H.-J., Gemsa, D., and Resch, K. (1987). The diacylglycerols dioctanoylglycerol and oleoylacetylglycerol enhance prostaglandin synthesis by inhibition of the lysophosphatide acyltransferase. *Biochem. J.* **247**, 773–777.

Gordon, E. L., Pearson, J. D., and Slakey, L. L. (1986). The hydrolysis of extracellular adenine nucleotides by cultured endothelial cells from pig aorta. *J. Biol. Chem.* **261**, 1053–1060.

Grabowski, E. F., Jaffe, E. A., and Weksler, B. B. (1985). Prostacyclin production by cultured endothelial cell monolayers exposed to step increases in shear stress. *J. Lab. Clin. Med.* **103**, 36–43.

Graier, W. F., Schmidt, K., and Kukowetz, W. R. (1990). Effect of sodium fluoride on cytosoloic free Ca^{2+}-concentrations and cGMP-levels in endothelial cells. *Cell. Signal.* **2**, 369–375.

Grimm, J., Keller, R., and de Groot, P. G. (1988). Laminar flow induces cell polarity and leads to rearrangement of proteoglycan metabolism in endothelial cells. *Thromb. Haemostas.* **60**, 437–441.

Grulich-Henn, J., and Müller-Berghaus, G. (1990). Regulation of endothelial tissue plasminogen activator and plasminogen activator inhibitor type 1 synthesis by diacylglycerol, phorbol ester, and thrombin. *Blut* **61**, 38–44.

Gupte, A., and Frangos, J. A. (1990). Effects of flow on the synthesis and release of fibronectin by endothelial cells. *In Vitro Cell. Devel. Biol.* **26**, 57–60.

Hallam, T. J., Pearson, J. D., and Needham, L. A. (1988). Thrombin-stimulated elevation of human endothelial-cell cytoplasmic free calcium concentration causes prostacyclin production. *Biochem. J.* **251**, 243–249.

Halldórsson, H., Kjeld, M., and Thorgeirsson, G. (1988). Role of phosphoinositides in the regulation of endothelial prostacyclin production. *Arteriosclerosis* **8**, 147–154.

Hanss, H., and Collen, D. (1987). Secretion of tissue-type plasminogen activator and plasmogen activator inhibitor by cultured human endothelial cells: Modulation by thrombin, endotoxin, and histamine. *J. Lab. Clin. Med.* **109**, 97–104.

Harlan, J. M., Thompson, P. J., Ross, R. R., and Bowen-Pope, D. F. (1986). α-Thrombin induces release of platelet-derived growth factor-like molecule(s) by cultured human endothelial cells. *J. Cell Biol.* **103**, 1129–1133.

Hollis, T. M., and Ferrone, R. A. (1974). Effects of shearing stress on aortic histamine synthesis. *Exp. Mol. Pathol.* **20**, 1–10.

Hong, S. L., and Deykin, D. (1982). Activation of phospholipases A_2 and C in pig aortic endothelial cells synthesizing prostacyclin. *J. Biol. Chem.* **257**, 7151–7154.

Hong, S. L., McLaughlin, N. J., Tzeng, C. Y., and Patton, G. (1985). Prostaglandin synthesis and deacylation of phospholipids in human endothelial cells: Comparison of thrombin, histamine and ionophore A23187. *Thromb. Res.* **38**, 1–10.

Hopkins, N. K., and Gorman, R. R. (1981). Regulation of endothelial cell cyclic nucleotide metabolism by prostacyclin. *J. Clin. Invest.* **67**, 540–546.

Hsieh, H.-J., Li, N.-Q., and Frangos, J. A. (1991). Shear stress increases endothelial platelet-derived growth factor mRNA levels. *Am. J. Physiol.* 29, H642–H646.

Iba, T., Shin, T., Sonoda, T., Rosales, O., and Sumpio, B. E. (1991). Stimulation of endothelial secretion of tissue-type plasminogen activator by repetitive stretch. *J. Surg. Res.* 50, 457–460.

Ignarro, L. J. (1990). Nitric oxide. A novel signal transduction mechanism for transcellular communication. *Hypertension* 16, 477–483.

Ingber, D. E. (1990). Fibronectin controls capillary endothelial cell growth by modulating cell shape. *Proc. Natl. Acad. Sci. USA* 87, 3579–3583.

Inyang, A. L., and Tobelem, G. (1990). Tissue-plasminogen activator stimulates endothelial cell migration in wound assays. *Biochem. Biophys. Res. Commun.* 171, 1326–1332.

Irvine, R. F. (1982). How is the level of free arachidonic acid controlled in mammalian cells. *Biochem. J.* 204, 3–16.

Ivarsson, B. L., Cambria, R. P., Megerman, J., and Abbott, W. M. (1989). Fibronectin enhances early shear stress resistance of seeded adult human venous endothelial cells. *J. Surg. Res.* 47, 203–207.

Jaffe, E. A., Grulich, J., Weksler, B. B., Hampel, G., and Watanabe, K. 1987. Correlation between thrombin-induced prostacyclin production and inositol trisphosphate and cytosolic free calcium levels in cultured human endothelial cells. *J. Biol. Chem.* 262, 8557–8565.

Jeremy, J. Y., and Dandona, P. (1988). Fluoride stimulates in vitro vascular prostacyclin synthesis: Interrelationship of G proteins and protein kinase C. *Eur. J. Pharmacol.* 146, 279–284.

Jirik, F. R., Podor, T. J., Hirano, T., Kishimoto, T., Loskutoff, D. J., Carson, D. A., and Lotz, M. (1989). Bacterial lipopolysaccharide and inflammatory mediators augment IL-6 secretion by human endothelial cells. *J. Immunol.* 142, 144–147.

Jo, H., Tarbell, J. M., Hollis, T. M., and Dull, R. O. (1991). Endothelial albumin permeability is shear-dependent, time dependent, and reversible. *Am. J. Physiol.* 260, H1992–H1996.

Johnson, G. S., Turrentine, M. A., and Sculley, P. W. (1986). Factor VIII coagulant activity and von Willebrand factor in post-exercise plasma from standard-bred horses. *Thromb. Res.* 42, 419–423.

Kamiya, A., and Togawa, T. (1980). Adaptive regulation of wall shear stress to flow change in the canine carotid artery. *Am. J. Physiol.* 239, H14–H21.

Kawaguchi, H., Iizuka, K., Sano, H., and Yasuda, H. (1990). Effect of streptokinase on prostacyclin synthesis and phospholipase activity in cultured pulmonary artery endothelial cells. *Biochim. Biophys. Acta* 1055, 223–229.

Keller, R., Silbert, J. E., Furthmayr, H., and Madri, J. A. (1987a). Aortic endothelial protoheparan sulfates I: Isolation and characterization of plasma membrane and extracellular species. *Am. J. Pathol.* 128, 286–298.

Keller, R., Pratt, B., Silbert, J. E., Furthmayr, H., and Madri, J. A. (1987b). Aortic endothelial protoheparan sulfates. II: Modulation by extracellular matrix. *Am. J. Pathol.* 128, 299–306.

Komuro, I., Kurihara, H., Sugiyama, T., Yoshizumi, M., Takaku, F., and Yazaki, Y. (1988). Endothelin stimulates c-fos and c-myc expression and proliferation of vascular smooth muscle cells. *FEBS Lett.* 238, 249–252.

Kooistra, T., van den Berg, J., Tons, A., Platenburg, G., Rijken, D. C., and van den Berg, E. (1987). Butyrate stimulates tissue-type plasminogen activator synthesis in cultured human endothelial cells. *Biochem. J.* 247, 605–612.

Kuchan, M. J., and Frangos, J. A. (1992). Fluid flow activates G-proteins that are coupled to calcium-independent EDRF production in cultured endothelial cells. *Am. J. Phys.* (Submitted for publication).

Kuhn, M., Otten, A., Frölich, J. C., and Förstermann, U. (1991). Endothelial cyclic GMP and cyclic AMP do not regulate the release of endothelium-derived relaxing factor/nitric oxide from bovine aortic endothelial cells. *J. Pharmacol. Exp. Ther.* **256**, 677–682.

Kurihara, H., Yoshizumi, M., Sugiyama, T., Takaku, F., Yanagisawa, M., Masaki, T., Hamaoki, M., Kato, H., and Yazaki, Y. (1989). Transforming growth factor-β stimulates the expression of endothelin mRNA by vascular endothelial cells. *Biochem. Biophys. Res. Commun.* **159**, 1435–1440.

Lambert, T. L., Kent, R. S., and Whorton, A. R. (1986). Bradykinin stimulation of inositol polyphosphate production in porcine aortic endothelial cells. *J. Biol. Chem.* **261**, 15288–15293.

Lang, D., and Lewis, M. J. (1991). Inhibition of inositol 1,4,5-trisphosphate formation by cyclic GMP in cultured aortic endothelial cells of the pig. *Br. J. Pharmacol.* **102**, 277–281.

Langille, B. L., and O'Donnell, F. (1986). Reductions in arterial diameter produced by chronic decreases in blood flow are endothelium-dependent. *Science* **231**, 405–407.

Lansman, J. B., Hallam, T. J., and Rink, T. J. (1987). Single stretch-activated ion channels in vascular endothelial cells as mechanotransducers? *Nature* **325**, 811–813.

Larsen, F. L., Katz, S., Roufogalis, B. D., and Brooks, D. E. (1981). Physiological shear stresses enhance the Ca^2 permeability of human erythrocytes. *Nature* **294**, 667–668.

Lawrence, M. B., Smith, C. W., Eskin, S. G., and McIntire, L. V. (1990). Effect of venous shear stress on CD18-mediated neutrophil adhesion to cultured endothelium. *Blood* **75**, 227–237.

Lawrence, M. B., Eskin, S. G., and McIntire, L. V. (1987). Effect of flow on polymorphonuclear leukocyte/endothelial cell adhesion. *Blood* **70**, 1284–1290.

Levine, E. G., and Santell, L. (1988). Stimulation and desensitization of tissue plasminogen activator release from human endothelial cells. *J. Biol. Chem.* **263**, 9360–9365.

Lückhoff, A., and Busse, R. (1990a). Calcium influx into endothelial cells and formation of endothelium-derived relaxing factor is controlled by the membrane potential. *Pflugers Arch.* **416**, 305–311.

Lückhoff, A., Mülsch, A., and Busse, R. (1990b). cAMP attenuates autacoid release from endothelial cells: Relation to internal calcium. *Am. J. Physiol.* **258**, H960–H966.

Lückhoff, A., and Busse, R. (1990c). Activators of potassium channels enhance calcium influx into endothelial cells as a consequence of potassium currents. *Naunyn-Scmiedeberg's Arch. Pharmacol.* **342**, 94–99.

Lückhoff, A., Pohl, U., Mülsch, A., and Busse, R. (1988). Differential role of extra- and intracellular calcium in the release of EDRF and prostacyclin from cultured endothelial cells. *Br. J. Pharmacol.* **95**, 189–196.

Lynch, J. J., Ferro, T. J., Blumenstock, F. A., Brockenhauer, A. M., and Malik, A. B. (1990). Increased endothelial albumin permeability mediated by protein kinase C activation. *J. Clin. Invest.* **85**, 1991–1998.

Magnússon, M. K., Halldórsson, H., Kjeld, M., and Thorgeirsson, G. (1989). Endothelial inositol phosphate generation and prostacyclin production in response to G-protein activation by AlF_4. *Biochem. J.* **264**, 703–711.

Makarski, J. S. (1981). Stimulation of cyclic AMP production by vasoactive agents in cultured bovine aortic and pulmonary artery endothelial cells. *In Vitro Cell. Devel. Biol.* **17**, 450–458.

Martin, T. W. (1988). Formation of diacylglycerol by a phospholipase D-phosphatidate phosphatase pathway specific for phosphatidylcholine in endothelial cells. *Biochim. Biophys. Acta* 962, 282–296.

Martin, T. W., Feldman, D. R., and Michaelis, K. C. (1990). Phosphatidylcholine hydrolysis stimulated by phorbol myristate acetate is mediated principally by phospholipase D in endothelial cells. *Biochim. Biophys. Acta* 1053, 162–172.

Martin, T. W., and Michaelis, K. (1988). Bradykinin stimulates phosphodiesteratic cleavage of phosphatidylcholine in cultured endothelial cells. *Biochem. Biophys. Res. Commun.* 157, 1271–1279.

Martin, T. W., and Wysolmerski, R. B. (1987). Ca^{2+}-dependent and Ca^{2+}-independent pathways for release of arachidonic acid from phosphatidylinositol in endothelial cells. *J. Biol. Chem.* 262, 13086–13092.

Martin, T. W., Wysolmerski, R. B., and Lagunoff, D. (1987). Phosphatidylcholine metabolism in endothelial cells: Evidence for phospholipase A and a novel Ca^{2+}-independent phospholipase C. *Biochim. Biophys. Acta* 917, 296–307.

Martin, W., White, D. G., and Henderson, A. H. (1988). Endothelium-derived relaxing factor and atriopeptin II elevate cyclic GMP levels in pig aortic endothelial cells. *Br. J. Pharmacol.* 93, 229–239.

Martinac, B., Adler, J., and Kung, C. (1991). Mechanosensitive ion channels of *E. coli* activated by amphipaths. *Nature* 348, 261–263.

Mawatari, M., Okamura, K., Matsuda, T., Hamanaka, R., Mizoguchi, H., Higashio, K., Kohno, K., and Kuwano, M. (1991). Tumor necrosis factor and epidermal growth factor modulate migration of human microvascular endothelial cells and production of tissue-type plasminogen activator and its inhibitor. *Exp. Cell Res.* 192, 574–580.

Milner, P., Bodin, P., Loesch, A., and Burnstock, G. (1990a). Rapid release of endothelin and ATP from aortic endothelial cells exposed to increased flow. *Biochem. Biophys. Res. Commun.* 170, 649–656.

Milner, P., Kirkpatrick, K. A., Ralevic, V., Toothill, V., Pearson, J., and Burnstock, G. (1990b). Endothelial cells cultured from human umbilical vein release ATP, substance P and acetylcholine in response to increased flow. *Proc. Roy. Soc. Lond. B* 241, 245–248.

Milner, P., Ralevic, V., Hopwood, A. M., Fehér, E., Lincoln, J., Kirkpatrick, K., and Burnstock, G. (1989). Ultrastructural localisation of substance P and choline acyltransferase in endothelial cells of rat coronary artery and release of substance P and acetylcholine during hypoxia. *Experientia* 45, 121–125.

Mitchell, J. A., Förstermann, U., Warner, T. D., Pollock, J. S., Schmidt, H. H. H. W., Heller, M., and Murad, F. (1991). Endothelial cells have a particulate enzyme system responsible for EDRF formation: Measurement by vascular relaxation. *Biochem. Biophys. Res. Commun.* 176, 1417–1423.

Mo, M., Eskin, S. G., and Schilling, W. P. (1991). Flow-induced changes in Ca^{2+} signaling of vascular endothelial cells: Effect of shear stress and ATP. *Am. J. Physiol.* 260, H1698–H1707.

Moncada, S., Gryglewsky, R., Bunting, S., and Vane, J. R. (1976). An enzyme isolated from arteries transforms prostaglandin endoperoxides to an unstable substance that inhibits platelet aggregation. *Nature* 263, 663–665.

Morris, C. E., and Horn, R. (1991). Failure to elicit neuronal macroscopic mechanosensitive currents anticipated by single-channel studies. *Science* 251, 1246–1249.

Moscatelli, D., and Rifkin, D. B. (1988). Membrane and matrix localization of proteinases: A common theme in tumor cell invasion and angiogenesis. *Biochim. Biophys. Acta* 948, 67–85.

Myers, P. R., Minor, R. L., Jr., Guerra, R., Jr., Bates, J. N., and Harrison, D. G. (1990). Vasorelaxant properties of the endothelium-derived relaxing factor more closely resemble S-nitrosocysteine than nitric oxide. *Nature* 345, 161–163.

Needham, L., Cusack, N. J., Pearson, J. D., and Gordon, J. L. (1987). Characteristics of the P$_2$ purinoreceptor that mediates prostacyclin production by pig aortic endothelial cells. *Eur. J. Pharmacol.* 134, 199–209.

Nevo, A. C., and Rikmenspoel, R. (1970). Diffusion of ATP in sperm flagella. *J. Theor. Biol.* 26, 11–18.

Newby, A. C., and Henderson, A. H. (1990). Stimulus-secretion coupling in vascular endothelial cells. *Annu. Rev. Physiol.* 52, 661–674.

Nollert, M. U., Diamond, S. L., and McIntire, L. V. (1991). Hydrodynamic shear stress and mass transport modulation of endothelial cell metabolism. *Biotechnol. Bioeng.* 38, 588–602.

Nollert, M. U., Eskin, S. G., and McIntire, L. V. (1990). Shear stress increases inositol trisphosphate levels in human endothelial cells. *Biochem. Biophys. Res. Commun.* 170, 281–287.

Nollert, M. U., Hall, E. R., Eskin, S. G., and McIntire, L. V. (1989). The effect of shear stress on the uptake and metabolism of arachidonic acid by human endothelial cells. *Biochim. Biophys. Acta* 1005, 72–78.

Olesen, S. P., Clapham, D. E., and P. F. Davies. (1988). Haemodynamic shear stress activates a K^+ current in vascular endothelial cells. *Nature* 331, 168–170.

Oliver, J. A. (1990). Adenylate cyclase and protein kinase C mediate opposite actions on endothelial junctions. *J. Cell. Physiol.* 145, 536–542.

Ovarfordt, P. G., Reilly, L. M., Lusby, R. J., Effeney, D. J., Ferrell, L. D., Price, D. C., Fuller, J., Ehrenfeld, W. K., Stoney, R. J., and Goldstone, J. (1985). Prostacyclin production in regions of arterial stenosis. *Surgery* 98, 484–491.

Parpart, A. K., and Hoffman, J. F. (1956). Flicker in erythrocytes. "Vibratory movements in the cytoplasm?" *J. Comp. Physiol.* 47, 295–303.

Pearson, J. D., and Gordon, J. L. (1979). Vascular endothelial and smooth muscle cells in culture selectively release adenine nucleotides. *Nature* 281, 384–386.

Pirotton, S., Erneux, C., and Boeynaems, J. M. (1987a). Dual role of GTP-binding proteins in the control of endothelial prostacyclin. *Biochem. Biophys. Res. Commun.* 147, 1113–1120.

Pirotton, S., Raspe, E., Demolle, D., Erneux, C., and Boeynaems, J.-M. (1987b). Involvement of inositol 1,4,5-trisphosphate and calcium in the action of adenine nucleotides on aortic endothelial cells. *J. Biol. Chem.* 262, 17461–17466.

Pober, J. S., and Cotran, R. S. (1990). Cytokines and endothelial cell biology. *Physiol. Rev.* 70, 427–445.

Podor, T. J., Curriden, S. A., and Loskutoff, D. J. (1988). The fibrinolytic system of endothelial cells. In *Endothelial Cells*, pp. 127–148, Ryan, U. S. (ed.). CRC Press, Boca Raton, FL.

Pytela, R., Pierschbacher, M. D., and Ruoslahti, E. (1985). Identification and isolation of a 140 kD cell surface glycoprotein with properties expected of a fibronectin receptor. *Cell* 40, 191–198.

Ralevic, V., Milner, P., Hudlická, O., Kristek, F., and Burnstock, G. (1990). Substance P is released from the endothelium of normal and capsaicin-treated rat hind-limb vasculature, in vivo, by increased flow. *Circ. Res.* 66, 1178–1183.

Reich, K. M., and Frangos, J. A. (1991). Effect of flow on prostaglandin E$_2$ and inositol trisphosphate levels in osteoblasts. *Am. J. Physiol.* 261, C428–C432.

Reich, K. M., Gay, C. V., and Frangos, J. A. (1990). Fluid shear stress as a mediator of osteoblast cyclic adenosine monophosphate production. *J. Cell. Physiol.* **143**, 100–104.

Reinders, J. H., de Groot, P. G., Dawes, J., Hunter, N. R., van Heugten, H. A. A., Zandbergen, J., Gonsalves, M. D., and van Mourik, J. A. (1985). Comparison of secretion and subcellular localization of von Willebrand factor with that of thrombospondin and fibronectin in cultured human vascular endothelial cells. *Biochim. Biophys. Acta* **844**, 306–313.

Resink, T. J., Grigorian, G. Y., Moldabaeva, A. K., Danilov, S. M., and Bühler, F. R. (1987). Histamine-induced phosphoinositide metabolism in cultured human umbilical vein endothelial cells associated with thromboxane and prostacyclin release. *Biochem. Biophys. Res. Commun.* **144**, 438–446.

Roach, M. R. (1977). The effects of bifurcations and stenoses on arterial disease. In *Cardiovascular Flow Dynamics and Measurements*, pp. 489–539. Wang, N. H., and Norrman, N. A. (eds.). University Park Press, Baltimore.

Rosen, L. A., Hollis, T. M., and Sharma, M. G. (1974). Alterations in bovine endothelial histidine decarboxylase activity following exposure to shearing stresses. *Exp. Mol. Pathol.* **20**, 329–343.

Ross, R., Raines, E. W., and Bowen-Pope, D. F. (1986). The biology of platelet-derived growth factor. *Cell* **46**, 155–169.

Russell, M. E., Quertermous, T., Declerck, P. J., Collen, D., Haber, E., and Homcy, C. J. (1990). Binding of tissue-type plasminogen activator with human endothelial cell monolayers. *J. Biol. Chem.* **265**, 2569–2575.

Saijonmaa, O., Ristimäki, A., and Fyhrquist, F. (1990). Atrial natriuretic peptide, nitroglycerine, and nitroprusside reduce basal and stimulated endothelin production from cultured endothelial cells. *Biochem. Biophys. Res. Commun.* **173**, 514–520.

Sato, Y., and Rifkin, D. B. (1988). Autocrine activities of basic fibroblast growth factor: Regulation of endothelial cell movement, plasminogen activator synthesis and DNA synthesis, *J. Cell. Biol.* **107**, 1199–1205.

Schmidt, K., Mayer, B., and Kukovetz, W. R. (1989). Effect of calcium on endothelium-derived relaxing factor formation and cGMP levels in endothelial cells. *Eur. J. Pharmacol.* **170**, 157–166.

Schröder, H., Machunsky, C., Strobach, H., and Schrör, K. (1990). Nitric oxide but not prostacyclin has an autocrine function in porcine aortic endothelial cells. *Adv. Prost. Thromb. Leuk. Res.* **21**, 671–674.

Shen, J., Luscinskas, F. W., Connolly, A., Dewey, C. F., Jr., and Gimbrone, M. A., Jr. (1992). Fluid shear stress modulates cytosolic free calcium in vascular endothelial cells. *Am. J. Phys.* **262**, C384–C390.

Sprague, E. A., Steinbach, B. L., Nerem, R. M., and Schwartz, C. J. (1987). Influence of a laminar steady-state fluid-imposed wall shear stress on the binding, internalization, and degradation of low-density lipoproteins by cultured arterial endothelium. *Circulation* **76**, 648–656.

Stamler, J., Mendelsohn, M. E., Amarante, P., Smick, D., Andon, N., Davies, P. F., Cooke, J. P., and Loscalzo, J. (1989). N-Acetylcysteine potentiates platelet inhibition by endothelium-derived relaxing factor. *Circ. Res.* **65**, 789–795.

Starksen, N. F., Harsh, G. R., Gibbs, V. C., and Williams, L. T. (1987). Regulated expression of platelet-derived growth factor A chain gene in microvascular endothelial cells. *J. Biol. Chem.* **262**, 14381–14384.

Stathopoulos, N. A., and Hellums, J. D. (1985). Shear stress effects on human embryonic kidney cells in vitro. *Biotechnol. Bioeng.* **27**, 1021–1026.

Steinberg, S. F., Bilezikian, J. P., and Al-Awqati, Q. (1987). Fura-2 fluorescence is localized to mitochondria in endothelial cells. *Am. J. Physiol.* **253**, C744–C747.

Steinchen, A., Gallez, D., and Sanfeld, A. (1981). A viscoelastic approach to the hydrodynamic stability of membranes. *J. Coll. Interfac. Sci.* **85**, 5–15.

Stelzner, T. J., Weil, J. V., and O'Brien, R. F. (1989). Role of cyclic adenosine monophosphate in the induction of endothelial barrier properties. *J. Cell. Physiol.* **139**, 157–166.

Stewart, D. J., Langleben, D., Cernacek, P., and Cianflone, K. (1990). Endothelin release is inhibited by coculture of endothelial cells with cells of vascular media. *Am. J. Physiol.* **259**, H1928–H1932.

Swan, D. C., McBride, O. W., Robbins, K. C., Keithley, D. A., Reddy, E. P., and Aaronson, S. A. (1982). Chromosomal mapping of the simian sarcoma virus *onc* gene analogue in human cells. *Proc. Natl. Acad. Sci. USA* **79**, 4691–4695.

Takagi, Y., Fusake, M., Takata, S., Yoshimi, H., Tokunaga, O., and Fujita, T. (1990). Autocrine effect of endothelin on DNA synthesis in human vascular endothelial cells. *Biochem. Biophys. Res. Commun.* **168**, 537–543.

Takamura, H., Kasai, H., Arita, H., and Kito, M. (1990). Phospholipid molecular species in human umbilical artery and vein endothelial cells. *J. Lipid Res.* **31**, 709–717.

Toothill, V. J., Needham, L., Gordon, J. L., and Pearson, J. D. (1988). Desensitization of agonist-stimulated prostacyclin release in human umbilical vein endothelial cells. *Eur. J. Pharmacol.* **157**, 189–196.

Unemori, E. N., Bouhana, K. S., and Werb, Z. (1990). Vectorial secretion of extracellular matrix proteins, matrix-degrading proteinases, and tissue inhibitor of metalloproteinases by endothelial cells. *J. Biol. Chem.* **265**, 445–451.

Vargas, F. F., Osorio, M. H., Ryan, U. S., and De Jesus, M. (1989). Surface charge of endothelial cells estimated from electrophoretic mobility. *Membrane Biochem.* **8**, 221–227.

Viret, J., Daveloose, D., and Leterrier, F. (1990). Modulation of the activity of functional membrane proteins by the lipid bilayer fluidity. *Membrane Transport Inform. Storage* **4**, 239–243.

Weast, R. C., and Astle, M. J. (1980). *CRC Handbook of Chemistry and Physics*, 61st ed., p. F-51. CRC Press, Boca Raton, FL.

White, D. G., and Martin, W. (1989). Differential control and calcium-dependence of production of endothelium-derived relaxing factor and prostacyclin by pig aortic endothelial cells. *Br. J. Pharmacol.* **97**, 683–690.

Whorton, A. R., Willis, C. E., Kent, R. S., and Young, S. L. (1984). The role of calcium in the regulation of prostacyclin synthesis by porcine aortic endothelial cells. *Lipids* **19**, 17–24.

Worthen, G. S., Smedley, L. A., Tonnesen, M. G., Ellis, D., Voelkel, N. F., Reeves, J. T., and Henson, P. M. (1987). Effects of shear stress on adhesive interaction between neutrophils and cultured endothelial cells. *J. Appl. Physiol.* **63**, 2031–2041.

Wright, J. G., Massop, D. W., Shyy, Y.-J., Smead, W. L., Jonasson, O. M., Kolattukudy, P. E., and Cornhill, J. F. (1990). Interleukin 6 secretion and gene expression by human aortic endothelial cells in response to laminar shear stress. *Arteriosclerosis*, **10**, A760.

Wysolmerski, R., and Lagunoff, D. (1985). The effect of ethchlorvynol on cultured endothelial cells. A model for the study of the mechanism of increased vascular permeability. *Am. J. Pathol.* **119**, 505–512.

Yanagisawa, M., Kurihara, H., Kimura, S., Tomobe, Y., Kobayashi, M., Mitsui, Y., Yazaki, Y., Goto, K., and Masaki, T. (1988). A novel potent vasoconstrictor peptide produced by vascular endothelial cells. *Nature* **332**, 411–415.

Yoshizumi, M., Kurihara, H., Sugiyama, T., Takaku, F., Yanagisawa, M., Masaki, T., and Yazaki, Y. (1989). Hemodynamic shear stress stimulates endothelin production by cultured endothelial cells. *Biochem. Biophys. Res. Commun.* **161**, 859–864.

Zavoico, G. B., Hrbolich, J. K., Gimbrone, M. A., Jr., and Schafer, A. I. (1990). Enhancement of thrombin- and ionomycin-stimulated prostacyclin and platelet-activating factor production in cultured endothelial cells by a tumor-promoting phorbol ester. *J. Cell. Physiol.* **143**, 596–605.

CHAPTER 6

Shear Stress Effects on the Morphology and Cytomatrix of Cultured Vascular Endothelial Cells

■ ■ ■ ■ ■

Peggy R. Girard, Gabriel Helmlinger, and Robert M. Nerem

I. INTRODUCTION

In recent years a number of laboratories have turned to cell culture to study the influence of flow on the vascular endothelial cell. In vivo, the vascular endothelium, as the interface between the underlying vessel wall and flowing blood, is exposed to hemodynamic forces. This hemodynamic environment not only is important to normal vascular biology but also may have a significant influence on pathobiologic mechanisms associated with disease processes.

The fluid-imposed hemodynamic force may be expressed as a stress, or force per unit area, and includes both a normal component—pressure—and a tangential component—shear stress. Pressure acts directly on the endothelium, but also serves to distend the underlying wall including the endothelium's basement membrane. Thus the endothelial cell both "sees" a time-varying pressure directly and "rides" on a basement membrane which is being cyclically stretched as the heart periodically ejects blood into the vascular system. On the other hand, the wall shear stress represents the frictional force per unit area and is associated with the local velocity

193

pattern. Although blood is not a Newtonian fluid, it still is instructive to note that, for a fluid exhibiting Newtonian behavior, there is a linear relationship between the shear stress exerted on the wall τ_w and the local velocity gradient at the wall, namely, the wall shear rate S_w such that

$$\tau_w = \mu \left(\frac{\partial u}{\partial y} \right)_w = \mu S_w \tag{1}$$

where u is the velocity component in the direction parallel to the wall, y is distance in the direction perpendicular to the wall, and μ is the viscosity coefficient. The wall shear rate S_w is a direct reflection of the local velocity pattern at the wall. For a non-Newtonian fluid such as blood, Eq. (1) does not strictly hold; however, even here one still would expect a direct dependence of the wall shear stress τ_w on the wall shear rate S_w.

As noted earlier, there are believed to be hemodynamic influences on vascular biology–pathobiology. Various in vivo evidence suggests differences in endothelial morphology and function in regions of differing hemodynamic environment. These are reviewed by Nerem and Girard (1990), and included have been studies of endothelial cell shape and orientation, F-actin localization, endothelial permeability, endothelial cell turnover rate, and monocyte recruitment into the intima (Flaherty et al., 1972; Silkworth and Stehbens, 1975; Nerem et al., 1981; Levesque et al., 1986; Kim et al., 1989; Bell et al., 1974a, 1974b; Caplan and Schwartz, 1973; Gerrity et al., 1985). Most compelling are experiments in which, through the introduction of an aortic stenosis, not only the blood flow pattern was altered, but there were concomitant changes in endothelial cell shape and orientation and in F-actin localization (Levesque et al., 1986; Kim et al., 1989). These studies demonstrate that in regions of high shear stress endothelial cells are more elongated, are aligned with the direction of flow, and contain stress fibers, also aligned with the direction of flow.

As seemingly convincing as such studies are, they suffer from the fact that it is extremely difficult to quantify the in vivo hemodynamic environment (Nerem et al., 1990; Nerem, 1981). For example, although a number of investigators have reported measurements of velocity profiles in large arteries using hot-film anemometry and pulsed Doppler ultrasound, such data cannot be obtained with the accuracy necessary to accurately determine the value of wall shear stress. Other investigators have tried to measure wall shear stress directly using a hot-film probe; however, these also suffer from a considerable uncertainty, associated primarily with probe positioning (Nerem, 1981). Some of the best information we have is in fact from in vitro glass- and plastic-tube studies where a model shaped as a vessel, such as the aorta or the carotid bifurcation (Deters et al., 1984; Ku et al., 1985), is used in conjunction with flow visualization and laser Doppler velocimetry techniques to measure the velocity pattern in considerable detail. However, here also various questions are raised, including the fact that in vivo the velocity pattern will be altered not only by vessel geometry and by the pulsatile nature of the waveform but also by longer-term time variations associated with changes in physiological condition.

The net result is that, although we know how pressure varies both spatially and temporarily in the vascular system and with this can deduce the extent of cyclic stretching of the vessel wall, our knowledge of the variation in wall shear stress is considerably more qualitative. Even so, there are estimates of wall shear stress available from the literature. These indicate a value of $15-20$ dyn/cm^2 for the mean shear stress in a straight section of artery, peak values that may exceed 100 dyn/cm^2, particularly on a flow divider, and still other regions where the mean shear stress may be close to zero, but where the flow is oscillatory in nature, thus exhibiting large excursions in shear stress, for example, as much as ± 20 dyn/cm^2 (Nerem and Girard, 1990).

With this as background and with the recognition that we can only describe the in vivo hemodynamic environment qualitatively, not quantitatively, a number of laboratories have turned to the use of cell culture in the investigation of how mechanical stresses alter endothelial cell structure and function. Although most of these studies have focused on fluid shear stress effects, there are a number of reports where endothelial cells grown on a compliant membrane have been exposed to cyclic stretch (Nerem and Girard, 1990). As important as these latter studies are, in this chapter we will focus on a review of experiments that have been designed to investigate the influence of a fluid-imposed shear stress on endothelial cell structure and function, including some comments on signaling processes and with an emphasis on studies from our own laboratory.

It should be noted that the first effort to study cultured vascular endothelial cells in the presence of flow was that of Dewey et al. (1981) who used a cone–plate viscometer to impose a shear stress on a cell monolayer. Other laboratories subsequently joined in this type of research, and in our own studies over the past 10 years we have employed a parallel-plate flow chamber which is briefly described in the next section. In such a device, as will be seen, the flow is well defined; that is, we know the wall shear stress and the pressure to which the cultured cells are exposed. One cautionary note, however, is as follows. Although the use of cell culture does allow one to study biological events under well-defined conditions of flow, it is not truly a physiologic model—it is a biologic model. Thus, a continuing agenda item must be to find ways to engineer the cell culture environment so as to make it more physiologic.

II. METHODS

In the studies to be reviewed here, multiply passaged bovine aortic endothelial cells (BAECs) were cultured on polyester substrates, specifically, Thermanox coverslips (Lux Culture Labware) or Mylar sheets, prior to being positioned in a parallel-plate flow chamber where they were exposed to a laminar flow. The chamber, a reservoir, and a circulation circuit were filled with culture medium [Dulbecco's modified Eagle's medium (DMEM) containing 25-mM Hepes buffer, 10–20% fetal bovine serum, and antibiotics]. The endothelial cells in the chamber were exposed to a specific shear stress condition defined by the dimensions of the flow chamber and the pressure drop across the chamber. The pressure drop was measured and controlled by adjusting the height of the upstream reservoir. A roller pump was used to return the outflow from the chamber to this feeding reservoir. The medium was kept at a constant temperature of $37 \pm 0.5°C$, and a gas mixture of 95% air and 5% CO_2 was provided to the reservoir.

The basic device being used in these studies to expose cultured endothelial cells to a known fluid mechanically imposed wall shear stress is a parallel-plate, channel-type flow system (Levesque and Nerem, 1985). It consists of a channel with a rectangular cross section whose height h is much less than either its length L or its width b. The test specimens of cultured cells are positioned on the bottom of this channel. By knowing the flow rate through the channel, which can be determined from a measurement of flow or by knowing the pressure drop, one can calculate the wall shear stress and thus will know the conditions to which the cells are being exposed.

This, of course, assumes that one has well-defined fluid mechanic conditions, and the simple geometry of a long channel with a rectangular cross section has been chosen for this reason. In such a channel, the velocity profile has the simple parabolic form

$$u = \frac{1}{2\mu} \frac{dp}{dx} \left(\frac{h^2}{4} - y^2 \right) \tag{2}$$

where y is measured in the vertical direction from the centerline both up and down, and the position of the upper and lower surfaces is at $y = \pm h/2$ where the velocity is zero. Using Eq. (2), and the definition from Eq. (1), it follows that

$$\tau_w = -\frac{h}{2} \frac{dp}{dx} \tag{3}$$

Thus, knowing the pressure gradient dp/dx, one can readily determine the wall shear stress τ_w.

It is of interest for the type of experiments discussed here to size the channel so that the flow conditions are well defined. To ensure a laminar flow it is necessary that the Reynolds number $Re = u(h)/v$, be less than 1000. For the flow to be two-dimensional in character, the channel width or span must be much greater than the channel height: $b \gg h$. Finally, for the velocity profile to be given by Eq. (2), the entrance length for the flow L_{ent} must be small compared to the total length of the channel L, where the entrance length L_{ent} may be approximated as

$$L_{ent} = 0.06(h)\,Re \qquad (4)$$

Practically speaking, Eq. (4) implies that the Reynolds number should be on the order of 100 or less.

It should be noted that one advantage of this type of flow chamber, as compared, for example, to a cone–plate viscometer, is being able to position the chamber in a light microscope for continuous observation during an experiment. This provides advantages not only in monitoring the progress of an experiment, but as part of that detecting possible problems, such as contamination, that might develop during the course of an experiment.

In the next few sections, results obtained in our laboratory as well as in others demonstrating how a cultured endothelial cell monolayer responds to a fluid-imposed shear stress will be reviewed. In each section the specific techniques used and the measurements reported will be briefly reviewed and referenced.

A. Cell Shape and Orientation

When a confluent BAEC monolayer is exposed to a steady, laminar flow, the cells elongate in shape and orient so as to align their major axis with the direction of flow as shown in Figure 1. This response to flow is dependent on many factors. Primary among these are the flow as characterized by the level of shear stress, the duration of exposure, and the substrate to which the cells are adherent (Nerem and Girard, 1990; Dewey et al., 1981; Levesque and Nerem, 1985; Eskin et al., 1984).

Figure 2 illustrates the influence of the level of shear stress on cell shape for BAEC cultured on a Thermanox coverslip. The parameter used as a measure of cell shape is the shape index SI, defined by the equation

$$SI = \frac{4\pi A}{p^2} \qquad (5)$$

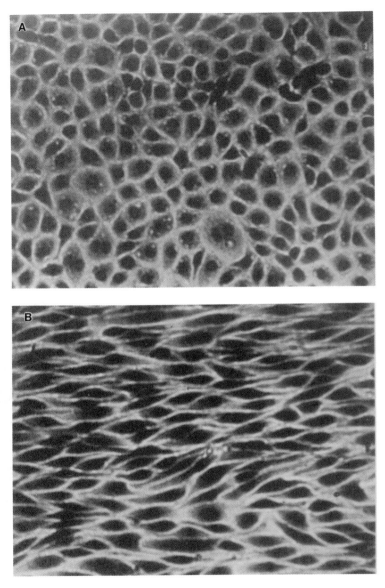

FIGURE 1   Photomicrographs of cultured BAEC grown on Thermanox under control
conditions (A) and under shear stress (85 dyn/cm^2) for 24 h (B); flow from left to right.
[Reproduced from Levesque and Nerem (1985), published by permission of The American
Society of Mechanical Engineers.

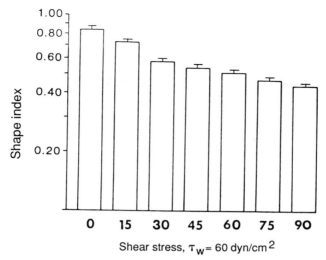

FIGURE 2   Shape index ($4\pi A/P^2$) for BAEC confluent monolayers grown on Thermanox exposed to steady laminar shear stress for 24 h.

Shape index is a nondimensional parameter defined such that a circular cell, or one which is round in shape, would have a shape index of SI = 1.0, while a highly elongated cell, in the limiting case of a straight line, would have a shape index of SI = 0.0. Thus, the more elongated a cell is, the smaller the value of its shape index. As may be seen in Figure 2, the higher the level of wall shear stress, the smaller the shape index and thus the more elongated the cells are. Figure 2 is based on calculating the shape index of all the cells in the field of view through a microscope and then doing population averages.

This effect of a steady laminar shear stress on the average cell shape for a confluent BAEC monolayer has been demonstrated for a variety of substrates, including glass, polyester-type materials (Thermanox coverslips and Mylar sheets), and extracellular matrix coated surfaces. In general, the effect is qualitatively the same, but quantitatively the influence of shear may be quite different (Levesque and Nerem, 1989).

In addition to measuring cell shape, one also can measure the average angle of orientation of the population of cells. In brief, the angle of orientation or degree of alignment of the cells is characterized by the deviation of the cell's major axis from the flow direction. In computing the mean deviation, only the absolute value of the angle between the cell's major axis and the flow direction is considered, and the mean of this quantity for all the cells in one image is calculated. If all cells in a picture

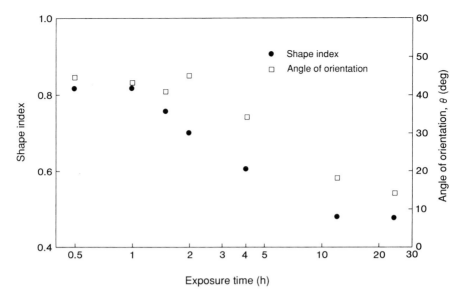

FIGURE 3   Graphical representation of the time–history effect of shear stress (85 dyn/cm²) on the shape index and angle of orientation of BAEC.

were exactly aligned with the flow, the mean would be equal to 0.0°. On the other hand, if the orientation of the cells were completely random, the mean would be equal to 45°. Just as is shown for cell shape in Figure 2, with increasing levels of steady, laminar flow as characterized by the shear stress, and for 24 h of exposure, there is a decrease in the average angle of orientation. Thus, for a given duration of exposure, the higher the level of wall shear stress, the more aligned the population of cells will be.

Even more interesting is the observation that there is a delay between the time at which changes in average cell shape are seen to be initiated and the time at which changes in the average angle of orientation are first seen (Levesque and Nerem, 1985). This is illustrated in Figure 3, and it is clear that the initiation of cell elongation precedes the initiation of cell alignment.

It should be noted that, although these effects on cell morphology have been fully documented for confluent monolayers, they also are observed to occur in subconfluent monolayers. Qualitatively, the effect of a steady, laminar flow on cell shape and orientation is the same for both subconfluent and confluent monolayers.

Of course, in vivo the flow is not steady; it is pulsatile in nature, although still laminar. Because the motivation for studying flow effects on endothelial cells in culture is to understand vascular biology, various attempts have been made to study the influence of pulsatile flow on the response of endothelial cells to shear stress. A variety of biological end points have been investigated, and these in general suggest that, whatever the effect of steady flow, this is exaggerated by adding pulsatility to the flow environment. This has been shown for cell shape (Levesque et al., 1989), for the mechanical stiffness of cells (Sato et al., 1988), for the proliferation of subconfluent monolayers (Levesque et al., 1990), and for PGI_2 synthesis (Frangos et al., 1985).

In all the investigations noted above, the pulsatile flow was of a very simple type; it was a nonreversing, pulsatile laminar flow, that is, one where the magnitude of the flow rate and thus shear stress were varying, but the direction was always in the forward direction. In contrast, in vivo one finds a wide variety of different pulsatile flow environments. In some regions the flow is nonreversing, in others reversing, and yet in others the mean shear stress may be very close to zero, but with large transient excursions during the cardiac cycle, what might be regarded as an almost pure oscillatory flow.

Recently the influence of various types of pulsatile flow on BAEC shape and orientation was investigated in our laboratory (Helmlinger et al., in press). Three types of pulsatile flow were examined: (I) a nonreversing pulsatile flow, with a shear stress of 40 ± 20 dyn/cm^2; (II) reversing pulsatile flows, including both a high-amplitude, reversing shear stress case of 20 ± 40 dyn/cm^2 and a low-amplitude, reversing shear stress case of 10 ± 15 dyn/cm^2; and (III) two cases of purely oscillatory flow, 0 ± 20 and 0 ± 40 dyn/cm^2. For all types of pulsatile flow, two controls were included, one with a wall shear stress corresponding to the peak pulsatile flow value and one corresponding to the mean value. For nonreversing pulsatile flow it was observed that, although BAEC shape initially changed less rapidly compared to cells in a steady flow, after $28–32$ h these cells were more elongated. This confirmed earlier findings from our laboratory (Levesque and Nerem, 1989; Levesque et al., 1989). For reversing pulsatile flow, BAEC shape also was found to change less rapidly, but in this case the cells were less elongated at all times compared to steady controls. Furthermore, for the large amplitude reversing case the BAEC monolayer invariably became detached within 24 h. Finally, for purely oscillatory flow, BAEC remained rounded, polygonal in shape, being virtually identical to that of static culture.

Even though reversal for the purely oscillatory flow was at least as great as that for the case of the large-amplitude reversing flow, no BAEC detachment was observed. This led us to initiate a new series of experiments in which BAEC monolayers were first exposed to steady laminar flow for 24 h, and then after this shear stress preconditioning, switched into a pulsatile flow environment. For the type II large-amplitude reversing flow, even with 24-h steady shear stress preconditioning at 20 dyn/cm^2, cell detachment still was observed within 24 h after the cells were switched to the pulsatile flow environment. When a BAEC monolayer was preconditioned for 24 h at 20 dyn/cm^2 and then switched into a purely oscillatory flow with a shear stress of 0 ± 20 dyn/cm^2, the cells were observed to change in morphology with time so as to become polygonal in shape. In fact, this return to a polygonal, rounded shape occurred more rapidly after preconditioning for cells placed in an oscillatory flow environment as compared to similarly preconditioned cells placed in static culture.

These experiments suggest that the preferred BAEC shape when exposed to a high mean shear stress is an elongated one that might involve some polarity, while for a purely oscillatory flow the preferred shape is a polygonal one, but perhaps without any polarity. Taken together and if one assumes that the large amplitude type II reversing flow is a combination of these two types of flow, that is, a steady flow and a purely oscillatory flow, this may represent a condition where the polarity in BAEC associated with a high mean shear stress is inconsistent with the lack of polarity favored by BAEC in a high-amplitude, reversing flow. Obviously, this whole question of cell polarity needs to be directly investigated. However, whatever the case, it is clear that BAEC in culture can not only discriminate between different types of pulsatile flow but also transduce that recognition into alterations in cell morphology.

There are other aspects of the cell culture environment that influence the elongation in shape and the alignment with flow direction that BAEC undergo in response to flow. One of these is the composition of the medium. It has been shown that BAEC on Thermanox in a medium supplemented with only 5% fetal calf serum (FCS) elongate more than BAEC exposed to the same level of shear stress, but in a medium supplemented by 20% (Levesque and Nerem, 1989). The difference in percent FCS could be important in terms of there being a different chemical milieu, such as the presence of growth factors, but it must also be recognized that there is an associated difference in the fluid mechanic environment. To be specific, the medium with only 5% FCS has a lower viscosity than the medium with 20% FCS; thus, with shear stress being the same, from Eq. (1) it may be seen that the wall shear rate S_w will be greater in the medium with the lower percent FCS.

There also could be differences associated with the cell; that is, one endothelial cell might not necessarily be the same as another endothelial cell. Even for endothelial cells from the same species and vessel, such as BAEC, there could be differences due to passage (i.e., early vs. late), and it has been demonstrated that there are differences due to time within passage, possibly due in part to a difference in the degree of confluency (Levesque et al., 1989).

Furthermore, there are species differences. A commonly used cell type is the human umbilical-vein endothelial cell (HUVEC). For HUVECs it has been reported that the time required for elongation is more than 5 days, or 120 h (Eskin et al., 1984; Ives et al., 1983), as compared to times on the order of 24 h for BAECs. Whether this is due to a species difference, a passage difference (HUVECs are normally used as primary or very early passage cultures), or a tissue difference (umbilical vein versus aorta) is not known.

Finally, although the emphasis here has been on laminar flow effects on cultured BAEC, there are rather isolated reports on the effect of a turbulent flow (Davies et al., 1986; Liu et al, 1990). These findings suggest that BAEC residing in a turbulent flow environment do not elongate, but rather retain a rounded, polygonal shape. In spite of this, there are flow-induced changes in function, such as cell turnover (Davies et al., 1986). This, together with our studies of various types of pulsatile flow, points to the fact that one should never consider flow as a single external stimulus. Just as there are a variety of chemical agonists, each with its own separate effect, there equally well are a variety of flow "agonists," or various types of flow environments, which, as a result of differences in frequency composition and amplitude as well as basic differences in the nature of the flow, will have a different "agonist" effect.

B. Cytoskeletal Localization

In the previous section we discussed the elongation in shape and the orientation of the cell's major axis with the direction of flow which endothelial cells (ECs) undergo as part of their response to flow. Associated with these changes, as part of this adaptive process, there are changes in cytoskeletal localization taking place; specifically, there is a rearrangement of cytoskeletal components (Eskin et al., 1985; Wechezak et al., 1987; Sato et al., 1987b).

The component that has been studied in most detail is the F-actin microfilament network (Wechezak et al., 1987; Sato et al., 1987b). For BAEC in static culture, the distribution of F-actin is characterized by dense peripheral bands (Wong and Gotlieb, 1986). However, with the onset of flow, a dramatic reorganization occurs. First, the dense peripheral bands begin to disappear. At a high shear stress of 85 dyn/cm^2 (Sato et al., 1987b), this is observed at a time as early as 30 min after the onset of flow, even before any change in cell shape can be detected. This disappearance of

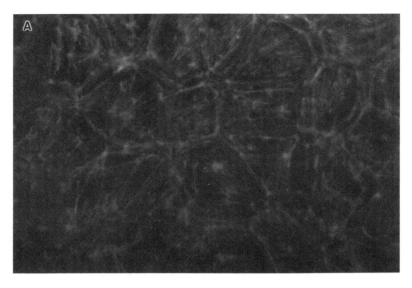

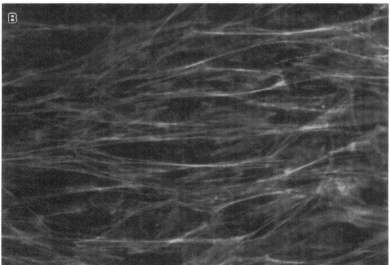

FIGURE 4 Photomicrographs of rhodamine-phalloidin-stained BAEC confluent monolayers grown on Thermanox for control conditions (A) and after exposure to shear stress (85 dyn/cm²) for 24 h (B); flow left to right.

the dense peripheral bands is followed by an increased number of centrally located stress fibers. Using the flow condition of 85 dyn/cm^2 again as an example, this change is observed as early as 4 h after exposure, and there is also a widening of intercellular spaces with few connecting microfilaments. At 12 h these stress fibers are mostly aligned with the direction of flow. Finally, as the cells reach a more complete state of elongation and alignment, intercellular spaces are no longer visible and a higher concentration of microfilaments at the cell's periphery is observed; there is a reappearance of the dense peripheral bands. This alteration in F-actin localization that occurs over a 24-h period is illustrated in Figure 4.

Obviously, this course of events will be influenced by the flow, including the level of shear stress, the exact cell type, and the substrate to which the cells are attached. In regard to the first of these, the level of shear stress, it thus is not surprising that for shear stress levels lower than 85 dyn/cm^2, the sequence of events is quite similar, but with an extended time course, commensurate with the increased time required for elongation and alignment. Also, K. Ookawa and M. Sato (University of Tsukaba, private communication) have quantified the rhodamine phalloidin florescence intensity in porcine aortic endothelial cells (PAEC) responding to the onset of flow. They have noted an increased intensity level at 24 h, which could be interpreted as representing an increased amount of F-actin.

It also should be noted that this dramatic reorganization in F-actin is directly linked to the elongation in cell shape that takes place. This is illustrated in Figure 5, where the time course in BAEC shape change is presented for different cases. If one compares the elongation observed over 48 h for BAEC exposed to flow with that for BAEC similarly exposed to flow, but in this latter case treated with cytochalasin B, which disrupts actin assembly, one can see a significant difference. Cells that have had actin assembly disrupted do not elongate in any major way. Their shape index only changes from a value of approximately 0.8 in static culture to a value of 0.7 with flow, a very small decrease. The cytochalasin-B-treated BAEC, even after 48 h of exposure to flow, are still round in shape, although with a slight eccentricity. It thus appears that the ability to assemble F-actin is critical to the cell elongation process.

The reorganization in cytoskeletal structure that occurs in cultured BAEC in response to shear stress also may be reflected in the cell's mechanical properties. In fact, it may be that the cell's ability to adapt its mechanical properties to the flow environment in which it resides is a key aspect of its response to the mechanical environment in which it finds itself. Thus, experiments were conducted in our laboratory to study the mechanical properties of endothelial cells. In these studies the deformation of BAEC was measured using the micropipette technique, as applied to

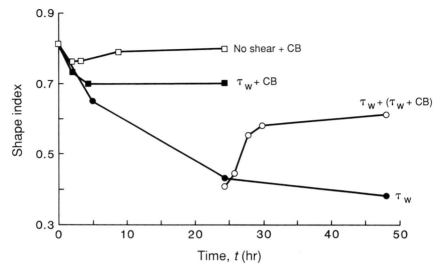

FIGURE 5   Shape index as a function of time for BAEC on Thermanox exposed to a steady shear stress (τ_w) of 60 dyn/cm^2 both with and without treatment with cytochalasin B (CB) = 2.0μm.

detached endothelial cells (Sato et al., 1987a, 1987b, 1988; Theret et al., 1988). A cell that has never been exposed to flow becomes spherical on detachment, as shown in Figure 6. F-Actin staining allows one to visualize a submembranous, cortical cytoskeletal layer, and we believe that it is this layer that represents the primary load-bearing component in a micropipette deformation, that is, stress–strain measurement. By using a membranelike model for this cortical layer, one can determine an effective shear modulus that represents the strength of this cortical cytoskeletal layer (Sato et al., 1987a). The importance of cytoskeletal elements in the determination of the effective cortical layer shear modulus can be demonstrated by using cytochalasin B to disrupt actin assembly. For BAEC so treated, the effective cortical layer shear modulus is found to be an order of magnitude less than that for control cells. Colchicine affects microtubule assembly, and when used there also is a reduction in the effective shear modulus, although not as great as that due to cytochalasin B.

   In contrast to cells detached from a static culture condition, an endothelial cell, which has been exposed to a high shear stress for a long period of time and subsequently mechanically detached, retains its elongated shape as shown in Figure 6 (Sato et al., 1987b). With increasing flow, as characterized by the level of wall shear stress, and/or increasing exposure time, there is an increase in cell stiffness. Our approach to the analysis of micropipette data obtained from BAEC exposed to shear stress has been

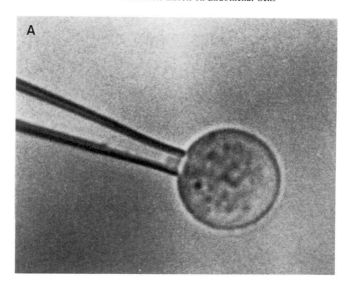

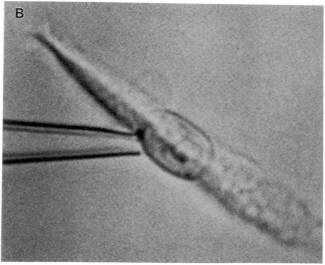

FIGURE 6   Examples of BAEC grown on to confluence on Thermanox and then detached for micropipette studies; control conditions (A), and after exposure to shear stress (85 dyn/cm^2) for 24 h (B).

to use a model that treats these elongated cells as a homogeneous material with an effective Young's modulus (Theret et al., 1988). Applying the results of this analysis to the micropipette data, characteristic values of Young's modulus E of the order of 10,000 dyn/cm^2 were calculated for BAEC exposed to 30 dyn/cm^2 for 24 h. This is as compared to values of E of approximately 1000 dyn/cm^2 for cells grown in static culture or on microcarrier beads under control, no-flow conditions. In this latter case, the measurements were performed with the cell both attached and detached to its microcarrier bead, with comparable results being obtained. Thus, our observations of F-actin and our measurements of endothelial cell mechanical properties all support the notion that endothelial cells, in response to shear stress, adapt their cytoskeletal structure to the flow environment in which they reside.

In more recent studies, it has been shown that endothelial cells are not only elastic in their mechanical nature but are viscoelastic (Sato et al., 1990). Furthermore, it is the F-actin that has been demonstrated to dominate the viscoelastic mechanical properties of endothelial cells. The microtubules also are important, but to a lesser extent, in terms of both elastic and viscoelastic properties.

Finally, in our investigation of the effects of various types of pulsatile flow as noted in the previous section, we also observed the distribution of F-actin (Helmlinger et al., in press). For types I and II pulsatile flows, specifically the nonreversing and reversing cases where cell elongation occurred, the localization of F-actin was very similar to that in a laminar, steady flow. Thus, even in the large-amplitude, reversing-flow case where cell detachment was observed, well-formed stress fibers aligned with the direction of flow were seen to develop and were present just prior to detachment. In contrast, in the oscillatory flow cases where the cell shape remained polygonal, the F-actin localization was concentrated in dense peripheral bands and thus appeared just as in cells in static culture. These pulsatile flow studies thus have reinforced our view that there is a direct relationship between cell shape and cytoskeletal localization. There must, however, be differences in the attachment of cells to the underlying surface since we observe for the large-amplitude, reversing-flow case cell detachment, and we do not for the case of a large-amplitude, purely oscillatory flow. The question of the influence of flow on properties related to cell attachment will be discussed next.

C. Extracellular Matrix and Cell-Surface Adhesion

As part of the adaptation of an endothelial cell in culture to its flow environment, mechanisms by which attachment is maintained would ap-

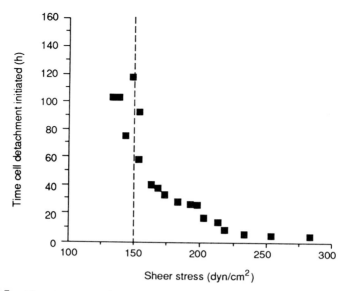

FIGURE 7 The time required for the initiation of BAEC detachment from a polyester substrate and in a steady laminar flow as a function of the level of shear stress; solid line represents the least squares fit, dashed line is the threshold value, or the critical denudative shear stress.

pear to be extremely important. In fact, there might be a certain threshold level of shear stress, above which cell detachment occurs. This is illustrated in Figure 7 for BAEC adherent to a Mylar sheet and exposed to a steady, laminar flow (Liu et al., 1990). For this case, the threshold appears to be in the neighborhood of 150 dyn/cm^2. Above this threshold value, there is an inverse relationship between the time required for cell detachment to be initiated and the level of wall shear stress. Of course, we would expect the threshold value to be dependent on a number of factors, including cell type (species, tissue origin, and passage), the nature of the flow environment, and most importantly the surface to which the cells are attached.

The mechanism of cell attachment is a complicated one involving the integration of the cell's cytoskeletal structure with its extracellular matrix (ECM) through focal contacts. These focal contacts, which are specialized regions of the plasma membrane where transmembrane cell adhesion receptors interact with ECM proteins and the actin microfilament network, are sites where cellular morphological changes may be modulated by alterations in focal contact-associated proteins or by modified contacts with the ECM (for a review, see Burridge et al., 1988). We have recently initiated studies to examine the role of ECM proteins and cytoskeleton-associated focal contact proteins in the modulation of shear-stress-related EC morphological changes. Using immunofluorescent staining techniques, vinculin, a member of the focal contact complex linking microfilaments to the plasma membrane, appeared to be located primarily at the periphery of each cell in static culture (Girard et al., 1990a). Following exposure of cells to 5 h of arterial levels of shear stress, vinculin staining appeared more streaked and seemed to undergo a reorganization concomitant with initial cell shape changes. Following 24 h of exposure to shear stress and corresponding to the formation of aligned stress fibers, we observed a reorganization of vinculin with a prominent localization at the "upstream" end of most BAEC. These data suggest that vinculin may play a role in directing flow-induced stress fiber assembly.

In addition to the regulation of actin microfilaments by focal contact-associated proteins, the stabilization of endothelial cells against fluid shear forces may be related to the cell's ability to synthesize, organize, and respond to the ECM proteins to which the cells adhere. The ECM of vascular endothelial cells involves a number of different types of adhesive proteins; however, in cell culture studies of flow effects on endothelial cells, the focus has been on fibronectin as a major ECM component. These experiments have shown an influence on fibronectin localization, that is, (Wechezak et al., 1987), as well as on the amount of fibronectin (Girard et al., 1990b; Gupte and Frangos, 1990).

In regard to an alteration in the pattern of localization of fibronectin (FN) associated with BAEC responding to the onset of flow, studies in our laboratory showed that, in static culture, cell-associated FN was organized in a diffuse network of fibers and mats as shown in Figure 8. When cells were exposed to relatively high levels of shear stress (30–50 dyn/cm^2) for 12 h, the FN appeared to be organized into a reticular pattern. At 24 h there was a diminished prominence of the FN reticular pattern and the FN was clearly aligned with the direction of flow, as is evident in Figure 8.

There also is a time variation in the amount of cell-associated fibronectin. Starting with the onset of flow, there initially is a decrease in the amount of fibronectin. This has been clearly demonstrated for HUVECs (Gupte and Frangos,1990), and this quite possibly is associated with the mobility required of endothelial cells if they are to change shape and align themselves with the direction of flow. Our own results for BAECs suggest that, after 12 h of flow exposure, the amount of cell-associated fibronectin may be slightly decreased, and that by 48 h, which is after the process of cell elongation and orientation has been completed, the amount of cell-associated fibronectin is comparable to control, perhaps even higher (Girard

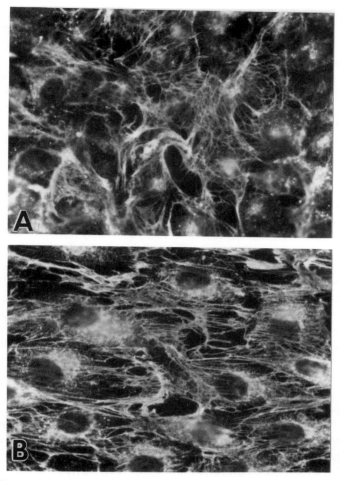

FIGURE 8   Immunofluorescent localization of FN associated with BAEC in static culture (A) and following 24 h of exposure to a shear stress of 30 dyn/cm² (B). Flow was from left to right.

et al., 1990b). At this point, the cells no longer need to be mobile and are becoming "glued" into place.

Critical to this entire issue of cell attachment is the interplay between the cell, particularly at focal contact sites, and the underlying ECM. We and others (Girard et al., 1990b; Gupte and Frangos, 1990) have shown that the ECM is a dynamic surface that changes during exposure to flow.

Since specific alterations in the ECM have been demonstrated to effect changes in cytoskeletal arrangement and in the organization of cell adhesion receptors in EC in static culture (Dejana et al., 1988; Herman, 1987; Fath et al., 1989; Semich and Robenek, 1990), it is possible that modifications of the ECM in response to flow may play a role in the observed cytoskeletal rearrangements that occur. The understanding of specific interactions between the cytoplasmic focal contact proteins, the cell adhesion receptors, and the ECM proteins, and how they are modified in response to flow, will be important in understanding the morphological changes that may influence endothelial monolayer integrity and adhesion to the basement membrane in vivo.

Clearly, there is much more needs to be done if we are to understand the nature of cell-surface attachment in a flow environment, the time course of events, and differences associated with the nature of the flow. Nothing points this out more dramatically than the pulsatile flow experiments described in the previous two sections. The fact that, for both the type I nonreversing and the type II high-amplitude reversing flows, one obtains very similar changes in cell shape and F-actin relocalization, but only in the latter type II flow does one see cell detachment, suggests that there may be major differences at the ECM–focal contact level. In this we need to recognize that there may be other contributors, such as other ECM components, which are important as well, and these also will need to be studied in the future.

D. Signal Recognition and Transduction

Although this chapter is not intended to provide a complete review of mechanisms involved in signal recognition and transduction, we would be remiss to not say something about this. Clearly, the types of changes that have been discussed in the previous sections must be "triggered" by a recognition event that is then transduced through second messengers.

In regard to this recognition event, little is still known. We are not even sure whether it is mediated by shear stress as opposed to shear rate or some other flow-related factor. Also, we do not know whether this sensing takes place on the lumenal or the ablumenal side of the cell or whether it is through a cell membrane receptor or channel or by way of the cytomatrix. There is a report on a shear-stress-activated K^+ channel (Olesen et al., 1987), as well as other reports on stretch-activated channels (Lansman et al., 1987), and several laboratories have measured a hyperpolarization of the cell due to the influence of flow (de Sousa et al., 1986; Nakache and Gaub, 1988). Alternatively, this recognition event could be associated with the transport of a small molecule, and in the last year considerable interest has developed in the possible role of ATP in the response of EC to flow, a molecule whose endogenous secretion rate itself may be flow-dependent (Milner et al., 1990). Whatever the case, it well may be that recognition is a multiple event, in other words, that there are several sensors of flow phenomena.

Whatever the recognition event, it is likely that the relevant signal pathways involve a linkage to the control of Ca^{2+} metabolism. The most widely studied biochemical pathway controlling Ca^{2+} mobilization is the phospholipase C-mediated hydrolysis of polyphosphoinositides. Thus, stimulation of a cell-surface receptor results in the activation of phospho-inositide specific phospholipase C to hydrolyze phosphatidylinositol 4,5 bisphosphate (PIP_2) to form inositol 1,4,5-trisphosphate (IP_3), resulting in a patent release of intracellular Ca^{2+}, and diacylglycerol (DAG), a stimula-tor of the Ca^{2+}, phospholipid-dependent enzyme, protein kinase C (Nishizuka, 1984). This signaling pathway is thought to be intimately involved in the dynamic control of cytoskeletal structure and function (Rasmussan et al., 1987). For example, gelsolin is a Ca^{2+}- and polyphos-phoinositide-regulated actin-binding protein that plays an important role in controlling the length of actin filaments within cells and as such is involved in regulating actin assembly (Yin et al., 1981). Another actin-binding protein, profilin, also modulates actin polymerization through interactions with phosphotidylinositol 4,5-bisphosphate (Lassing and Lindberg, 1988). Furthermore, data also exist suggesting that activation of PKC leads to actin polymerization, network assembly, and modulation of membrane–actin linkages (Schliwa et al., 1984; Burn et al., 1988; Ohto et al., 1987; Wong and Gotlieb, 1990).

In fact, a variety of investigators support the involvement of this signaling system in shear stress responses, if not explicitly, then at least implicitly. Such related observations include an elevation in intracellular Ca^{2+} within times of a minute or less after the onset of flow (Ando et al., 1988; Dull and Davies, 1991; Mo et al., 1992; Shen et al., 1992; Geiger et al., 1992), a flow-stimulation of phosphoinositide metabolism with a peak in IP_3 at times between 1 and 5 min (Prasad et al., 1989; Nollert et al., 1990), and a translocation of protein kinase C from cytosol to membrane (Girard and Nerem, 1990). Furthermore, studies utilizing the PKC inhibitor, staurosporine, have suggested the involvement of PKC in shear-stress-induced alterations in endothelial-cell cytoskeletal structure. BAECs were preincubated for one hour in 5-nM staurosporine, and then subjected to a shear stress of 30 dyn/cm² for 24 h in the continued presence of 5 nM staurosporine. These cells and control (static) cultures were then stained for F-actin using rhodamine phalloidin. BAEC in static culture showed F-actin organized in dense peripheral bands with some

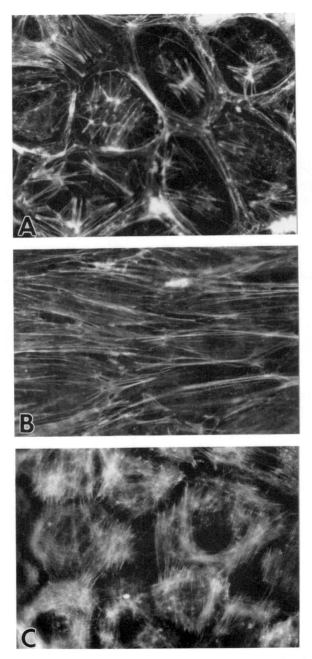

FIGURE 9 Effect of PKC inhibitor, staurosporine, on shear stress-induced F-actin microfilament reorganization. BAEC in static culture stained with rhodamine phalloidin showed F-actin organized in dense peripheral bands with some short, centrally located stress fibers evident (A). BAEC exposed to shear stress for 24 h but not staurosporine showed the appearance of long, prominent stress fibers aligned with the direction of flow (B), and BAEC subjected to flow in the presence of 5-nM staurosporine showed an inhibition of the reorganization of actin microfilaments (C); flow was from left to right.

short, centrally located stress fibers evident. BAEC exposed to shear stress, but not staurosporine, had developed long, prominent stress fibers aligned with the direction of flow. Cells subjected to flow in the presence of staurosporine show an inhibition of the reorganization of actin microfilaments, and this is illustrated in Figure 9.

Clearly, there are many "pieces of the puzzle" missing in terms of not only the recognition event but also signal transduction. In regard to the latter, although the second messengers that have been observed to be activated by flow are also generated in response to certain chemical agonists, what changes due to a flow signal are different from those associated with a chemical signal? The answer may lie, at least in part, with the fact that different receptors exist for each extracellular signal. The inositol phosphate signaling system involves many subtleties and complexities including multiple Ca^{2+} pools (various modes of entry into the cell of extracellular Ca^{2+}, and mobilization of intracellular pools), complex Ca^{2+} regulation by inositol phosphates, and a bewildering array of phosphotidylinositol species generated within the cell with a functional diversity so great that it is only now beginning to be understood. Subsequent to the formation of inositol phosphate second messengers is the activation of protein kinases, for example, the stimulation of PKC by DAG. As part of this, there is the question as to what substrates are specifically targeted for phosphorylation in response to a flow signal? Furthermore, are differences in the response to a flow signal as opposed to a chemical signal in any way associated with a pattern of events, either a pattern involving multiple second messengers, some of which are different, or the pattern of a single second messenger that may have differing spatial and/or temporal variations? Whatever the case, somehow the recognition and subsequent transduction of different external stimuli presents itself to the cell in the form of a differing content of information, which ultimately is read by the nucleus.

III. DISCUSSION

In concluding this chapter, it is important to integrate what has been presented in the previous sections as well as to place the studies reviewed here in perspective. In regard to the former, Table 1 summarizes the time course of events associated with the response of cultured EC to the onset of flow. Integrated into this table are the processes of cell elongation and orientation, cytoskeletal relocalization, ECM–focal contact modifications, and signal transduction. From the early studies that took place a decade ago, we clearly have increased our understanding considerably. Yet, there are still many holes in our knowledge. This is true in terms of some of the specifics of the events involved in the response and adaptation of endothe-

TABLE 1

Time Course of Events Following Onset of Flow for Bovine Aortic
Endothelial Cells Cultured on a Thermanox or Mylar Surface
Exposed to Moderate or High Shear Stress

Time after onset of flow (h)	Biological event
0.1	Second-messenger activation—Ca^{2+}, IP_3 [a,b]
0.2–0.5	Translocation of protein kinase C [c]
0.5	Disappearance of dense peripheral F-actin bands is initiated; [d] after peaking, IP_3 level by now has decreased to below initial, control value [a]
2	Cell shape changes first observed [e]
4	Cell orientation initiated [e]
6	Amount of cell-associated fibronectin decreased [f]
12	Extensive F-actin stress fibers aligned with the flow direction already are formed [d]
24	Dense peripheral F-actin bands reappear; [d] amount of cell-associated fibronectin increasing; [f] intracellular IP_3 level now returning to initial, control value [a]
36–48	Process of cell elongation–orientation complete; [e,g] extensive F-actin stress fibers as well as dense peripheral bands; [d] amount of cell-associated fibronectin comparable, perhaps even above initial control values [f]

[a] Prasad et al. (1989).
[b] Nollert et al. (1990).
[c] Girard and Nerem (1990).
[d] Sato et al. (1987b)
[e] Levesque and Nerem (1985).
[f] Girard et al. (1990b).
[g] Levesque and Nerem (1989).

lial cells to the onset of flow. The existence of "holes" is even more true when it comes to signal recognition and transduction as discussed in the last section.

Since most of the research reviewed here has been motivated by the possible role of flow in mediating vascular biologic and/or pathobiologic events, one might ask how the cell culture studies discussed here compare to results from in vivo investigations. In fact, there are a wide variety of observations from in vivo experiments suggesting differences associated with differing hemodynamic environments. The data here include endothelial cell shape (Flaherty et al., 1972; Silkworth and Stehbens, 1975; Nerem et al., 1981; Levesque et al., 1986), F-actin localization (Kim et al., 1989), transport across the endothelium into the intima (Bell et al., 1974a, 1974b), cell turnover (Caplan and Schwartz, 1973), and monocyte recruitment and intimal penetration (Gerrity et al., 1985). Of specific interest in the context of this review are the in vivo observations of endothelial cell shape and F-actin localization, and these suggest that in regions of high shear stress endothelial cells will be elongated, aligned with the direction of flow, and characterized by a cytoskeletal structure with microfilaments bundled into stress fibers, also aligned with the flow direction. Considering the difficulty in quantifying the detailed features of the hemodynamic environment in vivo, the good qualitative agreement is encouraging. However, it should be noted that endothelial cells in culture, even in the presence of a pulsatile flow, are not as elongated as endothelial cells in vivo, i.e. the shape index values of cultured BAEC exposed to flow (Levesque and Nerem, 1985; Helmlinger et al., in press) are larger than shape index values measured from in vivo endothelium (Nerem et al., 1981) for the same level of wall shear stress. In addition, cell turnover in a confluent endothelial-cell monolayer in culture (Davies et al., 1986) is much higher than that observed in vivo (Caplan and Schwartz, 1973). Our own studies have demonstrated that, for a BAEC monolayer exposed to a steady laminar flow, there is a reduced rate of cell proliferation (Levesque et al., 1990). This is supported by ^{3}H-thymidine incorporation and autoradiography measurements done postshear, that is, immediately after BAEC removal from the flow chamber, which demonstrate a reduction in DNA synthesis (Levesque et al., 1990). In a more recent study using flow cytometry, we have also shown that this inhibition of shear stress is due to a growth arrest in the G_0/G_1 phase of cell cycle (Mitsumata et al., in press). However, even with this inhibitory effect of flow, the level of cell turnover is still higher than that observed in vivo.

This simply suggests that, although adding flow to cell culture may represent an improved model for studying vascular biology in vitro, there is still much to be done in order to engineer the cell culture environment so as to make it more physiologic. There are many factors related to this including the engineering of the medium, the substrate, the presence of neighboring cells, and the flow environment itself, such as using an exact duplicate of a large artery flow waveform. However, a clear difference is that, whereas in vivo endothelial cells reside continually in a flow environment, albeit one that is changing considerably over a 24-h period, investigators have taken "naive" in vitro endothelial cells—ones that have never experienced the "pleasures" of fluid dynamics (after all, flow is the natural endothelial environment)—and then exposed them suddenly to the onset of flow. A much better model would be to (1) precondition the endothelial monolayer to flow and the associated level of wall shear stress and (2) then change the flow rate and the shear stress being imposed. In fact, several laboratories are initiating experiments of this type, and some of our own initial results were discussed in an earlier section.

Another difference between in vivo and cell culture experiments is that in vivo endothelial cells reside in a much more complicated mechanical stress environment. In vivo, endothelial cells, as the interface between blood and the underlying vessel wall, not only are exposed to the shear stress imposed by blood flow, but both "see" a pulsatile blood pressure and "ride" on a basement membrane that is being cyclically stretched by this time varying pressure. Herman et al. (1987) have shown, through in vitro experiments using perfused excised arteries, that pressure and flow may act in concert. In addition, from cell culture studies there are reports of efforts both due to hydrostatic pressure (Levesque and Nerem, 1983; Taylor et al., 1990) and cyclic stretch, in the latter case with experiments where endothelial cells are grown on a compliant membrane (Dartsch and Betz, 1989; Gorfien et al., 1989; Shirinsky et al., 1989; Sumpio and Banes, 1988; Sumpio et al., 1987, 1988). Clearly, the ultimate goal, from a mechanical stress perspective and in the engineering of the cell culture environment, has to be a device in which cultured endothelial cells can be exposed to the same complicated stress environment as in vivo. As important as such a goal is, however, it is equally important to continue the types of studies reviewed here. If we cannot fully understand the response of a "naive" cell to the simple onset of flow, we will have little chance of understanding cellular responses associated with a not-so "naive" cell and a more realistic mechanical stress environment.

ACKNOWLEDGMENTS

This work was supported by National Institutes of Health Grants HL-26890 and HL-41175 and NSF Grant ECS-8815656. The authors thank R. W. Alexander, B. C. Berk, M. J. Levesque, M. Sato, C. J. Schwartz, and E. A. Sprague for their contributions to the work and ideas reflected in this chapter and for their participation as collaborators in the study of this subject.

REFERENCES

Ando, J., Komatsuda, T., and Kamiya, A. (1988). Cytoplasmic calcium responses to fluid shear stress in cultured vascular endothelial cells. *In Vitro Cell. Devel. Biol.* 24, 871–877.
Bell, F. P., Admanson, I., and Schwartz, C. J. (1974a). Aortic endothelial permeability to albumin: Focal and regional patterns of uptake and transmural distribution of ^{125}I-albumin in the young pig. *Exp. Mol. Pathol.* 20, 57.

Bell, F. P., Gallus, A. S., and Schwartz, C. J. (1974b). Focal and regional patterns of uptake and the transmural distribution of ^{125}I-fibrinogen in the pig aorta *in vivo*. *Exp. Mol. Pathol.* **20**, 281.

Burn, P., Kupfer, A., and Singer, S. J. (1988). Dynamic membrane-cytoskeletal interactions: Specific association of integrin and talin arises in vivo after phorbol ester treatment of peripheral blood lymphocytes. *Proc. Natl. Acad. Sci. USA* **85**, 497–501.

Burridge, K., Fath, K., Kelly, T., Nuckolls, G., and Turner, C. (1988). Focal adhesions: transmembrane junctions between the extracellular matrix and the cytoskeleton. *Annu. Rev. Cell. Biol.* **4**, 487–525.

Caplan, B. A., and Schwartz, C. J. (1973). Increased cell turnover in areas of *in vivo* Evans Blue uptake in the pig aorta. *Atherosclerosis* **7**, 401.

Dartsch, P. C., and Betz, E. (1989). Response of cultured endothelial cells to mechanical stimulation. *Basic Res. Cardiol.* **84**, 268–281.

Davies, P. F., Remuzzi, A., Gordon, E. J., Dewey, C. F., Jr., and Gimbrone, M. A., Jr. (1986). Turbulent fluid shear stress induces vascular endothelial cell turnover *in vitro*. *Proc. Nat. Acad. Sci.* **83**, 2114–2117.

Dejana, E., Colella, S., Conforti, G., Abbadini, M., Gaboli, M., and Marchisio, P. C. (1988). Fibronectin and vitronectin regulate the organization of their respective Arg-Gly-Asp adhesion receptors in cultured endothelial cells. *J. Cell Biol.* **107**, 1215–1223.

de Sousa, P. A., Levesque, M. J., and Nerem, R. M. (1986). Electrophysiological response of endothelial cells to fluid-imposed shear stress. *Fed. Proc.* **45**, 471.

Deters, O. J., Mark, F. F., Bargeron, C. B., Hutchins, G. M., and Friedman, M. H. (1984). Comparison of steady and pulsatile flow near the ventral and dorsal walls of casts of human aortic bifurcations. *J. Biomech. Eng.* **106**, 79–82.

Dewey, C. F., Bussolari, S. R., Gimbrone, M. A., Jr., and Davies, P. F. (1981). The dynamic response of vascular endothelial cells to fluid shear stress. *ASME J. Biomech. Eng.* **103**, 177–181.

Dull, R. O., and Davies, P. F. (1991). Flow modulation of agonist (ATP)-response (Ca^{2+}) coupling in vascular endothelial cells. *Am. J. Physiol.* **261**, H149–H154.

Eskin, S. G., Ives, C. L., McIntire, L. V., and Navarro, L. T. (1984). Response of cultured endothelial cells to steady flow. *Microvasc. Res.* **28**, 87–94.

Eskin, S. G., Ives, C. L., Frangos, J. A., and McIntire, L. V. (1985). Cultured endothelium: The response to flow. *Am. Soc. Artif. Organs* **8**, 109–112.

Fath, K. R., Edgell, C.-J. S., and Burridge, K. (1989). The distribution of distinct integrins in focal contacts is determined by the substratum composition. *J. Cell Sci.* **92**, 67–75.

Flaherty, J. R., Pierce, J. E., Ferrans, V. J., Patel, D. J., Tucker, W. K., and Fry, D. L. (1972). Endothelial nuclear patterns in the canine arterial tree with particular reference to hemodynamic events. *Circ. Res.* **30**, 23–33.

Frangos, J. A., McIntire, L. V., Eskin, S. G., and Ives, C. L. (1985). Flow effects on prostacyclin production by cultured human endothelial cells. *Science* **227**, 1477–1479.

Geiger, R., Berk, B., Alexander, R. W., and Nerem, R. (1992). Spatial and temporal analysis of Ca^{2+} in single aortic endothelial cells (BAEC) exposed to flow. *Am. J. Physiol.*

Gerrity, R. G., Gross, J. A., and Soby, L. (1985). Control of monocyte recruitment by chemotactic factor(s) in lesion prone areas of swine aorta. *Arteriosclerosis* **5**, 55–56.

Girard, P. R., Nerem, R. M. (1990). Role of protein kinase C in the transduction of shear stress to alterations in endothelial cell morphology. *J. Cell Biol.* **14**, 238.

Girard, P. R., Keener, J. W., and Nerem, R. M. (1990a). Shear stress induces the modulation of vinculin and fibronectin and vitronectin receptors in cultured endothelial cells. *J. Cell Biol.* **111**, 300a.

Girard, P. R., Doty, S. D., Thoumine, O., and Nerem, R. M. (1990b). Fluid shear stress effects on fibronectin and cytoskeletal structure. Paper presented at First World Congress of Biomechanics, La Jolla, CA, Aug. 30–Sept. 4, 1990.

Gorfien, S. F., Winston, S. K., Thibault, L. E., and Macarak, E. J. (1989). Effects of biaxial deformation on pulmonary artery endothelial cells. *J. Cell. Physiol.* **139**, 492–500.

Gupte, A., and Frangos, J. A. (1990). Effects of flow on the synthesis and release of fibronectin by endothelial cells. *In Vitro Cell. Devel. Biol.* **26**, 57–60.

Helmlinger, G., Geiger, R. V., Schreck, S., and Nerem, R. M. (1991). The effect of a pulsatile laminar shear stress on cultured vascular endothelial cell morphology. *ASME J. Biomech. Eng.* **113**, 123–131.

Herman, I. M. (1987). Extracellular matrix-cytoskeletal interactions in vascular cells. *Tiss. Cell* **19**, 1–19.

Herman, I. M., Brant, A. M., Warty, V. S., Bonaccorso, J., Klein, E. C., Kormos, R. L., and Borvetz, H. S. (1987). Hemodynamics and the vascular endothelial cytoskeleton. *J. Cell Biol.* **105**, 291–302.

Ives, C. L., Eskin, S. G., McIntire, L. V., and Debakey, M. E. (1983). The importance of cell orgin and substrate in the kinetics of endothelial cell alignment in response to steady flow. *Trans. Am. Soc. Artif. Organs* **29**, 269.

Kim, D. W., Gotlieb, A. I., and Langille, B. L. (1989). *In vivo* modulation of endothelial F-actin microfilaments by experimental alterations in shear stress. *Arteriosclerosis* **9**, 439–445.

Ku, D. N., Giddens, D. P., Zarins, C. K., and Glagov, S. (1985). Pulsatile flow and atherosclerosis in the human carotid bifurcation. Positive correlation between plaque location and low and oscillating shear stress. *Arteriosclerosis* **5**(3), 293–302.

Lansman, J. B., Hallam, T. J., and Rink, T. J. (1987). Single stretch-activated on channels in vascular endothelial cells as mechanotransducers? *Nature* **325**, 811–813.

Lassing, I., and Lindberg, V. (1988). Specificity of the interaction between phosphatidylinositol 4,5-bisphosphate and the profilin:actin complex. *J. Cell Biochem.* **37**, 255–267.

Levesque, M. J., Liepsch, D., Moravec, S., and Nerem, R. M. (1986). Correlation of endothelial cell shape and wall shear stress in a stenosed dog aorta. *Arteriosclerosis* **6**, 220–229.

Levesque, M. J., and Nerem, R. M. (1983). Shear and pressure effects on cultured endothelial cells. *Proc. 36th Annual Conf. Eng. Med. Biol.*, p. 34.1. Alliance for Engineering in Medicine and Biology.

Levesque, M. J., and Nerem, R. M. (1985). The elongation and orientation of cultured endothelial cells in response to shear stress. *ASME J. Biomech. Eng.* **176**, 341–347.

Levesque, M. J., and Nerem, R. M. (1989). The study of rheological effects on vascular endothelial cells in culture. *Biorheology* **26**, 345–357.

Levesque, M. J., Sprague, E. A., Schwartz, C. J., and Nerem, R. M. (1989). The influence of shear stress on cultured vascular endothelial cells: The stress response of an anchorage-dependent mammalian cell. *Biotech. Prog.* **5**, 1–8.

Levesque, M. J., Sprague, E. A., and Nerem, R. M. (1990). Vascular endothelial cell proliferation in culture and the influence of flow. *Biomaterials* **11**, 702–707.

Liu, Z., Vanhee, C., Ziegler, T., Girard, P. R., Schreck, S., and Nerem, R. M. (1990). Hydrodynamic force effects on anchorage-dependent mammalian cells, *Bioprocess Engineering Symposium Proceedings*, Hochmuth, R. M. (ed.) ASME Winter Annual Meeting, Dallas, TX, November 25–30.

Milner, P., Bodin, P., Loesch, A., and Burnstock, G. (1990). Rapid release of endothelial and ATP from isolated aortic endothelial cells exposed to increased flow. *Biochem. Biophys. Res. Commun.* **170**, 649–656.

Mitsumata, M., Nerem, R. M., Alexander, R. W., and Berk, B. C. (1991). Shear stress inhibits endothelial proliferation by growth arrest in the G_0/G_1 phase of cell cycle. *FASEB J.* 5:A527.

Mo, M., Eskin, S. G., and Schilling, W. P. (1991). Flow-induced changes in Ca^{2+} signaling of vascular endothelial cells: effect of shear stress and ATP. *Am. J. Physiol.* **260**, H1698–H1707.

Nakache, M., and Gaub, H. E. (1988). Hydrodynamic hyperpolarization of endothelial cells. *Proc. Natl. Acad. Sci. USA* **63**, 1841–1843.

Nerem, R. M. (1981). Arterial fluid dynamics and interactions with the vessel wall. In *Structure and Function of the Circulation*, Vol. 11, pp. 719–835, Schwartz, C. J., Werthessen, N. T., and Wolf, S. (eds.). New York.

Nerem, R. M., and Girard, P. R. (1990). Hemodynamic influences on vascular endothelial biology. *Toxicol. Pathol.* **18**(4), Part 1.

Nerem, R. M., Levesque, M. J., and Cornhill, J. F. (1981). Vascular endothelial morphology as an indicator of blood flow. *ASME J. Biomech. Eng.* **103**, 172–176.

Nishizuka, Y. (1984). The role of protein kinase C in cell surface signal transduction and tumor promotion. *Nature* **308**, 693–698.

Nollert, M. U., Eskin, S. G., and McIntire, L. V. (1990). Shear stress increases inositol trisphosphate levels in human endothelial cells. *Biochem. Biophys. Res. Commun.* **170**, 281–287.

Ohto, Y., Akiyama, T., Nishida, E., and Sakai, H. (1987). Protein kinase C and cAMP-dependent protein kinase induce opposite effects on actin polymerizability. *FEBS Lett.* **222**, 305–310.

Olesen, S. P., Clapham, D. E., and Davies, P. F. (1987). Hemodynamic shear stress activates a K^+ current in vascular endothelial cells. *Nature* **331**, 168–170.

Prasad, A. R. S., Nerem, R. M., Schwartz, C. J., and Sprague, E. A. (1989). Stimulation of phosphoinositide hydrolysis in bovine aortic endothelial cells exposed to elevated shear stress. *J. Cell Biol.* **109**, 313a.

Rasmussan, H., Takwua, Y., and Park, S. (1987). Protein kinase C in the regulation of smooth muscle contraction. *FASEB J.* **1**, 177–185.

Sato, M., Levesque, M. J., and Nerem, R. M. (1987a). Application of the micropipette technique to the measurement of the mechanical properties of cultured bovine aortic endothelial cells. *ASME J. Biomech. Eng.* **109**, 27–34.

Sato, M., Levesque, M. J., and Nerem, R. M. (1987b). Micropipette aspiration of cultured bovine aortic endothelial cells exposed to shear stress. *Arteriosclerosis* **7**, 276–286.

Sato, M., Levesque, M. J., and Nerem, R. M. (1988). The effect of a fluid-imposed shear stress on the mechanical properties of cultured endothelial cells. *Proceedings of the International Symposium on the Role of Blood Flow in Atherogenesis*, pp. 189–194, Springer-Verlag, Tokyo.

Sato, M., Theret, D. P., Wheeler, L. T., Ohshima, N., and Nerem, R. M. (1990). Application of the micropipette technique to the measurement of cultured porcine aortic endothelial cell viscoelastic properties. *ASME J. Biomech. Eng.* **112**, 263–268.

Schliwa, M., Nakamura, T., Porter, K. R., and Euteneur, U. (1984). A tumor promotor induces rapid and coordinated reorganization of actin and vinculin in culture cells. *J. Cell Biol.* **99**,1045–1059.

Semich, R., and Robenek, H. (1990). Organization of the cytoskeleton and the focal contacts of bovine aortic endothelial cells cultured on type I and III collagen. *J. Histochem. Cytochem.* **38**, 59–67.

Shen, J., Luscinskas, F. W., Connolly, A., Dewey, C. F., Jr., and Gimbrone, M. A. (1992). Fluid shear stress modulates cytosolic free calcium in vascular endothelial cells. *Am. J. Physiol.* **262**, C384–C390.

Shirinsky, V. P., Anonov, A. S., Birukov, K. G., Sobolevsky, A. V., Romanov, Y. A., Kabaeva, N. V., Anonova, G. N., and Smirnov, V. N. (1989). Mechanochemical control of human endothelium orientation and size. *J. Cell Biol.* **109**, 331–339.

Silkworth, J. B., and Stehbens, W. E. (1975). The shape of endothelial cells in en face preparations of rabbit blood vessels. *Angiology* **26**, 474–487.

Sumpio, B. E., and Banes, A. J. (1988). Prostacyclin synthetic activity in cultured aortic endothelial cells undergoing cyclic stretch. *Surgery* **104**, 383–389.

Sumpio, B. E., Banes, A. J., Levin, L. G., and Johnson, G., Jr. (1987). Mechanical stress stimulates aortic endothelial cells to proliferate. *J. Vasc. Surg.* **6**, 252–256.

Sumpio, B. E., Banes, A. J., Buckley, M., and Johnson, G. (1988). Alternations in aortic endothelial cell morphology and cytoskeletal protein synthesis during cyclic tensional deformation *J. Vasc. Surg.* **7**, 130–138.

Taylor, W. R., Delafontaine, P., Grindling, K. K., Nerem, R. M., and Alexander, R. W. (1990). Pressure induces insulin-like growth factor-1 secretion by endothelial cells. Paper presented at 44th Annual Fall Conference and Scientific Sessions of the Council for High Blood Pressure Research, American Heart Association, Baltimore, MD, Sept. 12–15.

Theret, D. P., Levesque, M. J., Sato, M., Nerem, R. M., and Wheeler, L. T. (1988). The application of a homogeneous half-space model in the analysis of endothelial cell micropipette measurements. *ASME J. Biomech. Eng.* **110**, 190–199.

Wechezak, A. R., Viggers, R. F., and Sauvage, L. R. (1987). Fibronectin and F-actin redistribution in cultured endothelial cells exposed to shear stress. *Lab. Invest.* **53**, 639–647.

Wong, M. K. K., and Gotlieb, A. I. (1990). Endothelial monolayer integrity. Perturbation of F-actin filaments and the dense peripheral bandvinculin network. *Arteriosclerosis* **10**, 76–84.

Wong, M. K. K., and Gotlieb, A. I. (1986). Endothelial cell monolayer integrity. 1. Characterization of dense peripheral band of microfilaments. *Arteriosclerosis* **6**, 212–219.

Yin, H. L., Albrecht, J., and Fattoum, A. (1981). Identification of gelsolin, a Ca^{2+}-dependent regulatory protein of actin gel-sol transformation. Its intracellular distribution in a variety of cells and tissues. *J. Cell Biol.* **91**, 901–906.

CHAPTER 7

■ ■ ■ ■ ■

Fluid Shear-Stress-Dependent Stimulation of Endothelial Autacoid Release: Mechanisms and Significance for the Control of Vascular Tone

■ ■ ■ ■ ■

Rudi Busse and Ulrich Pohl

I. INTRODUCTION

Complex organisms need a convective transport system for substrates and metabolites of cells. Therefore, in the evolution, a circulatory system has emerged that not only allows such a transport by the streaming blood but also has a high capability to adapt rapidly to varying demands of the tissues. For this purpose the circulatory system in mammals exhibits central and local control mechanisms that act together to maintain, via a sufficient blood pressure level, (the energy of which is provided by the heart) an adequate tissue perfusion. The blood pressure is determined by both, the cardiac output and the peripheral vascular resistance, the latter parameter being a function of the degree of constriction of terminal arteries and arterioles. This constrictor level can be modulated by "systemic" (neural, humoral) as well as local signals. The obvious influence of systemic control, such as that of the sympathetic nervous system or the release of vasoactive hormones from the adrenal gland has been well recognized for a long time. Local control was thought to be mediated mainly by the tissue metabolites. In recent years we have learned that essential components of this control system are located within the vessel wall itself, particularly in the endothelium, which regulates vascular tone mainly by the release of endothelium-derived relaxant factor (EDRF) (Furchgott and Zawadzki, 1980). This factor has been identified as nitric oxide (NO) (Palmer et al., 1987). This chapter summarizes recent findings on the stimulation of endothelial cells to release EDRF in response to the hydrodynamic forces associated with the fluid flow. The concept emerges that the vascular endothelium acts as flow sensor and, by the release of EDRF, contributes decisively to the capability of the circulatory system to adapt rapidly to changes of flow.

II. FLOW-DEPENDENT CONTROL OF VASCULAR DIAMETER: OBSERVATIONS

More than 50 years ago, Schretzenmayr (Schretzenmayr, 1933) described a hemodynamic phenomenon in anesthetized cats: whenever blood flow to the leg was increased (by muscle work) there was a concurrent increase in the diameter of the feeding femoral artery. The author concluded that this flow-dependent dilator response, which improves conductivity of feeding vessels, was due to a tissue-derived signal transmitted along the vascular tree. Further studies led to the concept of "ascending dilation" in conduit arteries under conditions of high tissue oxygen demand (Fleisch, 1935; Hilton, 1959). However, distal transection did not impede the dilation of conduit arteries in response to flow (Lie et al., 1970), indicating that the sensing of blood flow changes and the subsequent vascular response were a strictly local phenomenon of the vascular wall instead of being elicited by an ascending signal. In 1975 Rodbard (Rodbard, 1975) presented a concept of vascular caliber adjustment to flow load initiated by elevation or reduction of flow rates in which, for theoretical reasons only, endothelial cells played a key role. Only some years later could essential parts of his predictions be proven experimentally. The potential role of endothelial cells in the flow-dependent dilation gained considerable interest after the pioneering observation by Furchgott and Zawadzki (Furchgott and Zawadzki, 1980) that the endothelium can actively induce changes of vascular tone by the release of a labile relaxant factor, EDRF. In fact, an obligatory role of endothelial cells in perceiving flow signals and transducing them into vasodilator responses was demonstrated in large conduit arteries of several organs and species in vivo (Hull et al., 1986; Pohl et al., 1986b; Smiesko et al., 1985): The removal or functional impairment of endothelial cells abolished (or suppressed) the flow-dependent dilation while the response to endothelium-independent

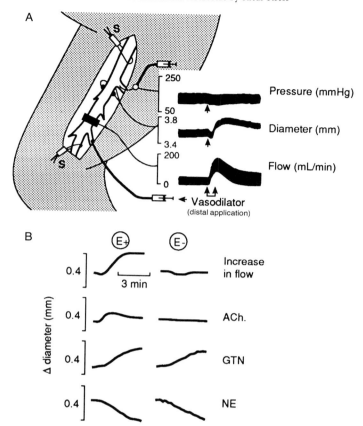

FIGURE 1   Flow-dependent dilation in the femoral artery of the dog is an endothelium-dependent response. (A) preparation of femoral artery in anesthetized dogs for analysis of flow-dependent dilation; injection of a vasodilator induced a transient increase in flow, followed by a slow, transient increase in arterial diameter, while femoral arterial pressure did not increase. (B) femoral artery diameter tracings from a typical experiment. Left-hand tracings: intact endothelium (E +). Right-hand tracings: endothelium mechanically removed (E −). Top tracings: the increase in flow (induced by distal application of a vasodilator) induces arterial dilation only in the presence of an intact endothelium, as did the EDRF stimulator acetylcholine (ACh). Removal of the endothelium did not affect the diameter response to nitroglycerin (GTN) and norepinephrine (NE), demonstrating well-preserved smooth-muscle responsiveness of the endothelium-denuded artery.

dilators such as nitroglycerin or adenosine was maintained (Fig. 1). Clinical studies revealed that flow-dependent dilation occurs also in human coronary and brachial arteries in vivo (Zeiher et al., 1991; Laurent et al., 1990; Heistad et al., 1984, 1986, 1987). This flow-dependent dilation was found to be reduced or even abolished in patients with hypercholesterolemia or atherosclerosis (Zeiher et al., 1991; Nabel et al., 1990; Creager et al., 1990; Cox et al., 1989). Both pathophysiologic states are associated with an impaired endothelial function, which shows indirectly that the endothelium plays a critical role in eliciting flow-dependent reactions in human arteries, too. In *isolated* canine femoral arteries (Rubanyi et al., 1986) it was shown that an increase in flow induced an augmented release of EDRF. This finding supported the concept that the flow-dependent dilation observed in large conduit arteries in vivo was due to an enhanced release of EDRF. Only recently could it be shown directly that a selective inhibitor of EDRF synthesis abolished the flow-dependent dilation in canine coronary arteries in situ (Bassenge, 1991).

Flow-dependent dilation, however, is not confined to large conduit arteries. In vivo observations of the microcirculation of the mesentery and cremaster muscle of the rat revealed dilator responses to increases in flow even in the smallest arterioles (Smiesko et al., 1989; Koller and Kaley, 1990a). When the endothelium was removed, this reaction was abolished, indicating its strict endothelial dependency (Koller and Kaley, 1990a). Experiments in isolated rabbit ear (Griffith and Edwards, 1990; Griffith et al., 1987) and mesentery (Pohl et al., 1991) showed, furthermore, that the flow-dependent dilation of small vessels could be suppressed by EDRF inhibitors, which argues again in favor of a critical role of this endogenous dilator in the flow-dependent control of vascular diameter.

III. SIGNALS AND SIGNAL RECOGNITION

It is generally assumed that flow or flow changes exert their stimulator effects on endothelial cells through the viscous drag, that is, the fluid-imposed shear stress acting on the luminal surface of the endothelium. It can be derived from the Navier–Stokes equations that in a rigid tube, under the conditions of a steady laminar flow, shear stress is a linear function of the viscosity of the fluid and of the flow rate and is inversely related to the third power of the vessel radius. If shear stress is the stimulus responsible for the flow-dependent dilation, then changes of not only flow but also fluid viscosity or vascular radius should result in vasodilation and/or release of EDRF from endothelial cells. A clear correlation of dilator responses and the magnitude of flow/flow changes has been

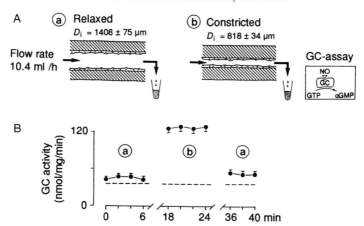

FIGURE 2   Reduction of vascular diameter increases EDRF release at constant flow rate. (A) experimental setup for measuring EDRF release from isolated saline-perfused arterial segments. EDRF was measured by activation of purified soluble guanylyl cyclase (GC) incubated with the effluent (Tyrode's solution) from the segments. (B) vasoconstriction leads to a more than 10-fold increase in the effluent-induced GC activation. Dotted line: GC activity in the absence of effluent. Mean value ± SEM (standard error of the mean) from six experiments.

demonstrated by several studies and in various types of vessels (Melkumyants et al., 1987; Smiesko et al., 1985; Pohl et al., 1986b). Likewise, an increase in flow in the intact coronary vascular bed was associated with an increase in EDRF release (Lamontagne et al., 1992). Moreover, dilations were also observed when instead of the flow the viscosity of the perfusate was increased (Melkumyants et al., 1987; Pohl et al., 1991; Tesfamarian and Cohen, 1988). Likewise, reductions of vascular diameter at constant flow led to an enhanced release of EDRF in isolated vessels (Busse et al., 1991) and intact hearts (Lamontagne et al., 1992) (Fig. 2). It is also consistent with a shear-stress-induced release of EDRF, that vasoconstrictions elicited by increases in the transmural pressure ("myogenic response") were significantly attenuated in the presence of an intact endothelium, indicating that shear-stress-induced increase of EDRF release counteracted in part the vasoconstriction. Moreover, a combined change of two of these parameters, including increase in flow and viscosity, further enhanced dilator responses of isolated arteries (Pohl et al., 1991). In summary, all these findings are consistent with the view that the shear stress imposed by the streaming fluid is the true stimulus that triggers a cascade of events that end up in the release of EDRF and endothelium-mediated, flow-dependent dilation.

Mechanoreception is a widespread sensory modality of living organisms. In general, local changes in the conductance of the cell membrane resulting in changes of the membrane potential have been described in several types of mechanoreceptor cells as primary response to the deforming forces and displacements in the process of sensory transduction. Although this mechanoelectrical transduction has been interpreted in many cells as the result of unspecific stretching of the membrane, it has been shown that in highly specialized mechanoreceptors, such as the hair cells of the inner ear, the minimal stimulus energy is in the range of $10-18$ W/s (comparable to that required for photoreceptors) (Thurm, 1983). This suggests that specific sensor molecules of the membrane are very closely and effectively coupled to ion channels. This process of sensory transduction has to be understood in endothelial cells, which as nonexcitable cells are not designed primarily as mechanoreceptors. However, very little is known about the shear-stress-sensing structures and the process of signal transduction in endothelial cells. One might speculate that the viscous drag is sensed by structures located at the outer leaflet of the cell membrane and coupled to regulating proteins (probably G proteins) at the level of membrane channels, the cytoskeleton, or membrane-attached calcium stores. This view is supported by the finding that exposure of the inner surface of blood vessels to low concentrations of glutaraldehyde, which denatured surface proteins without destroying the endothelial cells, resulted in an abolition of the flow-dependent dilation (Melkumyants et al., 1987). Recent experiments in which isolated, endothelium-intact vessels were treated with neuraminidase further points toward the existence of mechanotransducers at the luminal, endothelial surface (Pohl et al., 1991; Busse et al., 1991). This enzyme cleaves glycosidically bound sialic acids from the outer surface of the cell membrane. Removal of sialic acids from the membrane glycoproteins and glycolipids that form a network of anastomosing strands on the luminal surface of endothelial cells could impair the potential function of these structures as mechanoreceptors. There may be an additional role of sialic acids in the transduction of the flow signal at the cellular level (see Section IV). Sialic acids, which represent the terminal residues of the cell-coating glycocalix, provide the high density of negative surface charges on platelets, erythrocytes, and endothelial cells. There is some evidence that the electrostatic repulsion between endothelial cells and platelets is due to their high content of sialic acids and may account in part for the nonadherence of platelets to the endothelium (Sawyer and Srinivasan, 1972; Oka, 1983). Furthermore, it has been shown that the endothelial surface contains unusually high concentrations of sialic acids (Born and Palinski, 1985), the electronegative charge of which gives rise to a considerable binding of Ca^{2+} to the surface (Ludlam et al., 1988). Thus, this membrane-associated Ca^{2+} may represent a Ca^{2+} store, immediately available for the transmembrane Ca^{2+} influx into the cells and consequently for the triggering of Ca^{2+}-dependent EDRF synthesis. In this context, it should be mentioned however, that removal of sialic acid does not affect the net surface charge of endothelial cells estimated from electrophoretic mobility (Vargas et al., 1989).

IV. SIGNAL TRANSDUCTION

A. Formation of Endothelium Derived Relaxant Factor

Since the release of EDRF is a central part of the events leading to flow-dependent dilation, the cellular mechanisms of EDRF synthesis will be briefly summarized. EDRF that has been identified as nitric oxide (NO) (Palmer et al., 1987) is formed from one of the guanidino nitrogens of its precursor L-arginine (Schmidt et al., 1988; Palmer et al., 1988) in the presence of oxygen by a NADPH-dependent dioxygenase, the so-called NO synthase (Fig. 3). Consequently, the formation of EDRF can be blocked by means of analogues of L-arginine such as N^G-monomethyl-L-arginine (L-NMMA) (Iyengar et al., 1987; Sakuma et al., 1988; Rees et al., 1989) and N^G-nitro-L-arginine (L-NNA) (Mülsch and Busse, 1990; Moore et al., 1990), which inhibit in a stereospecific manner the NO-forming enzyme. The endothelial NO synthase is a constitutive enzyme and is similar to the enzyme found in neuronal tissues (Bredt and Snyder, 1990; Schmidt et al., 1991). Both forms differ from the NO synthase that is de novo synthetized in macrophages (Stuehr et al., 1989; Tayeh and Marletta, 1989), hepatocytes (Billiar et al., 1990), and vascular smooth-muscle cells (Busse and Mülsch, 1990b) on stimulation with endotoxin and/or cytokines. This inducible type of NO synthase does not need calcium-calmodulin for its activation (Busse and Mülsch, 1990b; Hauschildt et al., 1990). In contrast, endothelial NO synthase is directly activated by calcium-calmodulin in the presence of tetrahydrobiopterin (Busse and Mülsch, 1990a; Mülsch and Busse, 1991). Therefore, any elevation of intracellular free calcium in endothelial cells must result in an increased formation of EDRF. Increases in intracellular free calcium of endothelial cells are achieved by the activation of phospholipase C and the subsequent release of calcium from intracellular stores by IP_3 (Lambert et al., 1986; Pirotton et al., 1987; Freay et al., 1989). Simultaneously, a calcium influx from the extracellular space into the cytosol occurs through not yet defined but voltage-independent cation channels (Adams et al., 1989; Lückhoff and Busse, 1990b), which is responsible for the sustained elevation of Ca^{2+} within endothelial cells on receptor stimulation (Fig. 3).

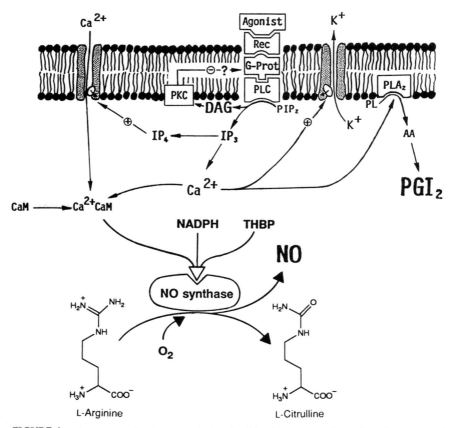

FIGURE 3   Signal transduction cascade involved in the receptor-dependent formation of nitric oxide and PGI_2 in endothelial cells. On binding of an agonist to its specific membrane receptor the activity of phospholipase (PLC) is enhanced via a GTP-binding protein (G-Prot.). PLC stimulation results in the formation of inositol-1,4,5-trisphosphate (IP_3) and diacylglycerol (DAG) from phosphatidylinositol-4,5-bisphosphate (PIP_2). IP_3 mobilizes Ca^{2+} from intracellular stores, whereas IP_4 may modulate the permeability of Ca^{2+} permeable cation channels in the plasma membrane, both of which increase the intracellular free Ca^{2+} concentration. Ca^{2+}, in turn, stimulates phospholipase A_2, resulting in the liberation of arachidonic acid (AA) and synthesis of PGI_2. Furthermore, Ca^{2+}-dependent K^+ channels are activated, which induces hyperpolarization and enhances transmembrane Ca^{2+} influx. The resulting Ca^{2+} increase activates in a calmodulin-dependent step the formation of NO from L-arginine in the presence of NADPH and tetrahydrobiopterin (THBP).

There is some evidence that simultaneous hyperpolarization as a consequence of the activation of calcium-activatable K^+ channel is a facilitating factor for the influx of calcium through these channels by enhancing the driving electrical force (Lückhoff and Busse, 1990a, 1990b). In contrast, depolarization inhibits Ca^{2+} influx into stimulated endothelial cells (Sauve et al., 1988; Adams et al., 1989; Lückhoff and Busse, 1990b; Schilling et al., 1989; Cannell and Sage, 1989; Laskey et al., 1990). Thus, the calcium influx into endothelial cells seems to be regulated by the membrane potential in a way opposite to that in vascular smooth-muscle cells where hyperpolarization inactivates and a depolarization activates voltage-gated calcium channels.

B. Effects of Shear Stress on Intracellular Free Calcium and Membrane Potential

Several studies have been performed in cultured endothelial cells investigating the effect of shear stress on cellular signal transduction pathways. A continuous increase of prostacyclin production (Bhagyalakshmi and Frangos, 1989; Frangos et al., 1985) as well as a rise in Ins-1,4,5-P_3 level (Nollert et al., 1990) have been described on exposure of cultured human endothelial cells to shear stresses. Using a fluorescent calcium indicator, a distinct increase in the concentration of intracellular free calcium being associated with the induction of shear stress on cultured bovine aortic endothelial cells has been demonstrated (Ando et al., 1988). Both findings have been questioned to be shear-stress-dependent recently, since superfusion of these cells has been performed with a medium containing micromolar amounts of ATP (M 199). Since the availability of ATP should increase with an increase in flow rate, it is possible that the observed effects are not caused primarily by the fluid-imposed shear stress but rather by an enhanced activation of P_{2y} receptors through higher amounts of ATP. It is consistent with this view that superfusion of cells with ATP-free medium induced only small or no changes in $[Ca^{2+}]_i$ (Mo et al., 1991; Dull and Davies, 1991). However, theoretical calculations of the flow-dependent change of ATP delivery to the cell surface reveal only minor changes in the range of shear stress discussed here (Berthiaume et al., 1991). Moreover, these findings are at variance with recent experiments in cultured human umbilical endothelial cells. In these cells imposed shear stresses in the range between 1.4 and 2.2 dyn/cm^2 induced a clear activation of the phospholipase C pathway irrespective of the presence or absence of ATP in the superfusate (Bhagyalakshmi et al., in press). Therefore, at least in human endothelial cells, shear stress can act as a stimulus per se. Since it is known that calcium is a necessary factor for the activation of NO synthase and that at least calmodulin and tetrahydrobiopterin are relatively constant within the cell, these experiments describe a very plausible way how calcium can act as the key messenger in the signal transduction of shear stress to EDRF formation.

Endothelial stimulation by ATP may, however, be a component of an additional activating pathway of shear stress-induced EDRF formation. It has been shown that high flow rates are able to release compounds from endothelial cells that are known to increase intracellular free calcium by receptor-dependent activation of phospholipase C. Increased perfusion of isolated rat hind legs as well as an enhanced superfusion rate of cultured endothelial cells grown on microcarrier beads and packed into columns, revealed a release of ATP, acetylcholine, and substance P (Milner et al., 1990a, 1990b; Ralevic et al., 1990). Apparently, these substances are produced in endothelial cells and released when the cells are exposed to shear stress.

Provided that the observed release was not an experimental artefact due to cell damage, this mechanism could lead indirectly (by receptor-mediated effects of the released compounds) to an increase in intracellular free calcium. However, it does not answer the primary question, how shear stress induces the release of the stimulator compounds. In this context, it is of particular interest that cultured endothelial cells release bradykinin in amounts sufficient to induce NO formation provided its degradation was inhibited by angiotensin-converting enzyme (ACE) inhibitors (Wiemer et al., 1991). If this release is enhanced or the activity of the ectoenzyme ACE is reduced by increased shear stress, then bradykinin might represent an essential link in the signal transduction cascade. The same might hold true with ATP provided that shear stress could reduce the activity of the ectonucleotidases.

It is not clear as yet whether the induction of K^+ currents and the resulting membrane hyperpolarization on stimulation of endothelial cells with shear stress (Olesen et al., 1988) is due to the action of these compounds on receptors or to a primary event occurring at the level of specific cation channels (Fig. 4). Stretch-activated cation channels in cell-attached patches from porcine aortic endothelial cells have been described (Lansman, et al., 1987) that could be involved in the response to the mechanical forces generated by the flow-induced viscous drag. These channels were nonselective and activated by suction in the patch pipette (Lansman et al., 1987; Popp et al., 1991). At resting membrane potential, activation of these channels would give rise for a depolarization. Whole-cell patch-clamp recordings in bovine aortic endothelial cells grown in capillary flow tubes revealed shear-stress-activated K^+ currents. These currents varied in magnitude and duration as a function of shear stress in a range

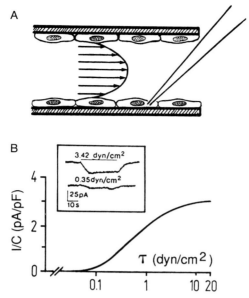

FIGURE 4   (A) experimental setup for whole-cell patch-clamp recordings in single endothelial cells grown on the inner surface of a perfused glass capillary tube. (B) relationship between fluid shear stress amplitude at the endothelial cell surface and whole-cell current. Insert: tracings showing the time course of whole-cell currents from one cell induced by two different flow rates. The whole-cell currents were normalized by cell capacity (C = 15 pF). [Modified from Olesen et al., 1988.]

between 0.1 and 20 dyn/cm^2 and represented a very fast and early event after initiation of shear stress increase (Olesen et al., 1988). While this may be due to a direct activation of a K$^+$-selective channel by a mechanical event, it cannot be excluded that the activation of the K$^+$ channel is secondary to a receptor-mediated increase in intracellular Ca^{2+}, followed by an activation of K$^+$-channels as demonstrated in patch-clamp studies on cultured endothelial cells (Sauve et al., 1988; Colden-Stanfield et al., 1990). As mentioned above, hyperpolarization facilitates the transmembrane influx of Ca^{2+} into endothelial cells preactivated with thimerosal (Lückhoff and Busse, 1990a). Therefore, this hyperpolarization may contribute to the maintenance of the intracellular Ca^{2+} elevation in response to shear stress. L-Arginine, the precursor of NO, is stoichiometrically converted to citrulline and NO (EDRF) in endothelial cells. Although in native and cultured cells, the intracellular concentration of L-arginine is surprisingly high [0.1–0.8 mM in cultured compared to 2–4 mM in freshly isolated endothelial cells (Hecker et al., 1990)], it is conceivable that under certain conditions the membrane transport rate of L-arginine may affect the overall NO formation. In various mammalian cells, L-arginine and other cationic amino acids are transported across the plasma membrane via a sodium-independent transporter (system $y+$) that operates as facilitated diffusion system driven by the membrane potential (Bussolati et al., 1989). Although a comparable study has not yet been performed in endothelial cells, the shear-stress-induced hyperpolarization might also be important in enhancing the cellular uptake of L-arginine. In fact, it has been reported that the uptake of L-arginine by cultured endothelial cells is enhanced under conditions of high shear stress (Sheriff et al., 1990). This is in line with the finding in cultured bovine aortic endothelial cells that high shear stress (30 dyn/cm^2) enhances the receptor-mediated internalization and degradation of low-density lipoproteins (LDL) by an increase in LDL receptor expression (Sprague et al., 1987).

C. Additional Endothelial Mediators Involved in Flow-Dependent Dilation

While there is good evidence for a role of EDRF as mediator of flow-dependent dilation in human vessels and in arteries of several other species, in rat arteries in vivo the mediator appears to be a prostaglandin. This derives from the observation that the flow-dependent dilation in rat cremaster arterioles could be suppressed by the treatment of the animals with the cyclooxygenase inhibitor indomethacin (Koller and Kaley, 1990a). Cyclooxygenase catalyzes the formation of prostaglandins from its precursor, arachidonic acid. Vascular-derived prostaglandins are likely of endothelial origin since (1) about 80% of the PGI$_2$ that is formed by the vessel wall is synthesized within endothelial cells (Moncada et al., 1977) and (2) not only cyclooxygenase inhibition but also a local disruption of the endothelium abolished the flow-dependent dilation in this preparation (Koller and Kaley, 1990a, 1990b). Since rat arteries exhibit a high sensitivity to PGI$_2$—in contrast to rabbit arteries (Förstermann et al., 1984)—it is conceivable that PGI$_2$, which is normally coreleased with EDRF, is an important dilator compound in this species. It has also been shown in isolated rabbit and canine arteries that pulsatile (Pohl et al., 1986a) or steady-state shear stress increases the release of PGI$_2$. The enhanced production of prostaglandin on shear stress may—similar to the formation of EDRF—be due to an increase of intracellular free Ca^{2+}. The subsequent activation of phospholipase A$_2$ releases arachidonic acid, the precursor of prostaglandins from membrane phospholipids (Lapetina, 1982).

When considering additional endothelial mechanisms that might be involved in eliciting flow-dependent dilation the shear stress-induced membrane hyperpolarization again comes into play. It has to be discussed with respect to not only the facilitation of Ca^{2+} entry but also the possibility that an electrical coupling exist between the endothelium and the underlying vascular smooth muscle. Admittedly, no detectable electrical connection between the endothelium and vascular smooth-muscle cells was found in the pig coronary arteries (Beny, 1990), although a transient hyperpolarization in response to bradykinin was observed in endothelial cells and neighboring smooth-muscle cells. In contrast, in small arteries and arterioles a high number of myoendothelial junctions have been found that offer the possibility that endothelial hyperpolarization is transmitted to the vascular smooth-muscle cells (Davies et al., 1988). Smooth-muscle hyperpolarization would result in an inactivation of voltage-operated Ca^{2+} channels, thereby reducing the available Ca^{2+} for contraction and inducing vasorelaxation. It must be emphasized however, that there is no direct experimental evidence for such a mechanism.

Much more substantiated is the tight electrical coupling of endothelial cells (Daut et al., 1988), which, is obviously the basis for the so-called propagated vasodilation that can be observed upstream of the application site of a vasodilator in small vessels (Segal et al., 1989; Segal and Duling, 1986, 1987; Gaehtgens et al., 1976). This response can be blocked by compounds that close gap junctions between endothelial cells but not by tetrodotoxin, which, by blocking specific Na^+ channels, inhibits signal conduction by neuronal structures. It remains to be established, however, whether "propagated vasodilation" is involved in the flow-dependent dilation. Such a concept would require that not all endothelial cells respond in the same manner to increases in flow but only some specialized cells—that spread the electrical signal to the adjacent cells similar to the situation described for the localization of vasoactive compounds in endothelial cells (Loesch and Burnstock, 1988; Milner et al., 1989).

V. FUNCTIONAL SIGNIFICANCE OF FLOW-DEPENDENT DILATION

Since the oxygen demand of the tissue varies considerably depending on the function and the work load of an organ an effective regulation of blood flow or oxygen transport capacity is required. The blood flow to organs is determined by the driving arteriovenous pressure difference generated by the action of the heart and by the peripheral vascular resistance of the organ, which according to Poisseuille's law is predominantly a fourth-power function of vascular radius. It is important to note that any increase in blood flow without a concomitant reduction in peripheral vascular resistance can be achieved only by a considerable increase in driving pressure and hence cardiac work. From this point of view it represents an optimizing principle that increased flow demands are associated with an essential vasodilation and consequently a decrease of vascular resistance.

According to the classic view, the resistance to blood flow and the control of tissue perfusion reside in the arteriolar section (range of vascular diameters $< 60 \; \mu$m) of the vascular bed. More recent studies on the distribution of intravascular pressure drop along the arterial tree indicate, however, that a major fraction of total vascular resistance resides in larger vessels that give rise to the vessels of the microcirculation. It can be calculated that up to 60% of the vascular resistance can be attributed to the segments proximal to the microcirculation. As a consequence, even maximal dilation of microvessels can result only in a twofold increase in flow as long as the resistance of the feed arteries remains unaltered (Segal et al., 1989), whereas the normal increase in blood flow is at least fivefold in most organs in vivo (Shepherd and Abboud, 1983). To satisfy this flow demand at an energy cost to the heart that is as low as possible, the diameter increases of arterial vessels during enhanced blood flow must be coordinated along the whole vascular tree. While the smallest vessels are thought to be under direct control of tissue metabolites, this is not necessarily the case in larger vessels. Therefore, tissue-derived signals cannot achieve the necessary coordination.

In the isolated saline-perfused rabbit ear with intact EDRF formation, X-ray microangiograms of the vascular tree revealed an optimal relationship between vascular diameters along the vascular tree and changes of blood flow. When the flow was altered, the relative changes of the diameter in vessels of different generations were the same. This "geometric similarity" leads to an approximate constant pressure drop along the vascular tree irrespective of the actual flow (Griffith et al., 1987). When, however, the effects of EDRF were blocked in this preparation, by infusion of the EDRF inhibitor hemoglobin, pressure drops along the vascular tree were significantly increased, thereby reducing the overall conductivity of the preparation (Griffith et al., 1987). These findings suggest that EDRF by eliciting flow-dependent dilation at any point of the vasculature in response to local stimulation by shear stress represents the mechanism that coordinates the vascular diameters.

Similar observations were made in isolated saline-perfused rabbit hearts. A sudden increase in perfusate flow (two-to-threefold) under flow-controlled conditions induced an initial increase in perfusion pressure due to the higher volume load at primarily unaltered resistance. This was followed by a secondary decrease in the pressure, indicating a reduced resistance of the coronary vascular bed due to flow-dependent dilation (Fig. 5). After inhibition of EDRF production by N^G-nitro-L-arginine (L-NNA), a stere-

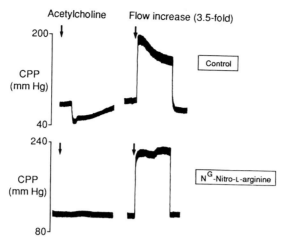

FIGURE 5   Coronary perfusion pressure in the isolated saline-perfused rabbit heart in response to acetylcholine (ACh, 1 μM) and a 3.5-fold increase in flow (from 20 to 70 mL/min) under control conditions and after treatment with N^G-nitro-L-arginine (30 μM; 30 min).

ospecific inhibitor of the EDRF generating enzyme, this flow-dependent dilation was completely abolished, resulting in a sustained augmentation of vascular resistance (Lamontagne et al., 1991). Since EDRF increases cyclic GMP in vascular smooth muscle and in platelets by the activation of soluble guanylyl cyclase (Busse et al., 1987; Pohl and Busse, 1989; Furlong et al., 1987; Rapoport et al., 1983), the cyclic GMP levels of platelets passing through the heart can be taken as an index of EDRF activity (Pohl and Busse, 1989). It was found that not only an increase in flow but also a pharmacologically induced vasoconstriction at constant flow, which must increase wall shear stress, was associated with an increased release of EDRF from the coronary vascular bed (Lamontagne et al., 1992) (Fig. 6). Similar results were obtained in isolated saline perfused arterial segments (Tesfamarian and Cohen, 1988; Busse et al., 1991).

These findings offer an interesting concept for EDRF acting as an attenuating principle of vasoconstriction that otherwise tends to impair tissue oxygen supply. As long as the flow is not proportionally reduced, a reduction in diameter would be associated with an increase in shear stress. The resulting release of EDRF would tend to oppose further vasoconstriction by two mechanisms. First, by a direct cyclic GMP-mediated inhibition of vasoconstriction at the level of the vascular smooth-muscle cell (Förstermann et al., 1986; Mülsch et al., 1987; Fiscus et al., 1983–84).

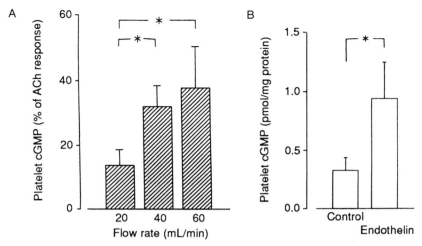

FIGURE 6   (A) Flow-induced increases in luminal NO release in the intact coronary bed of the rabbit heart as assessed by cGMP content of platelets after passage through the heart. Platelet cGMP content is expressed as percent of the platelet cGMP content measured during stimulation with acetylcholine (1 μM) at the respective flow rates. (B) Increase in luminal NO release elicited by endothelin-1-induced vasoconstriction at constant flow rate.

Second, by an inhibition of presynaptic release of the neurotransmitter norepinephrine (Tesfamarian and Cohen, 1988). This was shown in isolated saline-perfused rabbit carotid arteries exposed to electrical field stimulation in order to activate adrenergic nerves. Neurogenic vasoconstriction was significantly reduced with increases in shear stress (which were achieved by increasing perfusate viscosity at constant flow). On the contrary, neurogenic vasoconstriction was significantly enhanced following removal of the endothelium or blocking the target enzyme for EDRF, the soluble guanylyl cyclase, with the unspecific inhibitor methylene blue.

A relative fine-tuning of this opposing mechanism is possible, since at least in smaller vessels, the endothelium-mediated response to alterations in shear stress lies in the range of 5–10 s (Koller and Kaley, 1990a; Smiesko et al., 1989) and allows a very fast response. In larger vessels response times up to 40 s are reported (Pohl et al., 1986b; Hintze and Vatner, 1984; Smiesko et al., 1985; Hull et al., 1986). It is not clear whether these longer response times of large arteries are simply due to some surgical trauma (shorter response time appeared to occur in large arteries that were not acutely exposed for measurements) (Hintze and Vatner, 1984) or reflect lower shear rates in these vessels due to their much greater diameter. Using platelet velocity profiles as indicators of in vivo shear rates in microvessels

$(15-30 \ \mu m)$, shear stresses were found as high as about 120 dyn/cm^2 with a mean of around 40 dyn/cm^2 (Tangelder et al., 1988). Although these values were higher than those reported to be necessary for maximal activation of shear-stress-activated K$^+$ channels (Olesen et al., 1988), it is conceivable that the faster response in small vessels might also be due to the previously discussed electrotonic transmission of shear-induced hyperpolarization into vascular smooth muscle that certainly does not play any role in larger vessels (Beny, 1990). Shear-stress-induced counteraction of vasoconstriction is most important as a mechanism in the control of so-called "myogenic responses." When intravascular pressure is raised, a constriction of vascular smooth muscle occurs (myogenic response) (Johnson, 1981; Bayliss, 1902). This represents a positive-feedback mechanism which could lead to an instability of the circulation if no corresponding inhibitory mechanism is established. Studies in isolated pig coronary arterioles (diameters 40–80 μm) showed flow-dependent dilation that was endothelium-mediated and present even at high levels of myogenic activation although it was progressively reduced with higher pressures (Kuo et al., 1990). Similar results were obtained with pharmacologic activation of vascular smooth muscle in femoral arteries in vivo (Pohl et al., 1986b). Since in many physiological situations intravascular pressure and flow are elevated simultaneously (cf. muscle work), it is of interest to characterize the dynamic interaction of both mechanisms. In an isolated preparation of the rabbit mesentery the interaction of flow-dependent dilation and myogenic constriction was therefore studied in more detail. It could be shown that an increase in intravascular pressure resulted in a substantial vasoconstriction of small arteries (mean diameter 200–300 μm) (Pohl et al., 1991). When the increase in pressure was accompanied by a two- to threefold increase in flow or an increase in perfusate viscosity in the presence of intact EDRF production, a net dilation of the vessel resulted, indicating that flow-dependent dilation not only opposed myogenic responses but rather induced a significant reduction of vascular resistance. Interestingly, the vasodilation was never sufficient to compensate completely for the flow-induced increase of shear stress (Pohl et al., 1991). This is in accordance with experiments performed in small arterioles in situ (Koller and Kaley, 1990a; Smiesko et al., 1989). The still elevated shear stress might represent the stimulus for a maintained increase of vascular diameter as long as the vessel is being exposed to high flow.

Myogenic responses represent the crucial component in the autoregulation of blood flow that is observed in several organs. Autoregulation means, that the blood flow is independent of perfusion pressure over a wide pressure range (about 40–160 mm Hg) due to compensatory changes in vascular resistance of an organ (Johnson, 1981). If the shear-stress-

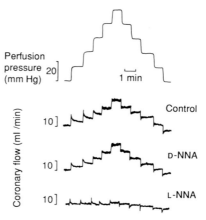

FIGURE 7   Reduced flow changes in response to changes in perfusion pressure in an isolated rabbit heart following inhibition of EDRF synthesis by N^{G}-nitro-L-arginine (L-NNA). Top tracing: stepwise changes of coronary perfusion pressures. Middle and bottom tracings: coronary flow (measured by electromagnetic flowmeter) under control conditions (control), after treatment with the ineffective stereoisomer N^{G}-nitro-D-arginine (D-NNA) and after treatment with N^{G}-nitro-L-arginine (L-NNA).

dependent inhibition of myogenic responses plays a functional role, it should be expected that autoregulatory responses are enhanced when the flow-dependent release of EDRF is no longer present. In isolated perfused rabbit hearts, stepwise increases in perfusion pressure from 40 to 120 mm Hg induced initial flow increases, which were followed by a secondary reduction of flow toward the previous control flow (see Fig. 7), indicating a tendency to autoregulate coronary flow. After inhibition of EDRF synthesis by N^{G}-nitro-L-arginine, the autoregulation was significantly enhanced (Pohl et al., 1990). This is consistent with the view that EDRF is a significant component controlling the extent of autoregulation of organ blood flow. Similar results were obtained in the isolated rabbit ear. Inhibition of EDRF release unmasked constrictor responses to increases in perfusion pressure that were not observed in a preparation with a functionally intact endothelium (Griffith and Edwards, 1990; Griffith et al., 1989).

    The functional consequences of an impaired flow-dependent dilation have also been studied in the isolated heart. Inhibition of EDRF release by L-NNA resulted in an increase of coronary vascular resistance (Lamontagne et al., 1991). Under pressure-control conditions this resulted in a nearly 50% decrease of resting coronary flow. This occurred in spite of a significant increase in the release of lactate indicating an oxygen deficit in

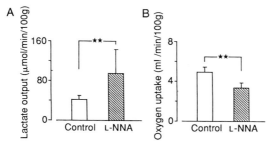

FIGURE 8   Lactate output (A) and oxygen consumption (B) measured in isolated, saline-perfused rabbit hearts before (white bars) and after treatment with N^G-nitro-L-arginine (L-NNA) (hatched bars). After inhibition of EDRF synthesis by L-NNA, the oxygen uptake of hearts perfused at constant pressure decreased significantly ($p < .01$) with signs of tissue hypoxia, as indicted by an enhanced release of lactate.

the myocardial tissue (Pohl et al., 1992) (Fig. 8). Direct measurements of tissue p_{O_2} distribution in rabbit skeletal muscle in vivo substantiate this explanation. After inhibition of EDRF release a significant reduction of tissue p_{O_2} was observed that could, however, be compensated by infusion of vasodilator substances that replaced endogenously released EDRF along the vascular tree (Pohl and Busse, 1988). Although tissue ischemia should lead to a significant dilation of small arterioles (Sparks, 1980) a compensatory increase was not achieved after inhibition of EDRF. This indicates not only an important role of larger feeding vessels in the control of organ vascular resistance but also a role of EDRF as main mediator in the control of nutritive blood flow at increased tissue oxygen demands.

REFERENCES

Adams, D. J., Barakeh, J., Laskey, R., and Van Breemen, C. (1989). Ion channels and regulation of intracellular calcium in vascular endothelial cells. *FASEB J.* 3, 2389–2400.

Ando, J., Komatsuda, T., and Kamiya, A. (1988). Cytoplasmic calcium response to fluid shear stress in cultured vascular endothelial cells. *In Vitro Cell. Devel. Biol.* 24, 871–877.

Bassenge, E. (1991). Endothelium-mediated regulation of coronary tone. *Basic Res. Cardiol.* 86 (Suppl. 2), 69–76.

Bayliss, W. M. (1902). On the local reaction of the arterial wall to changes of internal pressure. *J. Physiol. Lond.* 28, 220–231.

Beny, J. L. (1990). Endothelial and smooth muscle cells hyperpolarized by bradykinin are not dye coupled. *Am. J. Physiol.* 258, H836–H841.

Bhagyalakshmi, A., Berthiaume, F., Reich, K. M., and Frangos, J. A. (in press). Fluid shear stress stimulates membrane phospholipid metabolism in cultured human endothelial cells. *J. Vasc. Res.* (in press).

Bhagyalakshmi, A., and Frangos, J. A. (1989). Mechanism of shear-induced prostacyclin production in endothelial cells. *Biochem. Biophys. Res. Commun.* **158**, 31–37.

Billiar, T. R., Curran, R. D., Stuehr, D. J., Stadler, J., Simmons, R. L., and Murray, S. A. (1990). Inducible cytosolic enzyme activity for the production of nitrogen oxides from L-arginine in hepatocytes. *Biochem. Biophys. Res. Commun.* **168**, 1034–1040.

Born, G. V. R., and Palinski, W. (1985). Unusually high concentrations of sialic acids on the surface of vascular endothelia. *Br. J. Exp. Pathol.* **66**, 543–549.

Bredt, D. S., and Snyder, S. H. (1990). Isolation of nitric oxide synthetase, a calmodulin-requiring enzyme. *Proc. Natl. Acad. Sci. USA* **87**, 682–685.

Brunet, P. C., and Beny, J. L. (1989). Substance-P and bradykinin hyperpolarize pig coronary artery endothelial cells in primary culture. *Blood Vessels* **26**, 228–234.

Busse, R., Lückhoff, A., and Bassenge, E. (1987). Endothelium-derived relaxant factor inhibits platelet activation. *Nauyn-Schmiedebergs Arch. Pharmacol.* **336**, 566–571.

Busse, R., Fichtner, H., Lückhoff, A., and Kohlhardt, M. (1988). Hyperpolarization and increased free calcium in acetylcholine-stimulated endothelial cells. *Am. J. Physiol.* **255**, H965–H969.

Busse, R., Pohl, U., and Lückhoff, A. (1989). Mechanisms controlling the production of endothelial autacoids. *Z. Kardiol.* **78** (Suppl. 6), 64–69.

Busse, R., and Mülsch, A. (1990a). Calcium-dependent nitric oxide synthesis in endothelial cytosol is mediated by calmodulin. *FEBS Lett.* **265**, 133–136.

Busse, R., and Mülsch, A. (1990b). Induction of nitric oxide synthase by cytokines in vascular smooth muscle cells. *FEBS Lett.* **275**, 87–90.

Busse, R., Mülsch, A., and Bassenge, B. (1991). Shear stress-dependent nitric oxide release controls neuro- and myogenic vasoconstriction. In *Resistance Arteries, Structure and Function*, pp. 221–225 Mulvany, M. J., Aalkjaer, C., Heagerty, A. M., Nyborg, N. C. B., Strandgaard, S. (eds.). Elsevier Science Publishers, Amsterdam.

Busse, R., Mülsch, A., and Bassenge, E. (1991). Vasoconstriction elicits release of NO from isolated conduit and resistance-sized arteries by a shear stress-dependent mechanism. *Pflügers Arch.* **418** (Suppl. 1), R103.

Bussolati, O., Laris, P. C., Nucci, F. A., Dallasta, V., Franchigazzola, R., Guidotti, G. G., and Gazzola, G. C. (1989). Influx of L-arginine is an indicator of membrane potential in human fibroblasts. *Am. J. Physiol.* **256**, C930–C935.

Cannell, M. B., and Sage, S. O. (1989). Bradykinin-evoked changes in cytosolic calcium and membrane currents in cultured bovine pulmonary artery endothelial cells. *J. Physiol. Lond.* **419**, 555–568.

Colden-Stanfield, M., Schilling, W. P., Ritchie, A. K., Eskin, S. G., Navarro, L. T., and Kunze, D. L. (1987). Bradykinin-induced increases in cytosolic calcium and ionic currents in cultured bovine aortic endothelial cells. *Circ. Res.* **61**, 632–640.

Colden-Stanfield, M., Schilling, W. P., Possani, L. D., and Kunze, D. L. (1990). Bradykinin-induced potassium current in cultured bovine aortic endothelial cells. *J. Membr. Biol.* **116**, 227–238.

Cox, D. A., Vita, J. A., Treasure, C. B., Fish, R. D., Alexander, R. W., Ganz, P., and Selwyn, A. P. (1989). Atherosclerosis impairs flow-mediated dilation of coronary arteries in humans. *Circulation* **80**, 458–465.

Creager, M. A., Cooke, J. P., Mendelsohn, M. E., Gallagher, S. J., Coleman, S. M., Loscalzo, J., and Dzau, V. H. (1990). Impaired vasodilation of forearm resistance vessels in hypercholesterolemic humans. *J. Clin. Invest.* **86**, 228–234.

Daut, J., Mehrke, G., Nees, S., and Newman, W. H. (1988). Passive electrical properties and electrogenic sodium transport of cultured guinea-pig coronary endothelial cells. *J. Physiol. Lond.* **402**, 237–254.

Daut, J., Dischner, A., and Mehrke, G. (1989). Bradykinin induces a transient hyperpolarization of cultured guinea-pig coronary endothelial cells. *J. Physiol. Lond.* **410**, 48P.

Davies, P. F., Olesen, S. P., Clapham, D. E., Morrel, E. M., and Schoen, F. J. (1988). Endothelial communication. State of the art lecture. *Hypertension* **11**, 563–572.

Dull, R. O., and Davies, P. F. (1991). Flow modulation of agonist (ATP)-response (Ca^{2+}) coupling in vascular endothelial cells. *Am. J. Physiol.* **261**, H149–H154.

Fichtner, H., Fröbe, U., Busse, R., and Kohlhardt, M. (1987). Single nonselective cation channels and Ca^{2+}-activated K^+ channels in aortic endothelial cells. *J. Membr. Biol.* **98**, 125–133.

Fiscus, R. R., Rapoport, R. M., and Murad, F. (1983–84). Endothelium-dependent and nitrovasodilator-induced activation of cyclic GMP-dependent protein kinase in rat aorta. *J. Cycl. Nucleotide Protein Phosphorylation Res.* **9**, 415–425.

Fleisch, A. (1935). Les reflexes nutritifs ascendants producteurs de dilatation arterielle. *Arch. Int. Physiol.* **41**, 141–167.

Förstermann, U., Hertting, G., and Neufang, B. (1984). The importance of endogenous prostaglandins other than prostacyclin for the modulation of contractility of some rabbit blood vessels. *Br. J. Pharmacol.* **81**, 623–630.

Förstermann, U., Mülsch, A., Böhme, E., and Busse, R. (1986). Stimulation of soluble guanylate cyclase by an acetylcholine-induced endothelium-derived factor from rabbit and canine arteries. *Circ. Res.* **58**, 531–538.

Frangos, J. A., Eskin, S. G., McIntire, L. V., and Ives, C. L. (1985). Flow effects on prostacyclin production by cultured human endothelial cells. *Science* **227**, 1477–1479.

Freay, A., Johns, A., Adams, D. J., Ryan, U. S., and van Breemen, C. (1989). Bradykinin and inositol 1,4,5-trisphosphate-stimulated calcium release from intracellular stores in cultured bovine endothelial cells. *Pflügers Arch.* **414**, 377–384.

Furchgott, R. F., and Zawadzki, J. V. (1980). The obligatory role of endothelial cells in the relaxation of arterial smooth muscle by acetylcholine. *Nature* **288**, 373–376.

Furlong, B., Henderson, A. H., Lewis, M. J., and Smith, J. A. (1987). Endothelium-derived relaxant factor inhibits in vitro platelet aggregation. *Br. J. Pharmacol.* **90**, 687–692.

Gaehtgens, P., Benner, K. U., and Schickendantz, S. (1976). Nutritive and non-nutritive blood flow in canine skeletal muscle after microembolization. *Pflügers Arch.* **361**, 183–189.

Griffith, T. M., Edwards, D. H., Davies, R. L. I., Harrison, T. J., and Evans, K. T. (1987). EDRF coordinates the behavior of vascular resistance vessels. *Nature* **329**, 442–445.

Griffith, T. M., Edwards, D. H., Davies, R. L., and Henderson, A. H. (1989). The role of EDRF in flow distribution: A microangiographic study in the rabbit isolated ear. *Microvasc. Res.* **37**, 162–177.

Griffith, T. M., and Edwards, D. H. (1990). Myogenic autoregulation of flow may be inversely related to endothelium-derived relaxing factor activity. *Am. J. Physiol.* **258**, H1171–H1180.

Hauschildt, S., Lückhoff, A., Mülsch, A., Kohler, J., Bessler, W., and Busse, R. (1990). Induction and activity of NO synthetase in bone marrow-derived macrophages are independent of calcium. *Biochem. J.* **270**, 351–356.

Hecker, M., Sessa, W. C., Harris, H. J., Anggard, E. E., and Vane, J. R. (1990). The metabolism of L-arginine and its significance for the biosynthesis of endothelium-derived relaxing factor-cultured endothelial cells recycle L-citrulline to L-arginine. *Proc. Natl. Acad. Sci. USA* **87**, 8612–8616.

Heistad, D. D., Armstrong, M. L., Marcus, M. L., Piegors, D. J., and Mark, A. L. (1984). Augmented responses to vasoconstrictor stimuli in hypercholesterolemic and atherosclerotic monkeys. *Circ. Res.* **54**, 711–718.

Heistad, D. D., Armstrong, M. L., and Amundsen, S. (1986). Blood flow through vasa vasorum in arteries and veins: Effects of luminal p_{O_2}. Am. J. Physiol. 250, H434–H442.

Heistad, D. D., Mark, A. L., Marcus, M. L., Piegors, D. J., and Armstrong, M. L. (1987). Dietary treatment of atherosclerosis abolishes hyperresponsiveness to serotonin: Implications for vasospasm. Circ. Res. 61, 346–351.

Hilton, S. M. (1959). A peripheral arterial conducting mechanism underlying dilation of the femoral artery concerned in functional vasodilatation in skeletal muscle. J. Physiol. Lond. 149, 93–111.

Hintze, T. H., and Vatner, S. F. (1984). Reactive dilation of large coronary arteries in conscious dogs. Circ. Res. 54, 50–57.

Hull, S. S., Kaiser, L., Jaffe, M. D., and Sparks, N. V. (1986). Endothelium-dependent flow induced dilation of canine femoral and saphenous arteries. Blood Vessels 23, 183–198.

Iyengar, R., Stuehr, D. J., and Marletta, M. A. (1987). Macrophage synthesis of nitrite, nitrate, and N-nitrosamines: precursors and role of the respiratory burst. Proc. Natl. Acad. Sci. USA 84, 6369–6373.

Johnson, P. C. (1981). The myogenic response. In Handbook of Physiology. The Cardiovascular System. Vascular Smooth Muscle. Sect. 2, Vol. II, Chapter 15, pp. 409–442. American Physiological Society, Bethesda, MD, 1981.

Koller, A., and Kaley, G. (1990a). Endothelium regulates skeletal muscle microcirculation by a blood flow velocity-sensing mechanism. Am. J. Physiol. 258, H916–H920.

Koller, A., and Kaley, G. (1990b). Prostaglandins mediate arteriolar dilation to increased blood flow velocity in skeletal muscle microcirculation. Circ. Res. 67, 529–534.

Kuo, L., Davis, M. J., and Chilian, W. M. (1990). Endothelium-dependent, flow-induced dilation of isolated coronary arterioles. Am. J. Physiol. 259, H1063–H1070.

Lambert, T. L., Kent, R. S., and Whorton, A. R. (1986). Bradykinin stimulation of inositol polyphosphate production in porcine aortic endothelial cells. J. Biol. Chem. 261, 15288–15293.

Lamontagne, D., Pohl, U., and Busse, R. (1992). Mechanical deformation of vessel wall and shear stress determine the basal EDRF release in the intact coronary vascular bed. Circ. Res. 70, 123–130.

Lamontagne, D., Pohl, U., and Busse, R. (1991). N^G-nitro-L-arginine inhibits endothelium-dependent dilator responses in rabbit coronary resistance vessels. Pflügers Arch. 418, 266–270.

Lansman, J. B., Hallam, T. J., and Rink, T. J. (1987). Single stretch-activated ion channels in vascular endothelial cells as mechanotransducers? Nature 325, 811–813.

Lapetina, E. G. (1982). Regulation of arachidonic acid production: Role of phospholipases C and A2. Trends Pharmacol. Sci. 3, 115–118.

Laskey, R. E., Adams, D. J., Johns, A., Rubanyi, G. M., and Van Breemen, C. (1990). Membrane potential and Na^+-K^+ pump activity modulate resting and bradykinin-stimulated changes in cytosolic free calcium in cultured endothelial cells from bovine atria. J. Biol. Chem. 265, 2613–2619.

Laurent, S., Lacolley, P., Brunel, P., Laloux, B., Pannier, B., and Safar, M. (1990). Flow-dependent vasodilation of brachial artery in essential hypertension. Am. J. Physiol. 258, H1004–H1011.

Lie, M., Sejersted, O. M., and Kiil, F. (1970). Local regulation of vascular cross section during changes in femoral arterial blood flow in dogs. Circ. Res. 27, 727–737.

Loesch, A., and Burnstock, G. (1988). Ultrastructural localization of serotonin and substance P in vascular endothelial cells of rat femoral and mesenteric arteries. Anat. Embryol. 178, 137–142.

Lückhoff, A., and Busse, R. (1990a). Activators of potassium channels enhance calcium influx into endothelial cells as a consequence of potassium currents. *Naunyn-Schmiedebergs Arch. Pharmacol.* **342**, 94–99.

Lückhoff, A., and Busse, R. (1990b). Calcium influx into endothelial cells and formation of EDRF is controlled by the membrane potential. *Pflügers Arch.* **416**, 305–311.

Ludlam, C., Palinski, W., Görög, P., and Born, G. V. R. (1988). Concomitant release of sialic acids and calcium by neuraminidase from rat aorta in situ. *Proc. Roy. Soc. Lond.* **B235**, 139–144.

Melkumyants, A. M., Balashov, S. A., Veselova, E. S., and Khayutin, V. M. (1987). Continuous control of the lumen of feline conduit arteries by blood flow rate. *Cardiovasc. Res.* **21**, 863–870.

Milner, P., Kalevic, V., Hopwood, A. M., Feher, E., Lincoln, J., Kirkpatrick, K., and Burnstock, G. (1989). Ultrastructural localization of substance P and choline acetyltransferase in endothelial cells of rat coronary artery and release of substance P and acetylcholine during hypoxia. *Experientia* **45**, 121–125.

Milner, P., Bodin, P., Loesch, A., and Burnstock, G. (1990a). Rapid release of endothelin and ATP from isolated aortic endothelial cells exposed to increased flow. *Biochem. Biophys. Res. Commun.* **170**, 649–656.

Milner, P., Kirkpatrick, K. A., Ralevic, V., Toothill, V., Pearson, J., and Burnstock, G. (1990b). Endothelial cells cultured from human umbilical vein release ATP, substance P and acetylcholine in response to increased flow. *Proc. Roy. Soc. Lond.* **B241**, 245–248.

Mo, M., Eskin, S. G., and Schilling, W. P. (1991). Flow-induced changes in Ca^{2+} signaling of vascular endothelial cells: effect of shear stress and ATP. *Am. J. Physiol.* **260**, H1698–H1707.

Moncada, S., Herman, A. G., Higgs, E. A., and Vane, J. R. (1977). Differential formation of prostacyclin (PGX or PGI2) by layers of the arterial wall. An explanation for the antithrombotic properties of vascular endothelium. *Thromb. Res.* **11**, 323.

Moore, P. K., Alswayeh, O. A., Chong, N. W. S., Evans, R. A., and Gibson, A. (1990). L-ng-nitro arginine (L-noarg), a novel, L-arginine-reversible inhibitor of endothelium-dependent vasodilatation in vitro. *Br. J. Pharmacol.* **99**, 408–412.

Mülsch, A., Böhme, E., and Busse, R. (1987). Stimulation of soluble guanylate cyclase by endothelium-derived relaxing factor from cultured endothelial cells. *Eur. J. Pharmacol.* **135**, 247–250.

Mülsch, A., and Busse, R. (1990). N^G-nitro-L-arginine (N5-[imino(nitroamino)methyl]-L-ornithine) impairs endothelium-dependent dilations by inhibiting cytosolic nitric oxide synthesis from L-arginine. *Naunyn-Schmiedebergs Arch. Pharmacol.* **341**, 143–147.

Mülsch, A., and Busse, R. (1991). Nitric oxide synthase in native and cultured endothelial cells: calcium/calmodulin and tetrahydrobiopterin are cofactors. *J. Cardiovasc. Pharmacol.* **17** (Suppl. 3), S52–S56.

Nabel, E. G., Selwyn, A. P., and Ganz, P. (1990). Large coronary arteries in humans are responsive to changing blood flow—an endothelium-dependent mechanism that fails in patients with atherosclerosis. *J. Am. Colloid. Cardinol.* **16**, 349–356.

Nollert, M. U., Eskin, S. G., and McIntire, L. V. (1990). Shear stress increases inositol trisphosphate levels in human endothelial cells. *Biochem. Biophys. Res. Commun.* **170**, 281–287.

Oka, S. (1983). Physical theory of some interface phenomena in hemorheology. *Ann. N.Y. Acad. Sci.* **416**, 115.

Olesen, S. P., Clapham, D. E., and Davies, P. F. (1988). Haemodynamic shear stress activates a K$^+$ current in vascular endothelial cells. *Nature* 331, 168–170.

Palmer, R. M. J., Ferrige, A. G., and Moncada, S. (1987). Nitric oxide release accounts for the biological activity of endothelium-derived relaxing factor. *Nature* 327, 524–526.

Palmer, R. M. J., Ashton, D. S., and Moncada, S. (1988). Vascular endothelial cells synthesize nitric oxide from L-arginine. *Nature* 333, 664–666.

Pirotton, S., Raspe, E., Demolle, D., Erneux, C., and Boeynaems, J. M. (1987). Involvement of inositol 1,4,5-trisphosphate and calcium in the action of adenosine nucleotides on aortic endothelial cells. *J. Biol. Chem.* 262, 17461–17466.

Pohl, U., Busse, R., Kuon, E., and Bassenge, E. (1986a). Pulsatile perfusion stimulates the release of endothelial autacoids. *J. Appl. Cardiol.* 1, 215–235.

Pohl, U., Holtz, J., Busse, R., and Bassenge, E. (1986b). Crucial role of endothelium in the vasodilator response to increased flow in vivo. *Hypertension* 8, 37–44.

Pohl, U., and Busse, R. (1988). Reduced nutritional blood flow in autoperfused rabbit hindlimbs following inhibition of endothelial vasomotor function. In *Resistance Arteries*, pp. 10–16, Halpern, W., Pegram, B. L., Brayden, J. E., Mackey, K., McLaughlin, M., and Osol, G. (eds.). Perinatology Press, Ithaca, NY, 1988.

Pohl, U., and Busse, R. (1989). EDRF-induced increase of cGMP in platelets during passage through the coronary vascular bed. *Circ. Res.* 65, 1798–1803.

Pohl, U., Herlan, K., Huang, A., and Bassenge, E. (1991). EDRF-mediated, shear-induced dilation opposes myogenic vasoconstriction. *Am. J. Physiol.* 261, H2016–H2023.

Pohl, U., Lamontagne, D., Busse, R., and Bassenge, E. (1990). EDRF reduces coronary autoregulatory responses to increases in perfusion pressure. *Arch. Int. Pharmacodyn. Ther. Abstract.* 305, 273.

Popp, R., Hoyer, J., Meyer, J., Galla, H. J., and Gögelein, H. (1991). Stretch-activated nonselective cation channel in endothelial cells from porcine cerebral capillary. *Pflügers Arch.* 418 (Suppl. 1). R102 (abstract).

Popp, R., Hoyer, J., Meyer, J., Galla, H. J., and Gögelein, H. (1992). Stretch-activated nonselective cation channel in the antiluminal membrane of porcine cerebral capillaries. *J. Physiol.* (London) 454, 435–449.

Ralevic, V., Milner, P., Hudlicka, O., Kristek, F., and Burnstock, G. (1990). Substance-P is released from the endothelium of normal and capsaicin-treated rat hind-limb vasculature, in vivo, by increased flow. *Circ. Res.* 66, 1178–1183.

Rapoport, R. M., Draznin, M. B., and Murad, F. (1983). Endothelium-dependent relaxation in rat aorta may be mediated through cyclic GMG-dependent protein phosphorylation. *Nature* 306, 174–176.

Rees, D. D., Palmer, R. M. J., Hodson, H. F., and Moncada, S. (1989). A specific inhibitor of nitric oxide formation from L-arginine attenuates endothelium-dependent relaxation. *Br. J. Pharmacol.* 96, 418–424.

Rodbard, S. (1975). Vascular caliber. *Cardiology* 60, 4–49.

Rubanyi, G. M., Romero, J. C., and Vanhoutte, P. M. (1986). Flow-induced release of endothelium-derived relaxing factor. *Am. J. Physiol.* 250, H1145–H1149.

Sakuma, I., Stuehr, D. J., Gross, S. S., Nathan, C., and Levi, R. (1988). Identification of arginine as a precursor of endothelium-derived relaxing factor. *Proc. Natl. Acad. Sci. USA* 85, 8664–8667.

Sauve, R., Parent, L., Simoneau, C., and Roy, G. (1988). External ATP triggers a biphasic activation process of a calcium-dependent K$^+$ channel in cultured bovine aortic endothelial cells. *Pflügers Arch.* 412, 469–481.

Sawyer, P. N., and Srinivasan, S. (1972). The role of electrochemical surface properties in thrombosis at vascular interfaces: Cumulative experience of studies in animals and men. *Bull. N.Y. Acad. Med.* 48, 235–256.

Schilling, W. P., Rajan, L., and Strobl-Jager, E. (1989). Characterization of the bradykinin-stimulated calcium influx pathway of cultured vascular endothelial cells. Saturability, selectivity, and kinetics. *J. Biol. Chem.* 264, 12838–12848.

Schmidt, H. H. H. W., Nau, H., Wittfoht, W., Gerlach, J., Prescher, K. E., Klein, M. M., Niroomand, F., and Böhme, E. (1988). Arginine is a physiological precursor of endothelium-derived nitric oxide. *Eur. J. Pharmacol.* 154, 213–216.

Schmidt, H. H. H. W., Pollock, J. S., Nakane, M., Gorsky, L. D., Förstermann, U., and Murad, F. (1991). Purification of a soluble isoform of guanylyl cyclase-activating-factor synthase. *Proc. Natl. Acad. Sci. USA* 88, 365–369.

Schretzenmayr, A. (1933). Über kreislaufregulatorische Vorgänge an den großen Arterien bei der Muskelarbeit. *Pflügers Arch. Ges. Physiol.* 232, 743–748.

Segal, S. S., and Duling, B. R. (1986). Communication between feed arteries and microvessels in hamster striated muscle: Segmented vascular responses are functionally coordinated. *Circ. Res.* 59, 283–290.

Segal, S. S., and Duling, B. R. (1987). Propagation of vasodilation in resistance vessels of the hamster: Development and review of a working hypothesis. *Circ. Res.* 61 (Suppl. II), II-20–II-25.

Segal, S. S., Damon, D. N., and Duling, B. R. (1989). Propagation of vasomotor responses coordinates arteriolar resistances. *Am. J. Physiol.* 256, H832–H837.

Shepherd, J. T., and Abboud, F. M. (1983). *Handbook of Physiology*, Sect. 2: *The Cardiovascular System*. Vol. III, *Peripheral Circulation and Organ Blood Flow*. American Physiological Society, Bethesda, MD, 1983.

Sheriff, C. J., Bogle, R. G., Baydoun, A. R., Pearson, J. D., and Mann, G. E. (1990). Does increased fluid shear stress modulate L-arginine transport in perfused microcarrier cultures of vascular endothelial cells? *Arch. Int. Pharmacodyn. Ther.* 305, 279 (abstract).

Smiesko, V., Kozik, J., and Dolezel, S. (1985). Role of endothelium in the control of arterial diameter by blood flow. *Blood Vessels* 22, 247–251.

Smiesko, V., Lang, D. J., and Johnson, P. C. (1989). Dilator response of rat mesenteric arcading arterioles to increased blood flow velocity. *Am. J. Physiol.* 257, H1958–H1965.

Sparks, H.V., Jr. (1980). Effect of local metabolic factors on vascular smooth muscle. In *Handbook of Physiology*, Sect. 2: *The Cardiovascular System*. Vol. II, *Vascular Smooth Muscle*, pp. 475–513, Bohr, D. F., Somlyo, A. P., Sparks, H. V., Jr., and Geiger, S. R. (eds.). American Physiological Society, Bethesda, MD 1980.

Sprague, E. A., Steinbach, B. L., Nerem, R. M., and Schwartz, C. J. (1987). Influence of a laminar steady-state fluid-imposed wall shear stress on the binding, internalization, and degradation of low density lipoproteins by cultured arterial endothelium. *Circulation* 76, 648–656.

Stuehr, D. J., Kwon, N. S., Gross, S. S., Thiel, B. A., Levi, R., and Nathan, C. F. (1989). Synthesis of nitrogen oxides from L-arginine by macrophage cytosol—requirement for inducible and constitutive components. *Biochem. Biophys. Res. Commun.* 161, 420–426.

Tangelder, G. J., Slaaf, D. W., Arts, T., and Reneman, R. S. (1988). Wall shear rate in arterioles in vivo: Least estimates from platelet velocity profiles. *Am. J. Physiol.* 254, H1059–H1064.

Tayeh, M. A., and Marletta, M. A. (1989). Macrophage oxidation of L-arginine to nitric oxide, nitrite, and nitrate—tetrahydrobiopterin is required as a cofactor. *J. Biol. Chem.* **264**, 19654–19658.

Tesfamarian, B., and Cohen, R. A. (1988). Inhibition of adrenergic vasoconstriction by endothelial shear stress. *Circ. Res.* **63**, 720–725.

Thurm, U. (1983). Biophysics of mechanoreception (Sect. 15.2.2.1). Mechano-electric transduction (Sect. 15.2.2.1). In *Biophysics*, pp. 666–671, Hoppe, W., Lohmann, W., Markl, H., and Ziegler, H. (eds.). Springer-Verlag, Berlin, New York, 1983.

Vargas, F. F., Osorio, M. H., Ryan, U. S., and de Jesus, M. (1989). Surface charge of endothelial cells estimated from electrophoretic mobility. *Membr. Biochem.* **8**, 221–227.

Wiemer, G., Schölkens, B. A., Becker, R. M. A., and Busse, R. (1991). Ramiprilat enhances endothelial autacoid formation by inhibiting breakdown of endothelium-derived bradykinin. *Hypertension* **18**, 558–563.

Zeiner, A. M., Drexler, H., Wollschläger, H., and Just, H. (1991). Modulation of coronary vasomotor tone in humans. Progressive endothelial dysfunction with different early stages of coronary atherosclerosis. *Circulation* **83**, 391–401.

CHAPTER 8

■ ■ ■ ■ ■

Chronic Effects of Blood Flow on the Artery Wall

■ ■ ■ ■ ■

B. Lowell Langille

I. INTRODUCTION

Blood flow and artery size are intimately linked from the time the primordial vasculature first perfuses developing embryonic tissues. Vessel size greatly affects flow resistance, so growth of small resistance arteries must be finely tuned to the rapidly changing demands of these tissues. The growth of large vessels also follows changes in blood flow and this dependence plays a major role in determining their anatomy during development. Furthermore, the need for such control does not become redundant with maturation; instead, numerous physiological adjustments require chronic remodeling of blood vessels to accommodate altered blood flow requirements. While many factors participate in vascular growth control (Schwartz et al., 1990), there is much evidence that arteries are directly sensitive to intraluminal blood flow rates, such that they remodel to adjust their diameter in response to chronic blood flow changes. This sensitivity provides a potent negative-feedback mechanism for servocontrolling blood vessel growth according to tissue perfusion requirements.

II. BLOOD FLOW VERSUS BLOOD PRESSURE
IN ARTERIAL REMODELING

Available evidence indicates that intravascular pressures exert influences on vascular structure that are independent of those attributable to intraluminal blood flow. The wall tensions that both normal and hypertensive pressures produce correlate with medial thicknesses and have been implicated as a controller of vessel wall tissue proliferation from early life in utero to adulthood (Thoma, 1893; Folkow, 1982; Leung et al., 1977). Blood flow, on the other hand, correlates with internal vessel diameter and probably reflects vascular adaptation to the perfusion requirements of peripheral tissues (Thoma, 1893; Leung et al., 1977). It is important to recognize that these two hemodynamic variables can vary independently. Thus, alterations in total peripheral resistance cause arterial pressure and cardiac output to change in opposite directions, whereas changes in the level of cardiac function cause pressure and cardiac output to change in concert. This chapter focuses on the effect of blood flow on vascular structure and growth control, although some attention must be given to the interactive effects of blood pressure and wall tension.

III. BLOOD FLOW, SHEAR STRESS, AND DEFORMATION
OF VASCULAR TISSUES

Direct sensitivity of arteries to blood flow implies that vascular tissues detect and respond to shear stress, the frictional force that flow exerts on the vessel wall. Shear stresses arise because liquids adhere to, and remain stationary at, solid interfaces, so adjacent layers of blood must slide over each other in order to accommodate finite velocities at sites distant from the endothelial surface. Frictional forces (shear stresses) result and are transmitted to the vessel wall (Langille, 1991).

According to fundamental mechanical principles, there are two primary consequences of shear stress, but both require close scrutiny when applied to blood vessels. First, shear can exert a net longitudinal tension on vessels, since a unidirectional (downstream) force is applied to its internal surface. In practice, this tension is borne by the tethering provided by side branches and connective tissue attachments to contiguous tissues, and it is not transmitted along the vessel wall. Thus, a lengthwise, cyclic stretching of arteries from systole to diastole that is predicted by simple theory is virtually negligible in vivo (Womersley, 1957; Patel et al., 1964). However, arteries mounted for perfusion in vitro are freed from natural

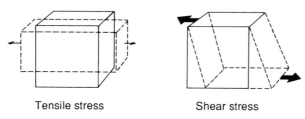

Tensile stress   **Shear stress**

FIGURE 1   Deformation of tissue elements exposed to tensile (left) or shear (right) stresses.

tethering and changes in longitudinal tension occur when flow is initiated. This effect is not eliminated by fixing the vessel at both ends since flow-induced shear stress then stretches the proximal portions of the vessel and compresses the distal portions. Consequently, artefacts may result when testing for physiologically relevant responses of the media to shear in vitro. These tension artefacts are small but, as described below, so is physiologic deformation of the media due to shear. Thus, efforts to isolate effects of the latter in vitro may be difficult.

The second classical deformation due to shear stress is shear strain of wall tissue (Fig. 1). Again, however, there is reason to question the biological significance of medial shear strain. The mechanical properties of arteries are complex (Patel and Vaishnav, 1980), but a simple and reasonable approximation indicates that bulk shear strain of artery walls is extremely modest, typically less than 1% (Langille and O'Donnell, 1986). Arteries stretch by 5–10% between diastole and systole (Patel and Vaishnav, 1980) and by over 25% with maximal physiological adjustments in arterial pressure, so it would appear unlikely that discrimination of shear strains below 1% by medial tissues is a viable mechanism for finely controlling arterial responses to blood flow changes.

If this argument is valid, it means that the medial smooth-muscle cells that mediate many flow-induced responses do not transduce the flow signal directly. Instead, most evidence now supports the contention that endothelium is sensitive to shear stress and that it controls both intimal and medial responses to shear (Langille, 1991). Although there is no theoretical reason to consider shear stresses as being substantially concentrated at the endothelial surface, shear deformation of these cells is not constrained by a surrounding solid matrix. Furthermore, the intimate contact between endothelium and flowing blood provides novel mechanisms by which these cells may respond to shear. Some of these are considered in Section VII. Consequently, it is not surprising that numerous endothelial cell functions, including replication rates (Langille et al., 1986; Davies et al., 1986), release of vasomotor substances (van Grondelle et al., 1984; Frangos et al.,

1985; Kaiser et al., 1986; Pohl et al., 1986) and release of tissue plasmino-gen activator (Diamond et al., 1989) are sensitive to shear. In contrast, only a few studies (Bevan et al., 1988) have reported flow-dependent responses from deendothelialized arteries.

IV. ARTERIAL REMODELING IN RESPONSE TO SHEAR FORCES

A. Endothelial Remodeling

Chronic changes in shear forces exerted on the endothelium elicit medial responses, but they also induce remodeling of the endothelial lining itself. Endothelial remodeling includes both cellular adaptations to the physical loads imposed by shear stress and compensatory adjustments to the changes in luminal surface area that result from medial adjustments.

In vivo, endothelial cells are long and thin and oriented in the direction of blood flow (see Fig. 2A). Manipulations that alter blood flow cause the cells to change shape and to reorient in a manner consistent with the altered flow pattern, or to lose any preferred orientation if flow is eliminated (Fig. 2B) (Langille et al., 1986; Reidy and Langille, 1980). More recently, there has been evidence of major intracellular adaptations of endothelium, particularly in the cell cytoskeleton. For example, endothelial F-actin undergoes profound redistribution when in vivo shear stresses are altered (Holtz et al., 1984; Kim et al., 1989). High shear induces the formation of very long, thick F-actin stress fibers that are aligned with shear (Fig. 2C). Elongation of cells in the direction of shear permits the formation of these lengthy intracellular structures, so the two adaptations may be functionally linked.

Stress fibers are thought to protect the integrity of the endothelial lining when it is exposed to high shear forces, and some evidence from tissue culture studies supports this hypothesis. Wechezak et al. (1989) showed that substrate adhesion of endothelial cells in a shear field is impaired if F-actin is disrupted with cytochalasin B. This finding probably reflects the participation of stress fibers in formation of substrate adhesion complexes; this group has also demonstrated reorganization of suben-dothelial fibronectin by endothelial cells under shear in vitro (Wechezak et al., 1985). The mechanisms by which actin participates in substrate adhesion have not been rigorously defined; however, Satcher et al. (1990) suggest that if stress fibers are associated with focal contacts with substrate at their upstream ends, as found by Wechezak et al. in tissue culture (Wechezak et al., 1989), then they may moor (anchor) cells that are exposed to shear forces. Alternatively, Sato et al. (1987) found that endothelial cells conditioned by shear stress exhibit decreased deformabil-

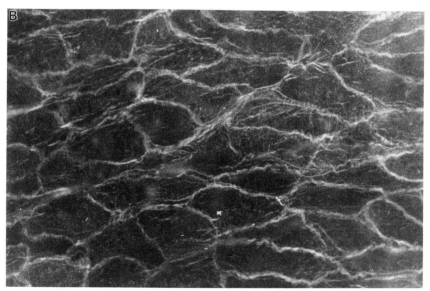

FIGURE 2   (A) F-Actin distribution in the rabbit aorta in a region of moderate shear stress. Actin is located at the peripheral regions of the cells (near cell junctions) and in short, central stress fibers. Peripheral staining reveals the elongated shape of cells and their orientation in the direction of shear stress. (B) F-Actin distribution in vivo under no-flow conditions (ligated rabbit carotid artery). Staining is concentrated at the cell periphery. Cells lose their elongate shape and preferred orientation. (C) F-Actin at high shear site. Peripheral staining is lost, and very long, thick stress fibers are formed. Some stress fibers appear to extend from one nucleus to another (arrowhead), indicating coordinated expression of stress fibers in adjacent cells. Marker is 10 μm. [Parts (A) and (C) are from Kim et al. (1989), by permission; (B) is from Walpola, Langille, and Gotlieb (unpublished data).] (*Figure continues.*)

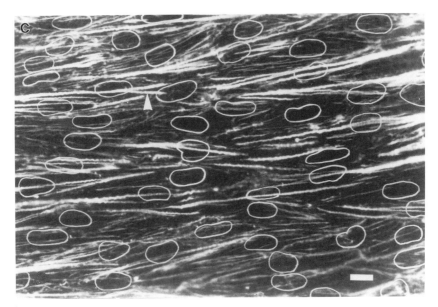

FIGURE 2   (*Continued*).

ity, whereas microfilament disrupting agents increase deformability. Presumably, decreased cell deformability may enhance resistance to shear damage by distributing stresses more evenly throughout the cell.

In high-shear regions in vivo, we have observed that many stress fibers in adjacent cells are in register, creating the illusion that single microfilament bundles extend from one cell to another (Fig. 2C). This finding suggests that stress fiber functions are integrated at the intercellular level. Most probably, these stress fibers emanate from adjacent nucleation sites at junctional complexes, and project upstream in one cell and downstream in the neighboring cell. If so, it is unlikely that both stress fibers associate with the basal surface at their upstream ends to "moor" the endothelium. However, if one function of stress fibers is to distribute stresses throughout the cell, then this concept may extend to intercellular stress distribution via intercellular F-actin organization. Thus, large stress fibers limit deformation of the endothelial monolayer over large areas, or they may tether cell junctions that are exposed to highly deforming stresses.

Remodeling of endothelium in response to flow also involves loss (decreased flow) or proliferation (increased flow) of cells as the vessel increases or decreases in size. Even in adult animals, where flow-induced diameter changes involve no detectable changes in amounts of major wall constituents (see Section IV,B,2), endothelial cell number is altered. We found that chronic reduction of carotid blood flow in rabbits caused loss of endothelial cells when diameter changes had gone to completion, and that endothelial cell density on the vessel surface was ultimately unchanged. Physiologically, adaptations that maintain endothelial cell density may characterize many adaptations that involve alterations in blood flow and blood vessel size. For example, Azmi and O'Shea (1984) describe deletion of endothelial cells during atrophy of the vascular supply to the corpus luteum during luteal regression. Endothelial deletion involved apoptosis of these cells, and these investigators suggested that it may result from the reduction in blood flow demand of luteal tissues at this time. At present, though, experimental evidence testing the role of flow in luteal vascular regression is lacking; nor do we know that apoptosis is the mechanism of endothelial cell deletion when blood flow is decreased experimentally.

Endothelial cell number also rises when flow increases. Interestingly, Masuda et al. (1989) found that this proliferative response preceded increases in arterial diameter when blood flows were increased by arteriovenous (A-V) anastomosis. However, this finding may be exceptional, since the anastomoses resulted in mean flow Reynolds numbers of about 1500, which almost certainly produces turbulent flow in arteries (McDonald, 1974). Turbulence stimulates endothelial cell replication even in cell culture (Davies et al., 1986) where remodeling of an underlying media is not a factor. This probably reflects injury to endothelium, and repair of endothelial injury can yield markedly elevated cell density, even long after repair has gone to completion (Reidy and Schwartz, 1981).

There are undoubtedly many other adaptive responses of endothelium to shear alterations that have gone undetected. This is partly because our current knowledge of what happens to these cells under shear is very primitive. For example, we do not know how stresses are distributed over the cell surface and through intracellular (cytoskeletal?) structures. It is even not known how much strain (deformation) these cells undergo when exposed to acute shear stress. Sizable strains (Fig. 1) imply increases in effective cell-surface area/volume ratios, but these could be accommodated by conformational changes of the highly interdigitated cell junctions, delivery to the cell surface of membrane stored in vesicles/caveolae, rapid loss of cell water, and/or simply by gross changes in cell shape. Hopefully, novel techniques will soon be developed to address these issues.

B. Medial Remodeling

Remodeling of the media in response to blood flow alterations is a two-phase response that involves acute vasomotion followed by chronic restructuring. The acute vasoconstriction caused by decreased flow predominates for 1–2 weeks (Langille et al., 1989), but the time course for responses to increased flow has not been established. The acute vasomotion is important as both a short-term vasoregulatory mechanism and a precursor to more permanent changes; however, it is considered here only briefly since Busse and Pohl deal with this subject in detail in Chapter 7 in this treatise.

1. Acute Vasomotor Responses

Shear stress mediates potent physiological responses of blood vessels to blood flow changes. Thus, arteries dilate when blood flow (shear stress) increases and constrict when flow decreases (Hintze and Vatner, 1984; Schretzenmayr, 1933; Smiesko et al., 1985) such that shear stresses exerted on the vessel wall return toward initial levels. Physiologically, these responses enhance local control of blood flow. For example, when peripheral tissues are metabolically active or hypoperfused, local vasodilators increase tissue perfusion. The vasodilators act only on the microcirculation since they have no access to feed arteries extrinsic to the tissue; however, the initial increase in flow they produce stimulates dilation of upstream arteries through the effects of shear on endothelium. Thus, vasodilation propagates upstream and the initial increase in blood flow is amplified. This response appears important in integrating vasomotion throughout the vascular tree to sustain normal intravascular pressure gradients (Griffith et al., 1987, 1989). It is probable that such mechanisms are important in flow-regulatory responses of many vascular beds.

Studies using large artery preparations indicate that flow-induced vasodilation is mediated through release of endothelium-derived relaxing factor (EDRF) (Kaiser et al., 1986; Rubanyi et al., 1986), an extracellular messenger responsible for many types of agonist-induced vasodilation (Furchgott, 1983). Actions of EDRF closely parallel those of the vasodilator nitrates. Palmer et al. (1987) produced strong evidence that EDRF is nitric oxide, although some investigators believe that it is a nitrosothiol that may release nitric oxide during assay procedures (Myers et al., 1990). Shear stress also induces endothelial cells to release prostacyclin (van Grondelle et al., 1984; Frangos et al., 1985). This agent appears to contribute little to flow-induced vasomotion in large arteries (Pohl et al., 1985), but a recent study implicates prostaglandins, possibly prostacyclin, in vasodilator responses to flow in the microcirculation. Koller and Kaley (1990) induced flow increases in cremaster muscle arterioles by occluding parallel vessels and found that the resulting vasodilation was abolished by indomethacin but not by an arginine analogue that blocks EDRF synthesis.

Arteries also constrict in response to reduced blood flow through an endothelium-dependent process (Langille et al., 1989), but apparently this is due, at least in part, to production of a vasoconstrictor, not reduction in tonic release of the vasodilators described above. We have found that deendothelialized arteries that are carrying reduced blood flow narrow as the endothelium regenerates, a finding inconsistent with a net vasodilator effect of these cells (Jamal et al., 1992). We have shown that prostanoids are not involved since constriction was not inhibited by indomethacin (Langille and Bendeck, 1990). Another attractive candidate mediator is angiotensin, since the artery wall contains a complete, local renin–angiotensin system (Dzau, 1989); however, we were unable to block the response with the converting enzyme inhibitor, captopril. We also assessed whether arteries carrying reduced flow express higher endothelin levels using radioimmunoassay (RIA) (Langille and Bendeck, 1990). Results were negative, although the nature of the study was such that only positive results would be conclusive. Arteries that had carried reduced or normal flows were incubated in media after excision, so positive results required that differential endothelin production would persist during incubation in vitro, after all flow-related stimuli had been removed.

2. Chronic Medial Remodeling in Adult Animals

When changes in blood flow rates through large arteries are chronic, acute vasomotor responses are followed by remodeling of the arterial wall that results in a net change in diameter (Langille et al., 1989; Langille and O'Donnell, 1986; Duling et al., 1987). These long-term adjustments in arterial diameters have important effects on vascular function in adult animals.

For example, the vascular supply to the corpus luteum grows and regresses during the reproductive cycle in concert with blood flow modulation, and it is thought that these changes are functionally linked (Azmi and O'Shea, 1984). Chew and Segal (1990) have shown that chronic reduction of load on skeletal muscle causes morphologic changes in supply arteries that result in reduced diameter, presumably in association with reduced blood flow. We have reported marked remodeling of renal arteries following nephrectomies that reduced total renal artery blood flow to adrenal artery flow (Duling et al., 1987), and we also have observed expansion of the contralateral renal artery following compensatory increases in flow to the remaining kidney (unpublished data). In addition, it is a common clinical observation that patients in whom dialysis shunts have been implanted because of renal failure develop a significantly enlarged radial artery on the shunted side.

Observations that arterial remodeling follows blood flow changes in a large number of physiological situations strongly suggest that the two are functionally linked. Often, however, it is difficult to establish cause and effect and to rule out independent, but synchronously, controlled mechanisms. For example, regression–regrowth of luteal tissue may be triggered by signals that also induce atrophy–regeneration of vascular supply, with changes in blood flow being a by-product rather than a controller of the vascular responses. Consequently, several laboratories have experimentally manipulated arterial blood flows to test for direct responses to shear stress. The clinical observation of arterial enlargement after implantation of A-V shunts is attractive in this regard. Kamiya and Togawa (1980) were particularly successful at exploiting this approach experimentally. They created carotid to jugular A-V shunts to increase blood flows by up to 10-fold and demonstrated that the artery carrying elevated flow increased its diameter under this stimulus. Diameter adjustments restored shear (as estimated using a Poiseuille flow approximation) to control levels over several months except when the highest flow changes were produced. It was not certain whether shears would have been restored in the latter cases if the experiments were carried for longer periods. Zarins and coworkers (Zarins et al., 1987) utilized iliac A-V anastomosis in cynomolgus monkeys to induce larger increases in blood flow (up to 30-fold) for 6 months and observed complete restoration of shear stress. Morphometric assessments in the latter study demonstrated that the diameter changes were not due to simple reorganization of preexisting wall constituents. Instead, significant increases in medial tissue mass were reported.

We have shown that chronic reductions in blood flow cause remodeling that reduces vessel diameter. The response was abolished when the endothelium was removed (Langille and O'Donnell, 1986) and restored when the endothelium regrew (Jamal et al., 1992). We also showed that the response was driven by the time-averaged, rather than the pulsatile component of blood flow (Langille et al., 1989). This observation is important because of evidence that some flow-dependent modulators of vessel tone are particularly responsive to pulsatile blood flow (Pohl et al., 1985). In this study (Langille et al., 1989), we showed that this response initially involved vasoconstriction (first 3–7 days) followed by structural remodeling that entrenched the response (7–14 days). Subsequently, the artery exhibited both a smaller maximally dilated and a smaller maximally constricted diameter; in other words, it functioned as a smaller artery not as a partially constricted, but otherwise normal artery. Unlike responses to increased flow, this remodeling occurred without detectable changes in vessel wall mass or DNA, elastin, or collagen contents (Fig. 3A). It is possible, however, that wall constituents change after a very long time. In adults, turnover of vascular tissues is very slow: elastin, collagen, and vascular cell half-lives are years to decades (LeFevre and Rucker, 1980; Schwartz and Benditt, 1977). Consequently, if flow affects only tissue synthesis, and not degradation, then long periods are required for even

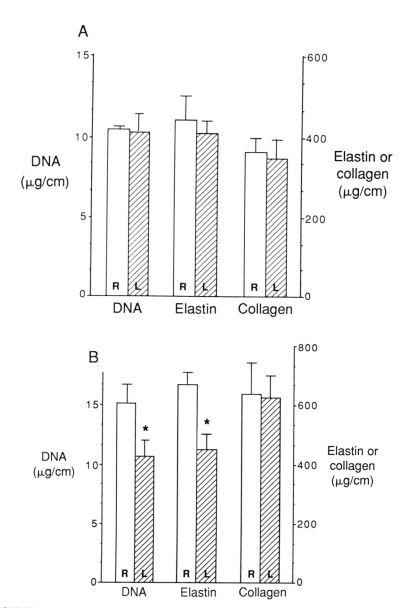

FIGURE 3   DNA, elastin, and collagen contents per centimeter of carotid artery 1 month after left external carotid artery was ligated. Ligation reduced left common carotid artery (L) blood flow by 70% without changing flow in right carotid artery (R). (A) Adult rabbits; (B) 6-week-old rabbits. [From Langille et al. (1989), by permission.]

total arrest of synthesis to alter content significantly. This could explain why increased flow causes changes in medial mass, but decreased flow does not (Langille et al., 1989; Zarins et al., 1987).

There is no direct evidence concerning how early vasomotor responses to altered blood flow are entrenched by structural remodeling. As stated above, our studies indicate that this is achieved without significant net changes in vessel wall elastin, collagen, or DNA contents. Furthermore, studies with ^{3}H-thymidine indicate that smooth-muscle turnover is not involved. We found that fewer than 0.03% of these cells were replicating per day, subsequent to flow reduction (unpublished data). It would appear inevitable, however, that remodeling involves some turnover of matrix constituents as the vessel establishes a new resting diameter. Recent studies implicate urokinase and tissue plasminogen activator (uPA and tPA) in reorganization of vascular cell interactions with matrix. Clowes and co-workers showed that both tPA and uPA are expressed as smooth muscle cells divide and then migrate to the intima following intimal-medial injury with a balloon catheter (Clowes et al., 1990). It is attractive to hypothesize that the same mechanisms are invoked during the structural remodeling that follows flow alterations.

An additional possibility is that intracellular turnover of contractile proteins in medial smooth-muscle cells ultimately produces a contractile unit with a resting length shorter than that encountered in control vessels. Our observations favor this hypothesis. Histological examinations of arteries after adaptation to reduced flow suggested that reduced arterial diameters became entrenched without restoration of initial muscle cell length (Fig. 4). Nevertheless, the vessels maximally contracted to 35–40% smaller diameters than control arteries in response to agonists. Thus, it appears that the smooth-muscle cells in these vessels could contract to much shorter cell lengths than those in control arteries. These vessels also exhibited smaller diameters (shorter cell lengths) when maximally dilated. These findings are consistent with a contractile apparatus that is at rest at shorter than normal cell lengths. Indeed, it is difficult to account for them without hypothesizing this type of intracellular remodeling.

3. Blood Flow-Induced Remodeling and Arterial Disease

Many vascular diseases exhibit clinical manifestations because blood flow is altered or compromised; therefore, flow-induced remodeling of arteries may result. In this case the stimulus for remodeling is abnormal by definition, and the remodeling process may or may not act to the advantage of the organism (Langille, 1991). In arterial occlusive disease, for example, the disease process can be affected at several levels. Initially, the

FIGURE 4   Histological cross sections of rabbit carotid arteries: (A) control artery; (B) carotid after one month of 70% blood flow reduction. (*Figure continues.*)

FIGURE 4   (*Continued*).

lesion narrows the vessel lumen and the resulting acceleration of blood flow through the lesion site elevates shear stress. Glagov et al. (1987) presented evidence that the vessel media subsequently expands to restore lumen diameter. Thus, flow-induced adaptations limit encroachment on the vessel lumen early in lesion development. Ultimately, however, growth of the lesion compromises blood flow and adjacent, healthy segments of the vessel wall experience reduced shear and may adapt by narrowing, a response that can exacerbate hypoperfusion. Finally, obstructive lesions partially depressurize the downstream vasculature and will therefore initiate flow through any collaterals arising from adjacent, pressurized vessels. If collateral vessels behave like other arteries that have been studied to date, they will expand in response to this flow stimulus. Consequently, collateral growth can be directly enhanced by the hemodynamic consequences of vessel occlusion.

Since pathological changes in blood flow induce vascular remodeling that may affect progression of the disease state, it is important to assess the reversibility of this remodeling if flow is restored to normal. We developed experimental models to reduce blood flow in rabbit carotid arteries for 2 months, and then restore these flows to normal (Langille and Brownlee, 1991). These studies demonstrated a rapid (within 1 week) return to normal structure after restoration of normal flows. Thus, it appears that successful surgical interventions directed at focal lesion sites in the clinical setting will reverse adverse adaptations of adjacent vasculature.

V. FLOW-RELATED REGULATION OF ARTERIAL GROWTH

Remodeling of arteries in adults may be limited by the slow turnover of mature vascular tissues, but such constraints do not apply to the developing vascular system. Instead, there is a potential for continuous modulation of vascular growth in accord with the blood flow requirements of developing tissues.

A. Blood Flow and Embryonic Vascular Development

The concept that blood flow affects the earliest vascular development is over a century old. Thoma (1893) conducted detailed examinations of the developing area vasculosa of the chick embryo and, on the basis of these studies, proposed four laws relating mechanical forces to vessel growth. These laws stated that increases in vessel diameter depend on blood flow rate, increases in vessel wall thickness depend on blood pressure, increases in vessel length depend on tension exerted by contiguous extravascular tissues, and angiogenesis can be initiated by increases in intravascular pressures. The first of these is the most relevant to the current discussion. (The second is widely accepted, and the third remains viable.) Thoma based his first law on observations that the development of supply vessels feeding regions of the area vasculosa could be predicted on the basis of which vessels within the plexus received the greatest initial flow. Experimental evidence to support this contention is sparse, but it is often dramatic. Thus Chapman (1918) observed that circulatory arrest in the early chick embryo halted or prevented the development of most (but not all) arteries. In addition, several studies showed that manipulating flows through embryonic aortic arches influences which vessels persist and which regress (Rychter, 1962; Ohlen et al., 1990). There is little other experimental support for Thoma's hypothesis; therefore, the alternate hypothesis, that changing blood flow rates and blood vessel growth reflect independent (but correlated) programs of cardiovascular development, cannot yet be discounted. At the least, an exclusive role for hemodynamics in the development of large arteries should be ruled out. For example, it appears inconceivable that the orderly development of the embryonic aortic arches could proceed solely on the basis of hemodynamic cues. Clearly, other factors stimulate the regression of the first arch and the growth of the second arch, because the former initially carries much greater flow than the latter from the heart to the dorsal aorta.

B. Perinatal Arterial Growth

If developmental changes in blood flow influence growth of arteries, then the perinatal period ought to be characterized by intense modulation of vascular growth. This is because blood flow through almost all arteries changes dramatically around the time of birth. Table 1 illustrates blood flow changes that occur in lambs between 140 days of gestation (about 5 days before birth) and 21 days of age. The large changes in perfusion of vascular beds after birth partly reflect the rise in p_{ao_2} (from 20–30 to 80–100 mm Hg), and hence the reduced perfusion needed to sustain a given oxygen delivery, and partly reflect changes in tissue metabolism.

The most striking cardiovascular adjustments at birth relate to the cessation of placental blood flow and the initiation of pulmonary gas exchange. The consequent separation of lung and systemic vasculature, triggered by closure of the ductus arteriosus and foramen ovale, allows

TABLE 1

Perinatal Blood Flow Changes

Organ	Blood flow change (%)[a]
Brain	−45
Kidneys	31
Adrenals	−46
Intestine (large)	−26
Muscle	−11
Skin	−54
Bone	−62

[a]Unpublished data from Bendeck and Langille. Blood flow (per gram tissue) measured using radioactive microspheres.

independent control of systemic and pulmonary arterial pressures. At the same time, nearly equal blood flows must enter the aorta and the pulmonary artery for the first time. Leung et al. (1977) assessed ascending aortic and pulmonary trunk growth and their relation to these hemodynamic changes. They found that the diameter of the two vessels, like the flows they carry, are tightly linked in young animals, whereas wall thickness follows pressure-induced changes in wall tension. These observations support the contention that Thoma's hypotheses concerning effects of pressure and flow on arterial growth apply at this time of life.

The loss of the fetal placental circulation at birth also has a dramatic effect on flow through the abdominal aorta, which directly supplies the umbilical arteries. In utero, abdominal aortic blood flow amounts to 40% of combined ventricular output, but after birth it falls by more than 90% (Langille et al., 1990). We have instrumented preterm fetal lambs to allow monitoring of abdominal aortic diameter and flow and arterial pressure through parturition. These experiments demonstrated that decreased abdominal aortic flow at parturition is accompanied by decreased diameter and inhibition of aortic tissue growth (Langille et al., 1990). These findings were particularly striking in that abdominal aortic diameter remained below in utero levels for several weeks (Fig. 5) despite large increases in the blood pressure distending the artery and substantial growth of the animal in the postpartum period. In contrast, flow through the thoracic aorta was only transiently decreased after birth, and this segment of the vessel exhibited rapid tissue proliferation and significant increases in diameter in the first weeks postpartum.

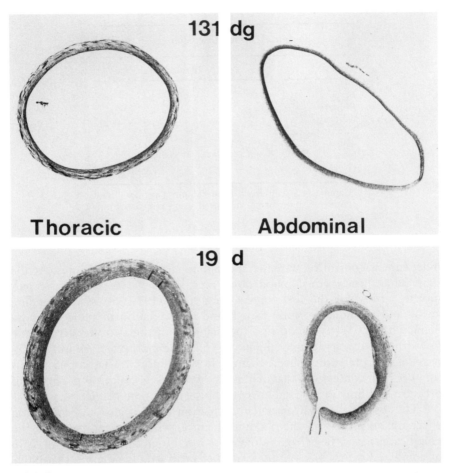

FIGURE 5   Thoracic (left) and abdominal (right) aortas from a 131-day-gestation (dg) fetal sheep (top) and a 3-week-old lamb (bottom). The thoracic aorta continually increases in diameter between these times, whereas the abdominal aorta exhibits a marked decrease in diameter after birth. Term in sheep is about 145 days. [After Langille et al., 1990, with permission.]

C. Postnatal Arterial Growth

These studies of pre- and perinatal development are largely correlative and do not prove a cause-and-effect relationship. After the transition to ex utero life is completed, there is a generalized increase in perfusion that accompanies growth of most peripheral tissues, and again a link between vessel perfusion and growth has been hypothesized. However, a significant amount of data supports the contention that flow influences arterial growth at this time. Thus, several laboratories have devised models to experimentally alter blood flow and assess whether subsequent growth and development were affected. Haworth and coworkers (Haworth et al., 1981) showed that development of the left lung vasculature was impaired after ductal and left pulmonary artery ligation in the newborn piglet, whereas development of collateral vessels from the bronchial circulation was greatly enhanced. Coyle (1984) has presented similar findings on expansion of collateral vessels following selective occlusion of cerebral vessels in rat pups.

A difficulty with these studies was that the flow changes induced by the experimental intervention could not be directly assessed. However, Guyton and Hartley (1985) used pulsed Doppler flowmetry to monitor experimental blood flow reductions in rat pup carotid arteries. They established that an experimentally induced change in blood flow affected accumulation of medial tissue mass. We used a similar preparation in rabbits and demonstrated that accumulation of elastin and DNA is significantly affected, while increases in collagen contents were not detectable (Fig. 3B). These studies were the first to demonstrate the effects of flow modulation on synthesis of specific wall constituents. The differential effects on elastin versus collagen synthesis are noteworthy given the strong link between synthesis of these two proteins in many developing tissues.

It is probable that growth modulation, like flow-induced remodeling in adults, is an endothelium-dependent process, and that it is controlled by soluble mediators. Vascular endothelial and smooth-muscle cells produce and/or respond to a variety of growth modulators including platelet-derived growth factor (PDGF) (Ross et al., 1986), epidermal growth factor (EGF) (Gospodarowicz et al., 1981), fibroblast growth factor (FGF) (Gospodarowicz et al., 1976), and transforming growth factor type b (TGF-b) (Heimark et al., 1986). Furthermore, interactions occur between different growth factors (Leof et al., 1986) and between these factors and other soluble agents (Pash and Bailey, 1988) or matrix constituents (Vlodavsky et al., 1987), at least in tissue culture systems. While there is evidence for a developmental role for these growth factors in blood vessel wall (Burgos, 1986; Moscatelli et al., 1986; Goustin et al., 1985), very little is known about their effects on perinatal arterial growth. One of the most informative studies in this area was that of Seifert et al. (1984). These investigators showed that smooth-muscle cells cultured from fetal and neonatal rat aorta produce 170 times the amount of PDGF, a potent smooth-muscle mitogen, that was produced by similar cultures of adult rat aortic smooth muscle. However, elevated PDGF mRNA was not observed in arteries from rat pups (Majesky et al., 1988), so the in vivo importance

of this observation remains in question, as does their sensitivity to hemodynamic changes during development.

VI. ARTERIAL REMODELING AND LONGITUDINAL ARTERIAL TENSION

Arterial remodeling in response to hemodynamic stresses is well established, but little is known about how specific modes of remodeling are
accomplished. Symmetrical growth implies proportional increases in circumference, length, and wall thickness; however, elevated blood pressure
induces arterial thickening, increased blood flow causes expansion of
arterial circumference, and both of these changes can occur without
changes in vessel length. How is preferential growth in one or two
dimensions accomplished?

Arterial growth without increases in length can be achieved if new wall
tissues are delivered by the cells that synthesize them only in the radial
and/or circumferential direction, but this seems highly unlikely. Alternatively, growth without vessel length change can be viewed as an inevitable
consequence of the longitudinal tension that is imposed on most arteries.
Most arteries retract by 20–40% when excised (Learoyd and Taylor,
1966), but the functional significance of lengthwise stretch has been little
explored. One of its major consequences is that new tissue that is integrated into the vessel media upstream or downstream of the cells that
produce them will not cause the vessel to lengthen. Instead, accumulation
of this tissue allows the remainder of the vessel to retract and reduce
longitudinal tension. Also, since vascular tissues are incompressible, other
dimensions must increase during this retraction. The end result is that all
newly synthesized tissue in mature, longitudinally stretched arteries contributes to wall thickening and/or increased circumference, regardless of
how it is deposited in the vessel wall. Ultimately, of course, a sufficient
growth response can totally eliminate longitudinal tension, so that further
growth may cause vessels to elongate.

Some experimental evidence supports this interpretation. First, unteth-
ered arteries in soft tissues do respond to growth stimuli by lengthening;
they take on a more convoluted path. This was observed by Coyle (1984),
who found that experimental increases in flow through small cerebral
arteries caused them to lengthen as well as increase in diameter. In
addition, hypertension-induced wall thickening is accompanied by reduced
longitudinal stretch on arteries (Cox, 1981), and we have observed the
same phenomenon when vessel growth is stimulated by chronic blood flow
elevation (Di Stefano and Langille, unpublished data).

It is less clear how growth can be preferentially directed to increase diameter when flow increases and preferentially directed to increase wall thickness in hypertension.

VII. TRANSDUCTION OF SHEAR STRESS

Although many endothelial functions are affected by mechanical stress, the mechanism through which the cells transduce physical forces remains unknown. Experiments from two laboratories have shed light on this question. Lansman et al. (1987) showed that stretch, exerted by applying suction to a microelectrode at the cell surface, activated ion channels permeable to calcium. In addition, Olesen et al. (1988) showed that shear stress activates endothelial cell potassium channels.

Activation of ion channels may represent the direct mechanism by which endothelial cells respond to shear, but it is also possible that ion channel activation is downstream of a more fundamental transduction process. Ingber and Folkman (1989) hypothesize that transmission of physical forces to cytoskeletal elements represents the primary means of mechanotransduction. The profound redistribution of cytoskeletal components under shear implies that any such transduction would probably be extremely complex. It is also possible that membrane-associated structures are redistributed by shear in a phenomenon similar to the "capping" of receptors that is driven by ligand binding in some cell types, and that this process elicits intracellular responses. Studies with red cells also clearly indicate the capacity of shear stress to mechanically redistribute cell-membrane components (Schmid-Schobein, 1981). A more basic process might involve shear-induced conformational changes of transmembrane proteins that drive intracellular responses in a fashion analogous to many receptor systems. However, these proposals are totally speculative, and evidence to date relates only to activation of ion channels.

VIII. SUMMARY

There is strong evidence that the structure of both growing and mature arteries is strongly influenced by a direct, local sensitivity to blood flow. Most data support the contention that structural responses represent negative feedback that serves to maintain a constant time-averaged shear stress exerted on the vessel wall, although this conclusion is subject to the caveat that evaluating shear stress is difficult and is often based on crude Poiseuille flow approximations. Structural responses of adult arteries involve endothelium-dependent remodeling of the media to adjust diameter in accord with the alteration in blood flow. These responses provide a mechanism for modifying the adult vasculature during a broad spectrum of physiological and pathological adjustments that have long-term effects on blood flow. During development, vessel growth rates are controlled by sensitivity of arteries to the changing amount of blood flow they deliver to peripheral tissues. Thus, the influence of systemic growth modulators can be tuned so that local vessel growth meets the physiological demands of developing peripheral tissues.

While these structural changes are becoming increasingly well characterized, very little is known about how shear stress is transduced by endothelium, and what role is played in the remodeling process by inter- and intracellular messengers that may be sensitive to shear stress. In addition, we know nothing about how remodeling processes are directed to achieve specific adjustments in diameter versus wall thickness. The roles of second messengers, growth factors, and so on may be addressed using straightforward approaches (e.g., measuring production of agonists by endothelium exposed to shear in vitro); however, defining transduction mechanisms and how arteries are reshaped by remodeling processes will probably await novel and imaginative experiments. However, the ubiquitous and diversified nature of vascular remodeling from embryonic to adult life underscores the importance of studies in these areas.

REFERENCES

Azmi, T. I., and O'Shea, J. D. (1984). Mechanism of deletion of endothelial cells during regression of the corpus luteum. *Lab. Invest.* 51, 206–217.

Bevan, J. A., Joyce, E. H., and Wellman, G. C. (1988). Flow-dependent dilation in a resistance artery still occurs after endothelium removal. *Circ. Res.* 63, 980–985.

Burgos, H. (1986). Angiogenic factor from human term placenta. Purification and partial characterization. *Eur. J. Clin. Invest.* 16, 486–493.

Chapman, W. B. (1918). The effect of the heart-beat upon the development of the vascular system in the chick. *Am. J. Anat.* 23, 175–203.

Chew, H. B., and Segal, S. S. (1990). Hindlimb unloading alters conduit artery morphology. *FASEB J.* 4, A722.

Clowes, A. W., Clowes, M. M., Reidy, M. A., and Belin, D. (1990). Smooth muscle cells express urokinase during mitogenesis and tissue-type plasminogen activator during migration in injured rat carotid artery. *Circl. Res.* 67, 61–67.

Cox, R. H. (1981). Basis for the altered arterial wall mechanics in the spontaneously hypertensive rat. *Hypertension* 3, 485–495.

Coyle, P. (1985). Diameter and length changes in cerebral collaterals after middle cerebral artery occlusion in the young rat. *Anat. Rec.* 210, 357–364.

Davies, P. F., Remuzzi, A., Gordon, E. J., Dewey, C. F., Jr., and Gimbrone, M. A., Jr. (1986). Turbulent fluid shear stress induces vascular endothelial cell turnover in vitro. *Proc. Nat. Acad. Sci. USA* **83**, 2114–2117.

Diamond, S. L., Eskin, S. G., and McIntire, L. V. (1989). Fluid flow stimulates tissue plasminogen activator secretion by cultured human endothelial cells. *Science* **243**, 1483–1485.

Duling, B. R., Hogan, R. D., Langille, B. L., Lelkes, P., Segal, S. S., Vather, S. F., Weigelt, H., and Young, M. A. (1987). Vasomotion control: Functional hyperemia and beyond. *Fed. Proc.* **46**, 251–263.

Dzau, V. J. (1989). Multiple pathways of angiotensin production in the blood vessel wall: Evidence, possibilities and hypotheses. *J. Hypertension* **7**, 933–936.

Folkow, B. (1982). Physiological aspects of primary hypertension. *Physiol. Rev.* **62** (2), 347–497.

Frangos, J. A., Eskin, S. G., McIntire, L. V., and Ives, C. L. (1985). Flow effects on prostacyclin production by cultured human endothelial cells. *Science* **227**, 1477–1479.

Furchgott, R. F. (1983). Role of endothelium in responses of vascular smooth muscle. *Circ. Res.* **53**, 557–573.

Glagov, S., Weisenberg, E., Zarins, C. K., Stankunavicius, R., and Kolettis, G. J. (1987). Compensatory enlargement of human atherosclerotic coronary arteries. *New Engl. J. Med.* **316**, 1371–1375.

Gospodarowicz, D., Hirabayashi, K., Giguere, L., and Tauber, J. P. (1981). Factors controlling the proliferative rate, final cell density, and life span of bovine smooth muscle cells in culture. *J. Cell. Biol.* **89**, 568–578.

Gospodarowicz, D., Moran, J., Braun, D., and Birdwell, C. (1976). Clonal growth of bovine vascular endothelial cells: FGF as a survival agent. *Proc. Natl. Acad. Sci. USA* **73**, 4120–4124.

Goustin, A. S., Betsholtz, C., Pfeifer-Ohlsson, S., Personn, H., Rydenert, J., Bywater, M., Holmgren, B., Heldin, C.-H., Westermark, B., and Ohlsomn, R. (1985). Co-expression of the sis and myc proto-oncogenes in developing human placenta suggests autocrine control of trophoblast growth. *Cell* **41**, 301–212.

Griffith, T. M., Edwards, D. H., Davies, R. L., Harrison, T. J., and Evans, K. T. (1987). EDRF coordinates the behavior of vascular resistance vessels. *Nature* (Lond.) **329**, 442–445.

Griffith, T. M., Edwards, D. H., Davies, R. L., and Henderson, A. H. (1989). The role of EDRF in flow distribution: A microangiographic study of the rabbit isolated ear. *Microvasc. Res.* **37**, 162–177.

Guyton, J. R., and Hartley, C. J. (1985). Flow restriction of one carotid artery in juvenile rats inhibits growth of arterial diameter. *Am. J. Physiol.* **248**, H540–H546.

Haworth, S. G., de Leval, M., and Macartney, F. J. (1981). How the left lung is perfused after ligating the left pulmonary artery in the pig at birth: Clinical implications for the hypoperfused lung. *Cardiovas. Res.* **15**, 214–226.

Heimark, R. L., Twardzik, D. R., and Schwartz, S. M. (1986). Inhibition of endothelial cell regeneration by type-beta transforming growth factor from platelets. *Science* **233**, 1078–1080.

Hintze, T. H., and Vatner, S. F. (1984). Reactive dilation of large coronary arteries in conscious dogs. *Circ. Res.* **54**, 50–57.

Holtz, J., Forstermann, U., Pohl, U., Giesler, M., and Bassenge, E. (1984). Flow-dependent, endothelium-mediated dilation of epicardial coronary arteries in conscious dogs: Effects of cycloogenase inhibition. *J. Cardiovas. Pharmacol.* **6**, 1161–1169.

Ingber, D. E., and Folkman, J. (1989). Tension and compression as basic determinants of cell form and function: Utilization of a cellular tensegrity mechanism. In *Cell Shape:*

Determinants, Regulation, and Regulatory Role, pp. 3–31, Stein, W., and Bronner, F. (eds.) Academic Press, San Diego, CA.

Jamal, A., Bendeck, M., and Langille, B. L. (1992). Structural changes and recovery of function after arterial injury. *Arteriosclerosis and Thrombosis* 12, 307–317.

Kaiser, L., Hull, S. S., Jr., and Sparks, H. V., Jr. (1986). Methylene blue and ETYA block flow-dependent dilation in canine femoral. *Am. J. Physiol.* 250, H974–H981.

Kamiya, A., and Togawa, T. (1980). Adaptive regulation of wall shear stress to flow change in the canine carotid artery. *Am. J. Physiol.* 239, H14–H21.

Kim, D. W., Gotlieb, A. I., and Langille, B. L. (1989). In vivo modulation of endothelial F-actin microfilaments by experimental alterations in shear stress. *Arteriosclerosis* 9, 439–445.

Koller, A., and Kaley, G. (1990). Prostaglandins mediate arteriolar dilation increased blood flow velocity in skeletal muscle microcirculation. *Circ. Res.* 67, 529–534.

Langille, B. L. (1991). Hemodynamic factors and vascular disease. In *Cardiovascular Pathology*, pp. 13–154. Silver, M. D. (ed.). Churchill Livingstone, New York.

Langille, B. L., Bendeck, M. P., and Keeley, F. W. (1989). Adaptations of carotid arteries of young and mature rabbits to reduced carotid blood flow. *Am. J. Physiol.* 256, H931–H939.

Langille, B. L., and Bendeck, M. P. (1990). Arterial responses to compromised blood flow. *Toxicol. Pathol.* 18, 618–622.

Langille, B. L., Brownlee, R. D., and Adamson, S. L. (1990). Perinatal aortic growth in lambs: Relation to blood flow changes at birth. *Am. J. Physiol.* 259, H1274–H1253.

Langille, B. L., and Brownlee, R. D. (1991). Arterial adaptations to altered blood flow. *Can. J. Physiol. Pharmacol.* 69, 978–983.

Langille, B. L., and O'Donnell, F. (1986). Reductions in arterial diameter produced by chronic decreases in blood flow are endothelium-dependent. *Science* 231, 405–407.

Langille, B. L., Reidy, M. A., and Kline, R. L. (1986). Injury and repair of endothelium at sites of flow disturbances near abdominal aortic coarctations in rabbits. *Arteriosclerosis* 6, 146–154.

Lansman, J. B., Hallam, T. J., and Rink, T. J. (1987). Single stretch-activated ion channels in vascular endothelial cells as mechanotransducers. *Nature* (Lond.) 325, 811–813.

LeFevre, M., and Rucker, R. B. (1980). Aortic elastin turnover in normal and hypercholesterolemic Japanese quail. *Biochim. Biophys. Acta* 630, 519–529.

Learoyd, B. M., and Taylor, M. G. (1966). Alterations with age in the viscoelastic properties of human arterial walls. *Circ. Res.* 18, 278–292.

Leof, E. B., Proper, J. A., Goustin, A. S., Shipley, G. D., DiCorleto, P. E., and Moses, H. L. (1986). Induction of c-sis mRNA and activity similar to platelet-derived growth factor by transforming growth factor b: A proposed model for indirect mitogenesis involving autocrine activity. *Proc. Natl. Acad. Sci. USA* 83, 2453–2457.

Leung, D. Y. M., Glasov, S., and Mathews, M. B. (1977). Elastin and collagen accumulation in rabbit ascending aorta and pulmonary trunk during postnatal growth. *Circ. Res.* 41, 316–323.

Majesky, M. W., Benditt, E. P., and Schwartz, S. M. (1988). Expression and developmental control of platelet-derived growth factor A-chain and B-chain/Sis genes in rat aortic smooth muscle cells. *Proc. Natl. Acad. Sci. USA* 85, 1524–1528.

Masuda, H., Kawamura, K., Tohda, T., Showaza, T., and Sageshima, M. (1989). Increase in endothelial cell density before artery enlargement in flow-loaded canine carotid artery. *Arteriosclerosis* 9, 812–823.

McDonald, D. A. (1974). *Blood Flow in Arteries*. Edward Arnold, London.

Moscatelli, D., Presta, M., and Rifkin, D. B. (1986). Purification of a factor from human placenta that stimulates capillary endothelial cell protease production, DNA synthesis, and migration. *Proc. Natl. Acad. Sci. USA* **83**, 2091–2095.

Myers, P. R., Minor, P. R., Jr., Guerra, R., Jr., Bates, J. N., and Harrison, D. G. (1990). Vasorelaxant properties of the endothelium-derived relaxing factor more closely resemble S-nitrosocysteine than nitric oxide. *Nature* (Lond.) **345**, 161–163.

Ohlen, A., Persson, M. G., Lindbom, L., Gustafsson, L. E., and Hedqvist, P. (1990). Nerve-induced nonadrenergic vasoconstriction and vasodilation in skeletal muscle. *Am. J. Physiol.* **258**, H1334–H1338.

Olesen, S. P., Clapham, D. E., and Davies, P. F. (1988). Haemodynamic shear stress activates a K channel current in vascular endothelial cells. *Nature* (Lond.) **331**, 168–170.

Palmer, R. M. J., Ferrige, A. G., and Moncada, S. (1987). Nitric oxide accounts for the biological activity of endothelium-derived relaxing factor. *Nature* (Lond.) **327**, 524–526.

Pash, J. M., and Bailey, J. M. (1988). Inhibition by corticosteroids of epidermal growth factor-induced recovery of cyclooxygenase afer aspirin treatment. *FASEB J.* **2**, 2613–2618.

Patel, D. J., Greenfield, J. C., and Fry, D. L. (1964). In vivo pressure-length-radius relationship of certain blood vessels in man and dog. In *Pulsatile Blood Flow*, pp. 293–305, Attinger, E. O. (ed.). McGraw-Hill, New York.

Patel, D. J., and Vaishnav, R. N. (1980). *Basic Hemodynamics and Its Role in Disease Processes*. University Park Press, Baltimore.

Pohl, U., Forstermann, U., and Busse, R. (1985). Endothelium-mediated modulation of arterial smooth muscle tone and PG12-release: Pulsatile versus steady flow. In *Prostaglandins and Other Eicosanoids in the Cardiovascular System. Proc. 2nd Int. Symp.*, pp. 553–558. Schror, K. (ed.). Karger, Basel.

Pohl, U., Holtz, J., Busse, R., and Bassenge, E. (1986). Crucial role of endothelium in the vasodilator response to increased flow in vivo. *Hypertension* **8**, 38–44.

Reidy, M. A., and Langille, B. L. (1980). The effect of local blood flow patterns on endothelial cell morphology. *Exp. Mol. Pathol.* **32**, 276–289.

Reidy, M. A., and Schwartz, S. M. (1981). Endothelial regeneration. III. Time course of intimal changes after small defined injury to rat aortic endothelium. *Lab. Invest.* **44**, 301–308.

Ross, R., Raines, E. W., and Bowen-Pope, D. F. (1986). The biology of platelet-derived growth factor. *Cell* **46**, 155–169.

Rubanyi, G. M., Romero, J. C., and Vanhoutte, P. M. (1986). Flow-induced release of endothelium-derived relaxant factor. *Am. J. Physiol.* **250**, H1145–H1149.

Rychter, Z. (1962). Experimental morphology of the aortic arches and heart loop in chick embryos. *Adv. Morphol.* **2**, 333.

Satcher, R., De Paola, N., Gimbrone, M. A., Jr., and Dewey, C. F., Jr. (1990). Endothelial cell structure resulting from shear stress. *First World Congress of Biomechanics* **II**, 243 (abstract).

Sato, M., Levesque, M. J., and Nerem, R. M. (1987). Micropipette aspiration of cultured bovine aortic endothelial cells exposed to shear stress. *Arteriosclerosis* **7**, 276–286.

Schmid-Schonbein, H. (1981). Factors promoting and preventing the fluidity of blood. In *Microcirculation*, pp. 249–266. Effros, R. M., Schmid-Schoebein, H., and Ditzel, J. (eds.). Academic Press, New York.

Schretzenmayr, A. (1933). Uber Kreislaufregulatorische Vorgange an den grossen Arterien bei der Muskelarbeit. *Pflugers Archiv. Ges. Physiol.* **232**, 743–748.

Schwartz, S. M., and Benditt, E. P. (1977). Aortic endothelial cell replication. I. Effects of age and hypertension in the rat. *Circulation* **41**, 248–255.

Schwartz, S. M., Heimark, R. L., and Majesky, M. W. (1990). Developmental mechanisms underlying pathology of arteries. *Physiol. Rev.* **70**, 1177–1209.

Seifert, R. A., Schwartz, S. M., and Bowen-Pope, D. F. (1984). Developmentally regulated production of platelet-derived growth factor-like molecules. *Nature* (Lond.) **311**, 669–671.

Smiesko, V., Kozik, J., and Dolezel, S. (1985). Role of endothelium in the control of arterial diameter by blood flow. *Blood Vessels* **22**, 247–251.

Thoma, R. (1893). Untersuchagen uberdie Histogenese und Histomechanik des Gefassystems. Enke, Stuttgart.

van Grondelle, A., Worthen, G. S., Ellis, D., Mathias, M. M., Murphy, R. C., Strife, R. J., Reeves, J. T., and Voekel, N. F. (1984). Increased prostacyclin production in endothelial cells during shear stress and in rat lungs at high flow. *J. Appl. Physiol.* **57**, 388–395.

Vlodavsky, I., Folkman, J., Sullivan, R., Fridman, R., Ishai-Michaeli, R., Sasse, J., Klagsbrun, M. (1987). Endothelial cell-derived basic fibroblast growth factor: Synthesis and deposition into subendothelial extracellular matrix. *Proc. Natl. Acad. Sci. USA* **84**, 2292–2296.

Wechezak, A. R., Viggers, R. F., and Sauvage, L. R. (1985). Fibronectin and F-actin redistribution in cultured endothelial cells exposed to shear stress. *Lab. Invest.* **53**, 639–647.

Wechezak, A. R., Wight, T. N., Viggers, R. F., and Sauvage, L. R. (1989). Endothelial adherence under stress is dependent on microfilament reorganization. *J. Cell. Physiol.* **139**, 136–146.

Zarins, C. K., Zatina, M. A., Giddens, D. P., Ku, D. N., and Glasov, S. (1987). Shear stress regulation of artery lumen diameter in experimental atherogenesis. *J. Vasc. Surg.* **5**, 413–420.

CHAPTER 9

■ ■ ■ ■ ■

Fluid Stress Effects
on Suspended Cells

■ ■ ■ ■ ■

Larry V. McIntire and Sridhar Rajagopalan

I. INTRODUCTION

This chapter presents a review of the effects of fluid stress on suspended cells, including effects on intracellular metabolism and aggregation. First we present an analysis of the cellular deformation induced by a simple shearing flow (a prototype "weak" flow). Some generalization of the simple shearing motion is the primary flow in all commonly used viscometric devices.

Cells suspended in shear flows are subjected to three types of stress: forces induced in the cell by the flowing fluid, forces arising from cell–cell collisions, and forces due to cell–wall collisions. Several general reviews of the theory involved in calculating these forces are available, including Goldsmith and Mason (1967) and Happel and Brenner (1986).

II. METHODS AND MATERIALS

A. Theory

Many cell types, including leukocytes, hybridomas, and tumor cells, can be accurately modeled as viscoelastic ellipsoids. We developed a

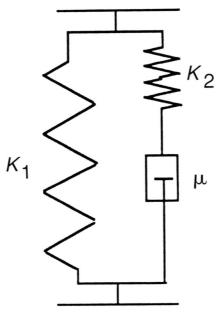

FIGURE 1   The linear viscoelastic solid model.

method to estimate the stresses and deformations induced in a cell by a shear flow. The forces induced were calculated using Roscoe's (1967) extension of the Jeffery's (1922) solution for particles in shear flows. Roscoe's procedure was followed except that the linear viscoelastic solid model shown in Figure 1 was employed for the particles, instead of the Voigt model employed by Roscoe. This model was used because Schmid-Schonbein et al. (1981) have shown that it describes the small deformation behavior of passive white blood cells quite well, and the formalism may apply, at least in a qualitative sense, to tumor cells and hybridomas. Other models have also been developed recently for leukocyte rheology (Yeung and Evans, 1989; Evans and Yeung, 1989). We do not intend to imply a preference, but rather use the earlier model since it is more easily incorporated into the Roscoe formalism. For small deformations, all models seem to give very good agreement, with appropriate choice of parameters. The procedure is described briefly below. [The notation used is similar to that of Roscoe (1967) and of Keller and Skalak (1982).]

The initially spherical cell of radius a_0 is assumed to be deformed into an ellipsoid with dedimensionalized semiaxes a_i $(i = 1, 2, 3)$. The ellipsoid is inclined to the external flow at an angle θ, and it rotates with a frequency ν. Assuming that the suspending fluid is an incompressible Newtonian fluid and inertial effects are negligible, the external flow is described by the Stokes approximation equations. Roscoe showed that the Jeffery solution can be modified to give a solution to this system, and from the solution the deviatoric stress traction vector $T_i'^d$, can be given as $T_i'^d = p_{ij}'^d \cdot n_j$, where $p_{ij}'^d$ is the deviatoric stress tensor induced on the particle by the flow given by Eqs. (16)–(21) of Roscoe (1967) [Note: The premultiplier in Roscoe's Eq. (16) should be $8/a_0^3$, not $a_0^3/8$].

The components of the deviatoric stress required to deform a viscoelastic particle in the manner described above can be calculated by using the constitutive equation for the particle and the imposed deformation. Because the linear viscoelastic model in Figure 1 is used, the constitutive equation is given by

$$p_{ij} + \frac{\mu}{K_2}\frac{dp_{ij}}{dt} = K_1 e_{ij} + \mu\left(1 + \frac{K_1}{K_2}\right)e_{ij}^{(1)} \tag{1}$$

where p_{ij} is the stress, e_{ij} the strain, and $e_{ij}^{(1)}$ the rate of strain tensor within the deformed particle and μ, K_1, and K_2 are the parameters of the linear viscoelastic model. The components of the strain tensor are

$$e_{ij} = (a_i - 1) \qquad \text{if } i = j \tag{2a}$$

$$e_{ij} = 0 \qquad \text{if } i \neq j \tag{2b}$$

and the (nonzero) components of the rate of strain tensor are

$$e_{12}^{(1)} = e_{21}^{(1)} = -\nu\frac{\left(a_1^2 - a_2^2\right)}{2a_1 a_2} \tag{3}$$

Using e_{ij} and $e_{ij}^{(1)}$ in the particle constitutive equation, one can calculate the stresses within the particle.

As in Roscoe (1967), the equilibrium condition (at equilibrium the stress p_{ij}' induced by the external flow equals the stress required for the equilibrium deformation of the particle p_{ij}), the conditions of symmetry of the stress tensor, and the condition of incompressibility of the particle yield five equations in the five unknown variables a_i, θ, and ν. Using the a_i, the components of the stress tensor, one can calculate the principal (deviatoric) stresses within the particle. For small deformations Schmid-Schonbein et al. (1981) have shown that for polymorphonuclear neutrophil leukocytes $K_1 = 275$ dyn/cm^2, $K_2 = 737$ dyn/cm^2, and $\mu = 130$ P. Because of lack of other data, we shall take these values to be representative of the rheological parameters for hybridoma and tumor cells also. When such a cell is suspended in a shear flow of rate $22,500 \text{ s}^{-1}$ in a fluid of viscosity 4 cP, the steady-state solution is

$$\nu \cong \frac{K}{2} = 11,250 \text{ s}^{-1}; \qquad \theta \cong 0.002° \tag{4}$$

$$a_1 \cong 1.00056; \qquad a_2 \cong 0.99944; \qquad a_3 \cong 1$$

These gives the stress at any point at any time as

$$\sigma \cong 2328 \sin(22,500t + \phi); \qquad \tau \cong 2328 \cos(22,500t + \phi) \tag{5}$$

where σ and τ give the normal and shear stresses.

The analysis given above is different from other similar analyses only in that a more realistic model for the cell response is used. Its utility is limited to understanding the qualitative nature of the stresses developed. The analysis suggests that large stresses that change with a high frequency may be developed, even though the maximum deformation induced will be very small. Thus, an observer rotating with the cell experiences a rapidly oscillating membrane stress but only a small maximum strain. Also, it can be easily proved that the magnitude of the developed stresses is approximately uniform throughout the particle.

B. Shear-Induced Damage of Mammalial Cells

There are many studies of shear-induced damage to blood cells. The erythrocyte was first examined because it is abundant and, in a biological sense, relatively uncomplicated. It has no nucleus and very little cytoskeletal structure, except for a two-dimensional spectin–actin network underlying the lipid bilayer. Shear-induced damage to human red blood cells has been extensively studied by several groups, including Leverett et al. (1972), Morris and Williams (1979), Williams (1973), and Sutera et al. (1972). Leverett et al. (1972) observed that when human blood was exposed to shear fields for two min, there was a threshold stress of 1500 dyn/cm^2 bulk shear flow, above which extensive red blood cell damage occurred. They also pointed out the importance of the interplay of time of exposure and fluid stress required to cause cell damage. Williams (1973) stressed red blood cell suspensions in dextran-containing media and observed that stresses above 700 dyn/cm^2 bulk shear for 5 min causes extensive red blood cell fragmentation. Morris and Williams (1979) showed that changing the medium viscosity with thickeners like dextran (or Ficol or polyvinyl pyrrolidone) affected cell deformation but did not directly affect damage. Sutera et al. (1972) also observed that there was not much effect of dextran on hemolysis. Much of the more recent work on stress effects on red cells has concentrated on modeling of membrane motion induced by fluid shear forces (e.g., Tran-Son-Tay et al., 1984).

Chittur et al. (1988) studied the effects of shear-induced stresses on human T cells, and observed that stresses of 200 dyn/cm^2 for 10 min could affect some of the T-cell functions, such as their response to lectins. Dewitz and coworkers have studied damage in human polymorphonuclear leukocytes caused by 2 or 10 min of exposure to shear fields (Dewitz, 1978; Dewitz et al., 1979; Martin et al., 1979). They observed that stresses above 600 dyn/cm^2 caused extensive morphologic destruction. There was a reduction in the number of leukocytes (measured with an electronic particle counter), and electron microscopic observation of stressed cells showed that the intact stressed leukocytes contained large vacuoles and condensed chromatin. At stresses of 75–150 dyn/cm^2 leukocyte functions like phagocytosis, chemotaxis, and adhesion were compromised.

There are many studies of shear-induced damage to hybridoma cells and other eukaryotic cells such as plant and insect cells. Schurch et al. (1988) observed that shearing hybridoma cells at stresses of up to 160 dyn/cm^2 for times of up to 10 min induced up to 4% damage per minute as measured by trypan blue exclusion. Smith et al. (1987) and Tramper et al. (1986) observed that there was extensive damage in insect and hybridoma cells when sheared at stress levels of 10–40 dyn/cm^2 for 3–15 h. Petersen et al. (1988, 1990) found that shearing CRL8018 hybridoma cells at stresses up to 50 dyn/cm^2 for 10 min induced up to 70% reduction in cell viability (measured by trypan blue exclusion). They also observed that the mode of cell culture (spinner flask versus T flask), culture age (lag or stationary phase in batch culture) and the concentration of metabolites such as ammonia, lactate, and pH could affect the shear sensitivity of the cells. Petersen (1989) observed that the organization of microfilaments, and not microtubules, affect cell deformability. Petersen (1989) also showed that modifications in mitochondrial respiration affected shear sensitivity more than glycolsis. Recent work by Bavarian et al. (1991) and Chalmers and Bavarian (1991) has demonstrated the importance of bubble rupture and jet formation in damage of insect cells in sparged cultures.

There are three studies involving shearing of tumor cell suspensions (Brooks, 1984; Koyama et al., 1987; Rajagopalan et al., 1992). Brooks (1984) suspended B16 melanoma cells in shear flows developed in a cone-and-plate viscometer. He observed that shearing the cells for 1 h at bulk shear stresses of 5.9–29 dyn/cm^2 damaged 20–70% of the cells (measured by erythrosin B staining). He also observed that cells from sublines of different metastatic potentials were damaged to essentially the same extent. Koyama et al. (1987) stressed KMT17 rat fibrosarcoma cells by suspending them in shear flows developed in a cone-and-plate viscometer. They observed that stressing the cells at 50 dyn/cm^2 for 1–2 h caused a 25–90% reduction in viability (trypan blue staining). They observed that stressing in the presence of calcium ion chelators such as ethylenediaminetetraacetic acid (EDTA) or calcium channel blockers like varapamil, intensified the stress-induced damage.

Rajagopalan et al. (1992) examined the effects of dynamic mechanical stresses on metastatic murine B16 melanoma and RAW117 large-cell lymphoma cells by suspending the cells in a laminar shear flow. The damage caused by the induced stresses was quantified using two measurements. Cell lysis in the shear field was used to measure short-term damage. The relative reduction in cell numbers in stressed cultures compared to unstressed controls at 48–72 h after stress exposure was used to asess long-term damage. Also, cell growth and lysis in stressed and unstressed cultures were monitored over time. The results could be modeled assuming that a subpopulation of stressed cells was lethally damaged and ultimately died, whereas a second subpopulation of stressed cells was only sublethally damaged and grew normally after a brief lag time.

Exposure of cells to bulk shear stresses did not result in a difference in the long-term damage between high- and low-metastatic-potential sublines. The minor differences in lytic damage caused in the shear field were insufficient to account for the differences in metastatic potential. Furthermore, the long-term damage coefficients indicated that a considerable fraction of the stressed cells was not lethally damaged, even at bulk shear stresses as high as 900 dyn/cm^2 for 5 min. Although the stresses applied were higher than normally seen in vivo, the results question the accepted view that mechanical stresses developed in the circulation directly cause the destruction of most of the tumor cells in the circulation during blood-borne metastasis and that highly metastatic cells can resist these stresses more effectively than poorly metastatic cells. They do not, however, rule out the possibility that mechanical stresses may be indirectly responsible for the destruction of circulating tumor cells through sublethal tumor cell damage, which might allow normal tumor cell clearing mechanisms to operate more efficiently.

C. Stress-Protein Synthesis in Cells

It has long been known that when cells are stressed by agents such as heat, they respond by synthesizing a small number of highly conserved proteins, called stress proteins (Lindquist, 1986; Subjeck and Shyy, 1986; Thomasovic, 1989; Carper et al., 1987; Burdon, 1986; Schlesinger, 1986). The most studied is the response to heat stress. When cultured cells (or whole organisms) are exposed to elevated temperatures, the synthesis of heat-stress proteins is induced. The major classes of heat-stress proteins are represented by the 70-, 90-, 110-, and 26-kD heat-stress proteins. This response is universal and highly conserved through evolution. Cells synthesize stress proteins in response to a wide variety of treatments, such as heat, hypoxia, glucose deprivation, and sodium arsenite. It is generally believed that all such treatments have the common feature of causing the accumulation of large quantities of denatured or damaged proteins within the call (Thomasovic, 1989). A hypothesis proposed to explain the mechanism of induction of stress protein synthesis (Munro and Pelham, 1985) suggests that when the levels of denatured or damaged proteins or structures in the cell increase, large amounts of the protein ubiquitin are required to target them for degradation. This binding of ubiquitin by the denatured proteins might indirectly induce the activation of stress proteins. Although their functions are not well understood, the stress proteins have protective effects (Thomasovic, 1989; Young and Elliot, 1989). The presence of stress proteins in cells has been shown to be critical to the survival of cells exposed to stresses, and in many instances the presence of stress proteins induces tolerance (defined as a transient increase in the resistance to stress damage) to a subsequent challenge. Pelham (1986) has hypothesized that the stress proteins (or at least some of them) help the cell to recover from stress-induced damage by helping the denatured proteins to renature and reassemble properly.

In our laboratory experiments were done to determine whether mechanical-stresses induced the synthesis of stress proteins and whether the presence of heat-stress proteins helped reduce mechanical-stress-induced damage. The cells used were RAW117-H10 cells, which were mechanically stressed by suspension in laminar shear flows for 5–10 min. The protein synthesis in stressed cells was monitored by culturing the stressed cells under regular conditions, in a medium containing [^{3}H]-leucine. The synthesized proteins were analyzed by using sodium dodecyl sulfate-polyacrylamide gradient gel electrophoresis and fluorography. To determine whether the presence of heat-stress protein protected cells from mechanical-stress-induced damage, synthesis was induced by exposing the cells to elevated temperatures (43°C) for 15–20 min. The heat-stressed cells were then cultured under normal conditions for 12 h to allow synthesis and accumulation of the stress proteins. These cells were then mechanically stressed in the viscometer and the long-term damage caused by the mechanical stress measured. The mechanical stress damage to preheated cells was compared with that in unheated cells to determine the protective effect, if any.

Results demonstrated that preexposure of RAW117-H10 cells to 43°C as described above did give some protection to damage from subsequent shear stress exposure (900 dyn/cm² for 5 min at 37°C). The relative amount of protection (as measured by the increase in cell number after 48 h of culture after mechanical stress exposure, compared to a matched control that had been sheared only) was more for cells that were heat-stressed more strongly or were allowed to recover longer.

The results of the stress protein metabolism studies were obtained as fluorographs of gels used to separate cellular proteins. Only the proteins with the radioactive leucine are detected in the fluorogram, and hence it represents only the proteins synthesized during the labeling period. The results of one experiment, where the heated sample was stressed at 43°C for 20 min and the sheared sample was stressed at 1200 dyn/cm² for 5 min, are shown in Figure 2. In this experiment an additional sample was

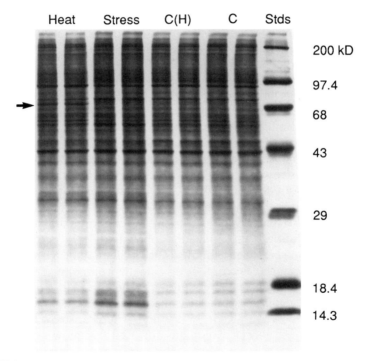

FIGURE 2 Stress protein metabolism in RAW117-H10 cells. Parameters are as follows: C = 37°C; (H) = 40.5°C, 20 min; Stress = 1200 dyn/cm², 5 min; Heat = 43°C, 20 min.

heated at 40.5°C for 20 min to determine whether the maximum temperature rise due to viscous heating could affect protein metabolism. Figure 2 indicates that the 70-kD heat-stress protein is strongly induced in RAW117-H10 cells heated at 43°C for 20 min. However, the metabolism of this heat-stress protein was *not* induced when the cells were shear-stressed. Synthesis of proteins in the 14–18-kD molecular weight range was induced in the sheared cells in this experiment. The data for the CH sample indicate that simply heating the cells at 40.5°C for 20 min did not induce stress protein metabolism. The experiment was repeated with slight variations: in three experiments the heat-stressed samples were heated for 15 min, in four experiments they were heated for 20 min, in two experiments the shear-stressed samples were sheared at 850 dyn/cm^2 for 10 min, and in four experiments they were sheared at 1100 dyn/cm^2 for 5 min. It was observed that increasing the heating time for 15 to 20 min increased the expression of heat-stress proteins. In all experiments heat-stress protein metabolism was not induced in shear-stressed cells.

Previous studies with mouse hybridoma cells (Passini and Goochee, 1989) have shown that mechanical stresses developed by agitation or sparging of batch cultures did not induce the synthesis of any known stress proteins.

D. Shear-Induced Platelet Aggregation

Early artificial-organ work indicated that platelet suspensions were probably induced to aggregate by mechanical stresses in the instruments. This phenomenon has been studied extensively, and some of the results are briefly reviewed below in this section. Most of the studies involved stressing platelets suspended in blood, plasma, or more defined solutions by subjecting the platelet suspension to laminar shear flow in tubes or in viscometric devices. Jen and McIntire (1984) observed that shearing anti-coagulated whole blood in a cone-and-plate viscometer at room temperature induced the formation of aggregates at stresses above a specific threshold stress. The threshold depended on the source of the blood and the anticoagulant used. For normal donors, the threshold was 2000–3000 s^{-1} for 1 min of stress exposure. Blood anticoagulated by heparin, a polysaccharide that inhibits thrombin, was more sensitive to stress-induced aggregation than was blood anticoagulated with sodium citrate, a calcium chelator. The sizes and stabilities of the aggregates increased with both shear rate and exposure time. Use of ADP antagonists reduced both sensitivity to and stability of the shear-induced aggregation. Dewitz et al. (1978) found that shearing blood in a modified Couette viscometer (Rice University's ROM viscometer) at 37°C induced aggregate formation at stress levels of the order of 75 dyn/cm^2 for 5 min. At higher stresses (about 400 dyn/cm^2) the aggregates were broken into smaller fragments. They also observed that leukocytes were associated with the aggregates. Subsequent work by Rhee et al. (1986) documented the role of leukocytes in aggregation and the interplay of shear-induced arachadonic acid metabolites from platelets and leukocytes in modulation of the aggregation process.

The presence of erythrocytes in blood increases the complexity of shearing experiments, and therefore, most studies have been done with platelet suspensions. (In some instances washed red blood cells have been added to such suspensions.) Colantuoni et al. (1977) subjected platelet suspensions in plasma, called *platelet-rich plasma* (PRP), to Poiseuille flow in narrow capillaries (100–200 μm inside diameter) for short times (1–5 ms). They observed that there was a decrease in the single platelet count for wall shear stresses above 20,000 dyn/cm^2. They also observed platelet serotonin secretion in the plasma above 7000 dyn/cm^2, whereas lysis did not occur below 15,000 dyn/cm^2. They concluded that the platelets were activated and released their dense granules at stress levels lower than that required for lysis. Rather high stresses were required in their studies, however, indicating either that for short exposure times rather high stress levels are required or that in Poiseuille flow some platelets may be in (or migrate to) the central region of the tube and hence be subjected to lower stresses. Using a flow-through Couette device, Wurzinger et al. (1985a, 1985b) have also observed that platelets in PRP treated with herparin required rather large stresses (570–2550 dyn/cm^2) for activation to aggregate when the shear exposure times were low (7–700 ms). Anderson et al. (1978) sheared PRP treated with citrate in a modified Couette flow viscometer. They observed that for 30 s stress exposure platelet aggregation began at 100 dyn/cm^2, and at least 600 dyn/cm^2 was required to induce lysis. For 5-min exposure they observed that 50–75 dyn/cm^2 was sufficient to induce aggregation, and stresses above 160 dyn/cm^2 were required for lysis. One potential problem in these studies is that lysis was measured using release of lactate dehydrogenase (LDH) as the marker of lysis, and the observed LDH may have been released from either the fragmentation of platelet aggregates or the shear-induced destruction of individual platelets. They also carried out their experiments at three different surface/volume ratios over a two- to three-fold range and observed that surface effects do not contribute significantly to platelet aggregation and lysis, and that aggregation (and lysis) resulted principally from bulk shear effects. Belval and co-workers have studied the kinetics of shear-induced platelet aggregation (Belval et al., 1984) and have described a model for the phenomenon (Belval and Hellums, 1986).

They observed that stable aggregation in PRP commenced at shear rates above 5000 s^{-1}, and they studied how the aggregation process evolved over time. The model describing this process involves a population balance equation [Eq. (1) in Belval and Hellums (1986)]. This equation is based on Smoluchowski's equation for gradient coagulation. The analysis has two adjustable parameters: the collision efficiency (ε) and the particle void volume fraction (ϕ). These were estimated using a best fit of the particle size distribution at the end of shearing. An interesting implication of their work was that only a very small fraction (1 in 1000) of the platelet collisions resulted in adhesion. Also, since the collision efficiency was lower for less concentrated platelet suspensions, platelet aggregation may be mediated by chemicals released actively/passively from platelets. The data (and analysis) due to Bell and Goldsmith (1984) and Bell et al. (1989a, 1989b) for platelet aggregation in Poiseulle flow also support the analysis of Beval and Hellums (1986). Goldsmith and co-workers (Bell and Goldsmith, 1984; Bell et al., 1989a, 1989b) also observed collision efficiencies of the same order of magnitude as those observed by Belval. They also observed that at lower shear rates the collision efficiency was higher, probably as a result of lower forces breaking up the aggregates. At very low shear rates, however, the physical forces do not initiate the platelets, and an additional agent like ADP had to be used to initiate aggregation. They observed that aggregation in platelets stimulated with 1 μM ADP increased with shear rate, probably due to an increase in the number of collisions.

There have been many studies that attempted to understand the mechanism of shear-induced platelet aggregation. Sutera and co-workers (Moritz et al., 1981) observed that when PRP is sheared in Couette flow, the platelet aggregation could be reduced by exposing the platelets to agents such a PGE$_1$ and theophyline, which reduce the extent of ADP secreted and released. Sutera and co-workers (Riemers et al., 1984) also performed studies with PRP to which ^{14}C-adenosine-labeled red blood cells or glutaraldehyde-fixed and ADP-depleted red blood cells were added. Their data suggest that ADP released from the lysis of a few red blood cells may be important for platelet activation in whole blood. (Their data also indicated that red blood cells also contribute to platelet aggregation by enhancing platelet convection.) Hellums and co-workers (Hellums and Hardwick, 1981; Hellums et al., 1986; Hardwick et al., 1980, 1981a, 1981b) have observed that some agents like aspirin have no effect on shear-induced platelet aggregation but can inhibit collagen-induced platelet aggregation of the same platelet preparations. Since aspirin inhibits thromboxane A$_2$ (TxA$_2$) synthesis, these results suggest that TxA$_2$ is not very important in shear-induced platelet aggregation. More recent studies (Rajagopalan et al., 1988) have shown that arachadonic acid metabolism in shear stressed platelets is shifted to produce essentially all 12 hydroxyeicos-apentanoeic (12-HETE) acid, with very little TxA_2 synthesis. Hellums and co-workers also observed that treating PRP with PGE_1 and theophyline reduced shear-induced platelet aggregation (although it also increased lysis above 150 dyn/cm^2 for 5 min). Hardwick et al. (1983) used colchicine to inhibit active secretion by disrupting platelet microtubules. They observed that serotonin was present in the external medium of PRP sheared at stresses greater than 40 dyn/cm^2 for 5 min. Hence Hardwick et al. (1983) concluded that the serotonin and consequently ADP might leak into the external medium in a passive manner. They also observed that plasma ADP concentration increased by 0.9 μM, a level sufficient to induce (reversible) platelet aggregation similar to that observed at this stress level. Wurzinger et al. (1985a) observed that on stressing PRP in a flow-through Couette device the relative amounts of β-thromboglobulin and LDH released were about the same. They concluded that any ADP present in the medium is probably due to cell lysis. Wurzinger et al. (1985a) further demonstrated that lysis of barely 1% of the available platelets could lead to an overall concentration of 0.1-μM ADP (0.1-μM ADP is sufficient to cause a reversible aggregation of platelet suspensions). Initially higher ADP concen-trations would be expected to be present in the immediate vicinity of a rupturing platelet. Thus Wurzinger et al. (1985a) argue that the other platelets might be stimulated by the ADP passively released from the lysed platelets. In another study, Wurzinger et al. (1985b) observed that the platelet aggregation in shear fields could be largely suppressed by adding ADP scavengers such as creatine phosphate–creatine phosphoki-nase, suggesting that most of the platelets are activated by ADP. The results of ultrastructural studies indicated that the source of the initial ADP could be the shear-destroyed platelets. Jen and McIntire (1984) documented similar reductions in aggregate volume using ADP scavengers in whole blood.

Moritz et al. (1981) observed that calcium in the external medium was required for platelet aggregation. Giorgio and Hellums (1986) observed that platelets that contained Quin-2, a calcium ion cheator that can penetrate and be retained within cells, ultimately behaved like untreated controls when stimulated with agents like thrombin, collagen, and ADP. However, shear-induced aggregation was significantly suppressed in Quin-2-loaded platelets. Since large intracellular concentrations of Quin-2 can act as a calcium sink, these results suggest that intracellular calcium mobilization is probably an important early step in shear-induced platelet aggregation. Their results also indirectly suggest that the platelets are probably activated in some way in the shear field.

Although there is consensus that shear-induced platelet activation, measured indirectly by measuring platelet-induced aggregation, may be mediated by ADP, there is still some dispute about the initial source of the ADP. Also there is a question of whether mechanical stresses can activate platelets directly, independent of ADP mediated events.

The role of specific bridging proteins in mediating shear-induced platelet aggregation has also been an area of significant recent progress. The first key observations by Moake et al. (1986, 1988) and Peterson et al. (1987) produced the concept that the von Willebrand factor (VWF) may play a crucial role in high-shear-induced platelet aggregation. In fact, VWF binding to the GPIb receptor was shown to be an essential step on the process. The unusually large multimers of VWF were particularly efficient in initiating stress induced aggregation. These forms do not normally circulate in plasma (Frangos et al., 1989) but may be released from granules stored in endothelial cells on activation of the endothelium. These findings have recently been independently confirmed (Ikeda et al., 1991). It is interesting that fibrinogen appears to play little or no role in this high-shear-induced aggregation.

REFERENCES

Anderson, G. H., Hellums, J. D., Moake, J. L., and Alfey, C. P. (1978). Platelet lysis and aggregation in shear fields. *Blood Cells* 4, 499–507.

Bavarian, F., Fan, L. S., and Chalmers, J. J. (1991). Microscopic visualization of insect cell-bubble interaction. I: Rising bubbles, air-medium interface and the foam layer. *Biotechnol. Prog.* 7, 140–150.

Bell, D. N., Spain, S., and Goldsmith, H. L. (1989a). Adenosine diphosphate induced aggregation of human platelets in flow through tubes, I: Effect of shear rate, donor sex, and ADP concentration. *Biophys. J.* 56, 829–843.

Bell, D. N., Spain, S., and Goldsmith, H. L. (1989b). Adenosine diphosphate induced aggregation of human platelets in flow through tubes, I: Measurement of concentration and size of single platelets and aggretates. *Biophys. J.* 56, 817–828.

Bell, D. N., and Goldsmith, H. L. (1984). Platelet aggregation in Poiseuille flow II: Effect of shear rate. *Microvasc. Res.* 27, 316–330.

Belval, T., and Hellums, J. D. (1986). Analysis of shear-induced platelet aggregation with population balance mathematics. *Biophys. J.* 50, 479–487.

Belval, T., Hellums, J. D., and Solis, R. T. (1984). The kinetics of platelet aggregation induced by fluid-shearing stress. *Microvasc. Res.* 28, 279–288.

Brooks, D. E. (1984). The biorheology of tumor cells. *Biorheology* 21, 85–91.

Burdon, R. H. (1986). Heat shock and the heat shock proteins. *Biochem. J.* 240, 313–324.

Carper, S. W., Duffy, J. J., and Gerner, E. W. (1987). Heat shock proteins in thermotolerance and other cellular processes. *Cancer Res.* 47, 5249–5255.

Chalmers, J. J., and Bavarian, F. (1991). Microscopic visualization of insect cell-bubble interaction. II: The bubble film and bubble rupture. *Biotechnol. Prog.* 7, 151–158.

Chittur, K. K., McIntire, L. V., and Rich, R. R. (1988). Shear stress effects on human T-cell function. *Biotech. Progress* 4, 89–96.

Colantuoni, G., Hellums, J. D., Moake, J. L., and Alfrey, C. P. (1977). The response of human platelets to shear stress at short exposure times. *Trans. Am. Soc. Artif. Int. Org.* 23, 626–631.

Dewitz, T. S. (1978). Fluid mechanical trauma in human blood leukocytes, Ph.D. thesis, Rice University, Houston, TX.

Dewitz, T. S., Martin, R. R., Solis, R. T., Hellums, J. D., and McIntire, L. V. (1978). Microaggregate formation in whole blood exposed to shear stress. *Microvasc. Res.* 16, 263–271.

Dewitz, T. S., McIntire, L. V., Martin, R. R., and Sybers, H. D. (1979). Enzyme release and morphological changes in leukocytes induced by mechanical trauma. *Blood Cells* 5, 499–510.

Evans, E., and Yeung, A. (1989). Apparent viscosity and cortical tension of blood granulocytes are determined by micropipete aspiration. *Biophys. J.* 56, 151–160, 1989.

Frangos, J. A., Moake, J. L., Nolasco, L., McIntire, L. V., and Phillips, M. (1989). Cyrosupernatent regulates accumulation of large multimeric forms of VWF from endothelial cells. *Am. J. Physiol.* 256, H1635–H1644.

Giorgio, T. D., and Hellums, J. D. (1986). A note on the use of Quin2 in studying shear-induced platelet aggregation. *Thromb. Res.* 37, 353–359.

Goldsmith, H. L., and Mason, S. G. (1967). The microrheology of dispersions. In *Rheology: Theory and Applications*, Vol. 4, pp. 84–250, Eirich, F. R. (ed.). Academic Press, New York.

Happel, J., and Brenner, H. (1986). *Low Reynolds Number Hydrodynamics*, 2nd ed. Martinus Nijhoff Publishers, Boston.

Hardwick, R. A., Gritsman, H. N., Stromberg, R. R., and Friedman, L. I. (1983). The biochemical mechanisms of shear-induced platelet aggregation. *Trans. Am. Soc. Artif. Int. Org.* 29, 448–453.

Hardwick, R. A., Hellums, J. D., Peterson, D. M., and Moake, J. L. (1981a). Effects of PGI_2 and theophylline on the response of platelets subjected to shear stress. *Blood* 58, 678–681.

Hardwick, R. A., Hellums, J. D., Peterson, D. M., Moake, J. L., and Olson, J. S. (1981b). Effects of PGI_2 and dibuutyryl cAMP on platelets exposed to shear stress. *Trans. Am. Soc. Artif. Int. Org.* 27, 192–196.

Hardwick, R. A., Hellums, J. D., Moake, J. L., and Peterson, D. M. (1980). Effects of anti-platelet agents on platelets exposed to shear stress. *Trans. Am. Soc. Artif. Int. Org.* 26, 179–184.

Hellums, J. D., and Hardwick, R. A. (1981). Response of platelets to shear stresses— A review. In *The Rheology of Blood, Blood Vessels and Associated Tissues*, pp. 160–183, Gross, D. R., and Hwang, N.H. C. (eds.). Sijhoff and Noordhoff Publishers, Amsterdam.

Hellums, J. D., Peterson, D. M., Stathopoulos, N. A., Moake, J. L., and Giorgio, T. D. (1986). Studies on the mechanisms of shear-induced platelet activation. Paper presented at the meeting on Cerebral Ischemia and Hemorheology, Rottach-Egern, Germany.

Ikeda, Y., Handa, M., Kawano, K., Ando, H., Sakai, K., and Ruggeri, Z. M. (1991). The role of von Willebrand factor and fibrinogen in platelet aggregation under varying shear stress *J. Clin. Invest.* 87, 1234–1240.

Jeffrey, G. B. (1922). The motion of ellipsoidal particles immersed in a viscous fluid. *Proc. Roy. Soc.* (Lond.) **A102**, 161–179.

Jen, C. J., and McIntire, L. V. (1984). Characteristics of shear-induced aggregation in whole blood. *J. Lab. Clin. Med.* **103**(1), 115–124.

Keller, S. R., and Shalak, R. (1982). Motion of a tank-treading ellipsoidal particle in a shear flow. *J. Fluid Mech.* **120**, 27–47.

Koyama, T., Arasio, T., Ishikawa, M., Okada, F., and Kobayahi, H. (1987). Intensification of rheological of rat fibrosarcoma KMT17 cells by elimination of divalent cations. *Biorheology* **24**, 775–782.

Leverett, L. B., Hellums, J. D., Alfrey, C. P., and Lynch, E. C. (1972). Red blood cell damage by shear stresses. *Biophys. J.* **12**, 257–273.

Lindquist, S. (1986). The heat-shock response. *Annu. Rev. Biochem.* **55**, 1151–1191.

Martin, R. R., Dewitz, T. S., and McIntire, L. V. (1979). Alterations in leukocyte structure and function due to mechanical trauma. In *Quantitative Cardiovascular Studies: Clinical Research Applications of Engineering Principles*, pp. 409–454, Hwang, N. H. C., and Patel, D. J. (eds.). University Park Press, Baltimore.

Moake, J. L., Turner, N. A., Stathopoulos, N. A., Nolasco, L. H., and Hellums, J. D. (1986). Involvement of large plasma von Willebrand factor (VWF) multimers and unusually large VWF forms from endothelial cells in shear stress induced platelet aggregation. *J. Clin. Invest.* **78**, 1456–1461.

Moake, J. L., Turner, N. A., Stathopoulos, N. A., Nolasco, L. H., and Hellums, J. D. (1988). Shear induced platelet aggregation can be mediated by VWF released from platelets, as well as by exogenous large or unusually large VWF multimers, requires adenosine diphosphate and is resistant to aspirin. *Blood* **71**, 1366–1374.

Moritz, M. W., Sutera, S. P., and Joist, J. H. (1981). Factors influencing shear-induced platelet alterations: Platelet lysis is independent of platelet aggregation and release. *Thromb. Res.* **22**, 445–455.

Morris, D. R., and Williams, A. R. (1979). The effects of suspending medium viscosity on erythrocyte deformation and hemolysis in vitro. *Biochim. Biophys. Acta* **550**, 289–296.

Munro, S., and Pelham, H. (1985). What turns on heat shock genes? *Nature* **317**, 477–478.

Passini, C. A., and Goochee, C. F. (1989). Response of a mouse hybridoma cell line to heat shock, agitation sparging. *Biotechnol. Prog.* **5**, 175–188.

Pelham, H. (1986). Speculations on the functions of major heat shock and glucose regulated proteins. *Cell* **46**, 959–961.

Petersen, J. F., McIntire, L. V., and Papoutsakis, E. T. (1988). Shear sensitivity of cultured hybridoma cells, CRL8018, depends on mode of growth, culture age, and metabolite concentration. *J. Biotechnol.* **7**, 229–246.

Petersen, J. F., McIntire, L. V., and Papoutsakis, E. T. (1990). Shear sensitivity of hybridoma cells in batch, fed batch and continuous cultures. *Biotechnol. Prog.* **6**, 114–120.

Peterson, D. A., Stathopoulos, N. D., Giorgio, T. D., Hellums, J. D., and Moake, J. L. (1987). Shear induced platelet aggregation, requires VWF and platelet membrane glycoproteins GPIb and IIb/IIIa. *Blood* **69**, 625–628.

Peterson, J. F. (1989). Shear stress effects on cultured hybridoma cells in a rotational Couette viscometer. Ph.D. thesis, Rice University, Houston, TX.

Rajagopalan, S., McIntire, L. V., Hall, E. R., and Wu, R. K. (1988). The stimulation of arachidonic acid metabolism in human platelets by hydrodynamic stresses. *Biochim. Biophys. Data* **958**, 108–115.

Rajagopalan, S., Updyke, T. V., Nicolson, G. L., and McIntire, L. V. (1992). Effects of dynamic stress on tumor cells of varying metastatic potentials. *Biophys. J.*

Rhee, B. G., Hall, E. R., and McIntire, L. V. (1986). Platelet modulation of polymorphonuclear leukocyte shear induced aggregation. *Blood* 67, 240–246.

Riemers, R. C., Sutera, S. P., and Joist, J. H. (1984). Potentiation by red blood cells of shear-induced platelet aggregation: Relative importance of chemical and physical mechanisms. *Blood* 64(6), 1200–1206.

Roscoe, R. (1967). On the rheology of a suspension of viscoelastic spheres in a viscous liquid. *J. Fluid Mech.* 28, 273–293.

Schmid-Schonbein, G. W., Sung, K. L. P., Tozeren, H., Skalak, R., and Chien, S. (1981). Passive mechanical properties of human leukocytes. *Biophys. J.* 36, 243–256.

Schleisinger, M. J. (1986). Heat shock proteins: The search for functions. *J. Cell Biol.* 103, 321–325.

Schurch, U., Kramer, H., Einsele, A., Widmer, F., and Eppenberger, H. M. (1988). Experimental evaluation of laminar shear stress on the behavior of hybridoma mass cell cultures producing monoclonal antibodies against mitochondrial creatine kinase. *J. Biotechnol.* 7, 179–184.

Smith, C. G., Greenfield, P. F., and Randerson, D. H. (1987). A technique for determining the shear sensitivity of mammalian cells in suspension culture. *Biotechnol. Techniques* 1(1), 39–44.

Subjeck, J. R., and Shyy, T.-T. (1986). Stress protein systems of mammalian cells. *Am. J. Physiol.* 250, C1–C17.

Sutera, S. P., Croce, P. A., and Mehrjardi, M. (1972). Hemolysis and subhemolytic alterations of human red blood cells induced by turbulent shear flow. *Trans. Am. Inst. Artif. Int. Org.* 18, 335–341.

Thomasovic, S. P. (1989). Fundamental aspects of the mammalian heat stress protein response. *Life Chem. Rep.* 7, 33–63.

Tramper, J., Williams, J. B., and Joustra, D. (1986). Shear sensitivity of insect cells in suspension. *Enz. Microb. Technol.* 8, 23–36.

Tran-Son-Tay, R., Sutera, S. P., and Rao, P. R. (1984). Determination of red blood cell membrane viscosity from rheoscopic observations of tank-treading motion. *Biophys. J.* 46, 65–72.

Williams, A. R. (1973). Shear-induced fragmentation of human erythrocytes. *Biorheology* 10, 303–311.

Wurzinger, L. J., Opitz, R., Blasberg, P., and Schmid-Schonbein, H. (1985a). Platelet and coagulation parameters following millisecond exposure to laminar shear stress. *Thromb. Hemost.* 54(2), 381–386.

Wurzinger, L. J., Opitz, R., Wolf, M., and Schmid-Schonbein, H. (1985b). Shear induced platelet activation: A critical re-appraisal. *Biorheology* 22, 399–413.

Yeung, A., and Evans, E. (1989). Cortical shell-liquid core model for passive flow of liquid like spherical cells into micropipetes. *Biophys. J.* 56, 139–149.

Young, R. A., and Elliot, T. J. (1989). Stress proteins, infection, and immune surveillance. *Cell* 59, 5–8.

CHAPTER 10

■ ■ ■ ■ ■

Physical Forces in Mammalian Cell Bioreactors

■ ■ ■ ■ ■

Eleftherios T. Papoutsakis and James D. Michaels

I. INTRODUCTION

The effects of physical forces on mammalian cells are numerous, as can be seen from the discussions in other portions of this volume. Results of these studies can be used to help us gain a further understanding of these effects in mammalian cell bioreactors since mammalian cell culture technology has become an important tool in the production of numerous high-value products used to satisfy the needs for advanced therapies and transplantation (Culliton, 1989; Hansen and Sladek, 1989; Rosenberg et al., 1985, 1988), novel diagnostic and testing methodologies, and new therapeutic proteins. However, several of these cells are difficult to grow in large quantities while maintaining their specialized biological activity. These include various fetal cells, malignant cells, and a large variety of hemopoietic, blood, and related cells (Muul et al., 1986, 1987; Rosenberg et al., 1985, 1988; Yssel et al., 1984). Cells that are used in transplantation therapies include the tumor-infiltrating lymphocyte (TIL) cells and the lymphokine-activated killer (LAK) cells that are used in experimental cancer therapies (the so-called adoptive immunotherapies) by Rosenberg's group at NIH (National Institutes of Health, Bethesda, MD) (Aebersold et al., 1988; Jadus et al., 1988; Muul et al., 1986, 1987; Rosenberg et al., 1985, 1988; Rosenberg, 1990; Yssel et al., 1984). TIL and LAK cells are obtained from the patients tumors or peripheral blood, respectively. They need to be grown quickly to very high numbers (10^{11}–10^{12} live, biologically active cells) and are then administered back to the patient. Growth of such large numbers of TIL and LAK cells for adoptive immunotherapies constitutes a formidable task. Choosing and scaling-up the appropriate culture system (i.e., bioreactor), depending on the characteristics of the cell, demands that we understand how the complex fluid-mechanical, nutritional, and physicochemical environment in bioreactors affects the cells. While a number of cells can be grown in various static-type bioreactors (e.g., hollow-fiber and other perfusion bioreactors), in this review we are concerned only with the bioreactors where the cells are subjected to substantial fluid forces. A cell growing in these bioreactors might find itself in any of a large variety of fluid dynamic environments that may differently influence the responses of the cell to the various forces discussed in this text.

Most normal and many transformed animal cells are anchorage-dependent for growth; that is, they require a surface for attachment. One of the most efficient means (large surface area per unit volume) to provide surface for cell growth is the use of microcarriers (small polymeric or glass beads, typically 120–400 μm in diameter) in stirred bioreactors. On the other hand, many transformed cells, insect cells, and blood cells can be readily grown in free suspension. Growth of anchorage-dependent and freely suspended animal cells in *mixed* bioreactors has a variety of advantages such as scalability, ease of controlling and monitoring important bioreactor parameters such as pH, nutrient concentrations, relatively uniform bioreactor conditions, and use of existing industrial capacity. Mixing in these bioreactors is necessary in order to increase oxygen transfer and provide a homogeneous environment for cell growth.

Aside from applications for producing recombinant or naturally excreted glycoproteins, microcarriers will also find applications in the emerging field of tissue or organ engineering. In this case, it is desirable to grow and/or maintain complex and inhomogeneous populations of cells that retain the functional and differentiation characteristics of the tissues or organs from which they were extracted. Examples include the growth of neuronal, muscle, liver, pancreas, and bone-marrow cells (Erickson et al., 1983; Glasgow et al., 1984; Rhee et al., 1986; Rhee and McIntyre, 1986). The effect of fluid-mechanical forces is particularly important in these systems since such cells are typically more fragile, and also because fluid events vastly affect the interactions and adherence of cells to/with substrates and other cells. Such interactions and adherence are very important for retaining the functional characteristics and differentiation properties of these cells.

Examples of freely suspended cells that are used in various applications include transformed mammalian and insect cells for production of recombinant proteins, hybridoma cells for production of monoclonal antibodies, and certain bone-marrow and blood cells for production of cell components (needed for diagnostic purposes) or as final products (such as for transfusion therapies). The last category includes the aforementioned TIL and the LAK cells that are used in experimental adoptive cancer immunotherapies. Finally, a variety of suspension cells are used for virus production, diagnostic purposes, and possibly for vaccine production in the future. For example, the human immunodeficiency (HIV) virus is produced by infecting a variety of normal or neoplastic blood cells grown in large scale (Greenaway and Farrar, 1990).

Freely suspended cells can be grown in a variety of bioreactor configurations such as microencapsulation reactors, hollow-fiber reactors, airlift reactors, and stirred-tank reactors (Prokop and Rosenberg, 1989). While most monoclonal antibody production can be accomplished in static perfusion-type bioreactors (such as hollow-fiber reactors), growth of cells for the other aforementioned applications will require the use of suspension reactors, with or without perfusion and/or cell recycle. The mass production of cells for autologous and heterologous transfusion therapies, including the expected future development of culturing systems for massive production of blood cells and components (Golde and Glasson, 1988) will amplify the currently underappreciated difficulty of fluid-mechanical cell damage. Indeed, it has been shown that even low-level fluid forces can cause injury-related cell alterations without cell death, and these alterations can result in dramatic adverse patient responses (Martin et al., 1979) or in reduced biological effectiveness. The latter has been documented for the LAK cells, whereby cells grown in suspension spinner cultures were found to have substantially lower biological activity compared to cells grown in static or roller-bottle cultures (Muul et al., 1986). It is evident that the need to effectively and expediously culture various difficult cell types will require cell culture technologies far more advanced than those available today. Nutrient and oxygen supply and the associated problems of agitation and/or aeration are at the core of this difficulty. Furthermore, in view of the central role of agitation and/or aeration in many cell culturing systems, it is an engineering anomaly that no basic understanding or engineering correlations have been developed for this fundamental problem.

From the above discussion it becomes clear that understanding how fluid-mechanical forces due to agitation and/or aeration in bioreactors affect the viability and biological functionality of freely suspended cells is a problem of both practical and fundamental importance. A better qualitative and quantitative understanding of this problem would benefit the design, operation, and productivity of many bioreactor types.

II. PROBLEM OVERVIEW

The problem of fluid-mechanical effects on cultured animal cells has two components: one fluid mechanical and the other biological. The fluid-mechanical component of the problem addresses the following questions, including which of the following interactions or events is most damaging to the cells in reactors or other processing devices:

1. For freely suspended cells is it due to cell–fluid, cell–bubble, or cell–solid surface interactions?
2. For microcarrier culture is it due to bead–fluid, bead–bead, bead–solid surfaces, or bead–bubble interactions?
3. What forces or events affect the cells in a flow environment and how?
4. Is the effect on the cells due to the intensity or the frequency of the forces or both?

Biological questions include the following:

1. Do fluid-mechanical stresses reduce cell growth, or just cause cell death?
2. Do fluid-mechanical forces affect the physiology (e.g., the cell cycle), product expression, molecular processes, and the cytoskeleton of the cells? If so, how?
3. Do cells react to fluid-mechanical forces? If so, how?

A number of reports in the biomedical engineering literature have shown that well-defined, laminar shear stresses affect the shape, physiology, cytoskeletal structure, membrane structure and processes, and the formation of certain metabolites and proteins of a number of cells. It is well established in the cell biology literature that the shape, surface attachment and spreading of anchorage-dependent cells, and the shape of freely suspended cells affect DNA and protein synthesis, RNA metabolism, and cytoskeletal protein production in a profound and specific way (Ben-Ze'ev et al., 1980; Folkman and Greenspan, 1975; Folkman and Moscona, 1978; Reiter et al., 1985; Wittelsberger, 1981). There is sufficient evidence to support the hypothesis that fluid forces may affect cultured cells in bioreactors in profound specific and nonspecific ways, although this has not yet been documented very effectively (see discussion in Section V,C). First, we will review the various flow environments encountered by cells in bioreactors. Then we will discuss specific and general responses of a variety of cells to various flow and bioreactor environments, and finally we will discuss the use of various "shear" protectants (additives) in free-suspension bioreactors.

III. FLOW ENVIRONMENTS EXPERIENCED BY CELLS IN BIOREACTORS

A. Classification

We are considering cells either freely suspended or on microcarrier beads in agitated, airlift, and bubble-column reactors. So far, it has not been practical to use microcarriers in bubble-column or airlift reactors because the upward motion of bubbles results in bead aggregation and accumulation at the free air-medium surface. Similarly, in agitated micro-carrier bioreactors direct air sparging may result in aggregation and accumulation of the beads at the free surface. Although it has been suggested that with proper sparging this difficulty can be eliminated, we have decided not to discuss effects on cells from bead–bubble interactions, primarily for the lack of any information on this issue. A cell in a bioreactor may be freely suspended, part of a larger cell aggregate, attached on a single microcarrier, or part of a multibead aggregate (Fig. 1). For an agitated or otherwise mixed bioreactor, a cell may experience a variety of flow environments which may be subdivided into four broad categories.

1. The Bulk Liquid without Interference of Any Bubbles, Free Liquid Surfaces, or Nearby Walls

In this environment, we have some of the most important interactions that affect cells on microcarrier beads, namely bead–fluid and bead–bead (Cherry and Papoutsakis, 1986, 1990; Croughan et al., 1987) interactions. This environment does not play a major role in the damage of freely suspended cells (Kunas and Papoutsakis, 1990b). This first environment corresponds to most of the contents of a stirred or airlift vessel, and can be compared to turbulent flow near the center of a pipe. The flow of the liquid is mostly turbulent.

2. Flow Regimes near Moving and Rearranging Gas–Liquid Interfaces

This category includes interfaces at the free surface or near the moving-bubble surface in a bubble-column bioreactor with or without macroscopic foaming (Handa et al., 1987; Handa-Corrigan et al., 1989; Tramper and Vlak, 1986, 1988; Tramper et al., 1986, 1988). Similar regimes are encountered in agitated bioreactors with or without direct sparging. They

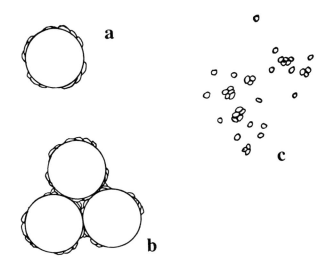

FIGURE 1   The magnitude, frequency, and type of forces experienced by cells in agitated and/or aerated bioreactors depends strongly on the size of the suspended entity; (a) cells attached on a single microcarrier bead; (b) cells attached on a bead aggregate; (c) freely suspended cells either as single cells or as part of an aggregate.

include flows near the unstable vortex surface, and flows that result from the associated gas entrainment, and bubble coalescence and breakup (Kunas and Papoutsakis, 1990b). As is discussed in more detail in Sections V,B and C below, this is the flow regime where damage of freely suspended cells is most likely to occur in most bioreactors.

3. Fluid Flows near a Solid Surface

This includes boundary layers near moving or stationary surfaces in reactors and pipe flows and is characterized by a relatively ordered local flow pattern. Such flows are unlikely to cause any detrimental effects on cells freely suspended or on microcarriers (Cherry and Papoutsakis, 1986, 1990), except for the case of turbulent pipe flows (McQueen et al., 1987). It is unlikely that such flows are of any practical significance unless cell suspensions need to be piped very quickly from one vessel to another or during very fast filtration in membrane channels.

4. Flow-Induced Collisions of a Suspension Cell or Microcarrier with Solid Objects in the Reactor

With freely suspended cells, such collisions are very unlikely (because of the small size and inertia of the cells). These collisions are low energy events if they occur at all. Therefore, they cannot be expected to cause cell damage.

B. Turbulent Flow in Agitated Bioreactors in the Absence of Bubbles

In moderately or intensely agitated bioreactors in the absence of bubbles, liquid flow is mostly or entirely turbulent, depending on the prevailing Reynolds number (Nagata, 1975). Kolmogorov's theory has been used to understand cell damage in turbulent microcarrier bioreactors (Croughan et al., 1987, 1988; Cherry and Papoutsakis, 1988, 1990) and in suspension cell bioreactors (Kunas and Papoutsakis, 1990b). Kolmogorov's isotropic turbulence relationships can be used to predict several bioreactor phenomena in a unified manner, and the theory may be used as a guide for practical design since the analysis is relatively simple even though many complex phenomena are involved (Kawase and Moo-Young, 1990). This has been discussed extensively in the literature of mixing of one and two-phase (gas–liquid, liquid–liquid) systems [see, e.g., Clift et al. (1978); Placek et al. (1986); Shinnar and Church (1960)] for bioreactors (Kawase and Moo-Young, 1990; Prokop and Rosenberg, 1989) and, more specifically, for particle or microcarrier reactors (Cherry and Papoutsakis, 1986, 1990; Croughan et al., 1987; Batchelor 1980). Thus, it is appropriate and convenient to discuss the turbulent environment that cells and beads will experience in agitated reactors in the absence of bubbles in terms of the properties of the small eddies in the viscous–dissipation range of the spectrum of isotropic turbulence (Nagata, 1975; Hinze, 1975; Cherry and Papoutsakis, 1990). It was suggested by Cherry and Papoutsakis (1986) that a major source of cell damage in turbulent microcarrier reactors was the interaction of microcarriers with turbulent eddies of a size comparable to the microcarrier size or the distance between the microcarriers. They used the eddy size correlation from Kolmogorov's isotropic turbulence theory for agitation rates where they found cell damage. Similarly, Kunas and Papoutsakis (1990b) used the correlation in free-suspension bioreactors to model cell damage due to hydrodynamic shear in the bulk turbulent liquid and found that the predicted eddy size is approaching the size of single hybridoma cells at these agitation rates. In Kolmogorov's isotropic turbulence theory, the Kolmogorov-scale eddy size η is calculated from [Hinze, 1975]

$$\eta = \left(\frac{\nu^3}{\varepsilon} \right)^{1/4} \tag{1}$$

where η is the Kolmogorov eddy length (cm), ν is the kinematic viscosity

(cm^2/s), and ε is the specific turbulent power dissipation rate (cm^2/s^3) given by

$$\varepsilon = \frac{N_p n^3 d_i^5}{V_d} \tag{2}$$

where d_i is the impeller diameter (cm), N_p the dimensionless power number $(P/\rho_f n^3 d_i^5)$, n the agitation rate (rev/s), and V_d the power dissipation volume, or liquid volume for energy dissipation (cm^3). N_p is a function of impeller geometry, impeller Reynolds number

$$Re = \frac{n d_i^2}{\nu} \tag{3}$$

and whether the vessel is baffled. Turbulence is taken to prevail in such a reactor at Re > 1000. The value of N_p can be estimated from the widely known diagrams of N_p versus Re for various impeller designs (Nagata, 1975), or predicted from direct experimental measurements [see, e.g., Aunins et al. (1989)]. A major difficulty in predicting the Kolmogorov-scale eddy size is the calculation of a representative value for ε. The problem is that ε near the impeller can be vastly different from ε in the bulk liquid. Thus, the choice for the power dissipation volume V_d is critical, and this choice depends on several reactor parameters (e.g., impeller type and impeller diameter to tank diameter ratio) (Nagata, 1975). But even though the ambiguity and lack of information on the variation of ε in a stirred-tank reactor do not allow for a precise determination of the Kolmogorov-scale eddy size, these calculations may still be used as approximate criteria for correlating cell death to the bioreactor agitation intensity. Presently, the level of sophistication in the analysis of the system and the accuracy of the experimental data do not justify the introduction of additional complexities such as the residence time of a particle in the various regions of the agitated reactor. Instead, one may obtain an estimate of the *average high* ε (i.e., in the area of the most severe turbulence around the impeller) by taking the dissipation volume V equal to d_i^3 (Cherry and Papoutsakis, 1988, 1990) to give

$$\varepsilon = \frac{P}{\rho_f V} = \frac{P}{\rho_f d_i^3} = N_p n^3 d_i^2 \tag{4}$$

where ρ_f is density of the fluid (g/m^3). With the understanding that some very small regions near the impeller will have a larger ε and that other

TABLE 1

Range of Eddy Sizes Calculated for a Single Agitation Rate of 800 rpm for a Completely Filled 2-liter Bioreactor [a]

Power number	Kolmogorov eddy size (μm)
V_d = 2000 cm^3	
1.5	22.2
2.0	20.7
2.5	19.6
$V_d = d_i^3$ = 343 cm^3	
1.5	14.3
2.0	13.3
2.5	12.6

[a]The eddy size is given dependent on the value used for the power number and the volume available for power dissipation, V_d in the reactor. Parameter d_i is the impeller diameter (cm). Taken from Kunas and Papoutsakis (1990b).

regions in the reactor will have smaller ε values, the expression of Eq. (4) can be used. To exemplify how calculations of η vary from using either total reactor volume or the volume in the vicinity of the impeller for V_d, Table 1 shows calculations of the average Kolmogorov eddy size for the 2-L bioreactor employed by Kunas and Papoutsakis (1990b). The eddy size shown is determined at agitation rates of 800 revolutions per minute (rpm) using power numbers of 1.5, 2.0, and 2.5. Using either expression for power dissipation volume, the prediction of Kolmogorov-scale eddy size approaches the size of the hybridoma cells (10–15 μm). It was suggested that eddies of sizes less than the cell diameter cause pressure differences across the cell and deformation of the surface that could cause lysis. McQueen et al. (1987) measured cell death rates for several freely suspended animal cells in turbulent capillary flows. They found that cell death becomes evident when the Kolmogorov-scale eddies are smaller than 3–5 μm or one-third to one-half the cell diameter.

For the case of bubble-column and airlift bioreactors, one may calculate P for Eq. (4) from the power that is necessary to expand the gas isothermally from the pressure at the sparger to the pressure at the free surface (headspace pressure) (Cherry and Papoutsakis, 1990). The final result recast in the form of the superficial gas velocity (cm/s) U_G is

$$P = \rho_f g U_G V_L \tag{5}$$

where V_L is the liquid volume. For an airlift reactor the expression of

Eq. (5) must be multiplied by the ratio of the riser cross section to the total cross section because in this case U_G is typically U_{GR} [i.e., the superficial velocity based on the riser cross section (cm/s) only]. In deriving Eq. (5), the kinetic energy of the gas at the sparger was ignored, and the length of the reactor was taken to be less than 3 m. If necessary, corrections for both assumptions can be immediately made (Cherry and Papoutsakis, 1990). Are the necessary assumptions and calculations for the small-scale (Kolmogorov) eddy size [Eq. (3)] valid in the case of bubble columns and airlift reactors? This has been addressed at some length by Kawase and Moo-Young (1990). They concluded that with some degree of caution, the calculations are relevant at least for large-scale, vigorously aerated (i.e., for high U_G values) reactors. But this is unlikely to occur in the bubble columns or airlift reactors that have been used so far to culture animal and plant cells (see Section V,A). However, as we shall discuss in Section V,A, all the available evidence so far suggests that stresses in the bulk liquid and stresses due to bubble motion away from the free surface do not harm cells in such reactors. The main damage appears to come from stresses during bubble breakup and draining foams at the free-reactor surface. So, at this point the Kolmogorov theory is not necessary for modeling or understanding call damage in bubble columns or airlift reactors. It may, however, become necessary at some point in the future when more is understood about cell-damage mechanisms.

C. What Fluid–Bead or Fluid–Cell Interactions May Cause Cell Injury in Turbulent Flows?

We will first consider interactions between a freely suspended "particle" (i.e., a free cell, a microcarrier bead, a cell aggregate, or a bead aggregate) and the surrounding fluid in the absence of bubbles that are detrimental to the cell. We will ignore bead–bead interactions and bead–internals interactions for now.

The experimental support for the existence of such detrimental interactions is consistent over a wide size range of biological "particles." In agitated bioreactors, protozoa cells [*Tetrahymena pyriformis* of 80-μm average diameter (Midler and Finn, 1966)] were severely damaged in an agitated vessel at high agitation intensities. It was assumed that bubble entrainment and breakup was not the predominant mechanism of cell damage. (A careful examination of the experimental protocol employed leaves us suspicious about the validity of this assumption; if this assumption is not valid, one must completely disregard the evidence coming from these experiments.) Croughan et al. (1987) and Cherry and Papoutsakis (1988) calculated that cell damage occurs when the eddy size calculated from Eq. (4) becomes approximately equal to the microcarrier bead site. For cells on microcarriers, Croughan et al. (1988) have used increasing bead concentrations to show that FS-4 cells on Cytodex 1 beads (average diameter of 185 μm) are damaged by forces due to bead–fluid interactions in addition to bead–bead interactions. The implication is that cells on microcarriers are damaged even at very low bead concentrations (where the bead–bead interactions become negligible; see Section III, D) when the Kolmogorov eddy size becomes approximately equal to or less than the bead size (Cherry and Papoutsakis, 1986, 1988, 1989; Croughan et al., 1987, 1988). A variety of freely suspended animal cells were shown to be damaged in turbulent pipe flows in the absence of any bubbles (Augenstein et al., 1971; McQueen et al., 1987), and it was calculated that cell damage becomes severe when the Kolmogorov-eddy size becomes smaller than the cell size of approximately 10 μm (McQueen et al., 1987). Finally, Kunas and Papoutsakis (1990a) have recently shown that in agitated bioreactors under conditions that carefully avoid the presence of bubbles and all other gas–liquid interfaces, damage of the hybridoma CRL8018 cells occurs at very high agitation rates (700 rpm and higher in their 2-L bioreactor) when the Kolmogorov-scale eddy becomes approximately equal to the cell size of 10–12 μm. It becomes clear that interactions of biological "particles" with eddies may result in detrimental effects, but what is the nature of these interactions? What stresses does a cell experience and at what frequencies during these interactions? What factors affect these interactions?

To address these questions, one needs a detailed description of the shear and normal forces a particle experiences in a turbulent-flow field. This is a formidable problem. Ultimately, some assumptions have to be made regarding the properties of the eddies that interact with the "particles." Cherry and Papoutsakis (1986, 1990) have discussed the forces that affect particles in such flows and have provided several literature references on the subject. We would like to briefly discuss four additional and more recent references. These do not merely strengthen the earlier analysis, but also provide direct experimental evidence about some key assumptions and a more detailed picture of the interactions between beads and eddies.

Kuboi et al. (1974) have carried out a detailed theoretical and computational analysis of the relative particle-to-fluid motion in a turbulent dispersion. In order to determine important constants for the expressions they derived, they used experimental data whereby the motion of both the fluid and particles were continuously recorded by a motion-picture method. The data was then treated by a Fourier analysis. The experimental data in conjunction with the theoretical analysis show that neutrally buoyant particles follow the motion of eddies of size larger than that of the particles. The implication here is that eddies smaller than the particle size may be finally dissipated on the surface of the particles when they collide with a particle. Thus these eddies may release all their energy on the particles on collision. On the other hand, eddies larger than the particles are responsible for little or no relative motion between the particle and the turbulent fluid.

Lee et al. (1988) summarized theoretical arguments from several investigators and from his own research to show that the important dynamic interactions between (spherical) particles and the fluid in a turbulent suspension is governed by the simple Stokes law of drag applied for the large values of the turbulent particle Reynolds number, where instead of the molecular viscosity, a turbulent equivalent viscosity must be used. He also presented detailed correlations for the calculation of this turbulent viscosity in terms of the particle size and concentration, the local flow turbulence Reynolds number, and the ratio of particle to fluid densities. In essence, one may use this turbulent equivalent viscosity to estimate shear forces and stresses on the surface of the spherical particle as in the case of a small particle Reynolds number [i.e., creeping flow; see Bird et al. (1960)].

Batchelor (1980) analyzed the relative motion between a small particle and the fluid in a turbulent flow in order to calculate the rate of mass transfer from the particle to the fluid. He assumed that the suspension is dilute so that there are no significant interactions among particles. He showed that the flow around the particle is a superposition of (1) the flow due to the velocity gradient in the ambient fluid and (2) a streaming flow due to a translational motion of the particle relative to the fluid, with a velocity proportional to the density difference between the particle and the fluid. For neutrally buoyant particles (as in the present case) the second contribution is zero. In his analysis, Batchelor justified and first used the Stokes equations for the velocity distribution near the particle (because of the locally small particle Reynolds number). Second, he used the properties of small-scale isotropic (statistically steady) turbulence (although the flow does not have to be either isotropic or homogeneous). If we use his Eq. (4.2) together with his equations $\langle V_\omega \rangle$ (parameter relating the mean motion of fluid elements relative to the particle size) $= 0$ and $\langle E_\omega \rangle$ (parameter of the turbulent motion in which the particle is immersed) $= 0.18 \ (\varepsilon/\nu)^{1/2}$ [for the notation, see Batchelor (1980)], we obtain the following expression for the tangential stress τ (dyn/cm^2) in the polar direction (θ is the polar angle) $\tau(\theta)$ on the spherical particle:

$$\tau(\theta) = 0.675 \rho_f (\varepsilon\nu)^{1/2} \sin(2\theta) \tag{6}$$

The maximum value for $\tau(\theta)$ is obviously obtained for $\sin(2\theta) = 1$.

Batchelor's expression poses no restrictions on the "particle" size, as long as the aforementioned assumptions are valid. His assumptions are apparently valid for dilute suspensions of both microcarriers and freely suspended cells.

Finally, Cherry and Kwon (1990) have presented an analysis to calculate the magnitude and frequency of shear stresses acting on a freely suspended animal cell in a turbulent flow field. They calculate that the maximal shear stress is given by

$$\tau_{max} = 5.33 \rho_f (\varepsilon \nu)^{1/2} \qquad (7)$$

They assumed that the eddies that are responsible for the shear stresses must completely surround the particle, thus questioning the validity of their expression for microcarrier systems. We note that the τ_{max} calculated from Eq. (6) is identical in functional form to the expression of Eq. (7) and differs by a factor of only about 8. For estimating the stresses on a cell or microcarrier under these agitation conditions, an order of magnitude calculation is all that can be reasonably expected. The results of Eqs. (6) and (7) can therefore be viewed as equivalent. We note that $\rho_f(\varepsilon/\nu)^{1/2}$ is the Kolmogorov-scale [i.e., corresponding to the expressions of Eqs. (1) and (2)] shear stress, which is calculated as the product of the viscosity and the shear rate. The shear rate can be obtained by dividing the expression of Eq. (2) by the length η of Eq. (1), as has already been pointed out by Papoutsakis and Kunas (1989).

For nondilute suspensions of microcarriers, the approach of Lee (1988) could possibly give a more accurate estimation, compared to Eq. (6) or (7), of the shear stresses acting on a microcarrier or particle.

D. Bead–Bead Interactions

The experimental evidence for the importance of bead–bead interactions in cell damage is very strong (Hu, 1983; Croughan et al., 1987, 1988, 1989), but modeling and predicting cell damage due to the interaction between beads is difficult. Even though we know that both the frequency and the severity of bead-to-bead and bead–fluid interactions are important determinants of cell injury, we do not know which of the two predominates under various agitation and bead concentration conditions. For example, Croughan et al. (1988) showed that intense agitation reduces the growth rates of cultures even at very low microcarrier concentrations. Cherry and Papoutsakis (1988) decided to correlate their data based on the turbulent collision severity (TCS) per bead (g cm^2/s^3) to characterize cell damage due to bead-to-bead interactions. The TCS was defined as

$$\text{TCS} = \frac{(\text{kinetic energy of interaction})\,(\text{interaction frequency/volume})}{\text{bead concentration}}$$

(8)

and represents the interaction energy per bead per unit time. A TCS expression was obtained using equations that estimate the relative velocity of the beads and the kinetic energy that characterizes all possible interactions between the beads. The relative velocity of the beads can be predicted by two methods. One is by using the velocity of the smallest eddies in turbulence, which is valid if the beads have nearly the same size and density of those eddies. This will give an "eddy-based" TCS. If the eddies are much larger than the beads, the relative velocity between neighboring beads can be predicted by a shear-based mechanism. Given two beads in a shear field, the relative velocity between the beads will equal the distance between the streamlines along which the beads are moving multiplied by the local velocity gradient [shear rate γ^* (s^{-1})] across the streamlines. With beads moving on streamlines less than one bead diameter apart, a collision can occur with the velocity of the collision on order $(\gamma^* d)$. Using a shear rate based on Kolmogorov-size eddies, a "shear-based" TCS can be calculated. Experimental data of growth rates and death rates using bovine embryonic kidney cells can be correlated quite well using either an "eddy-" or "shear-based" TCS expression. Specifically, the apparent growth rate decreased and the death rate increased with an increase in TCS. It has been established that both bead–bead collisions and eddy–bead interactions are important and that the former interactions contribute more to cell damage at higher agitation intensities (Croughan et al., 1988; Cherry and Papoutsakis, 1990).

Cells in microcarrier bioreactors are exposed to forces due primarily to the interaction of beads with individual small eddies and bead–bead collisions. The beads may also collide with the internal parts of the bioreactor (probes and impellers). Bead–internals interactions occur much less frequently than do bead–bead collisions, but with potentially higher severity. Available data, however, indicate that the bead–internal collisions do not cause substantial cell damage since the microcarriers appear to follow the fluid streamlines around objects with severe stagnation points (e.g., probes and impellers).

IV. FLUID-MECHANICAL EFFECTS IN MICROCARRIER CULTURES

A. Fluid-Mechanical Considerations in Nonporous Microcarrier Bioreactors

Considering the types of cell responses to different levels and frequencies of fluid forces, a variety of effects on cells in microcarrier reactors can be expected due to fluid-mechanical forces. Other than macroscopic cell death or reduction of cell growth, such effects have not been thoroughly investigated. Hu (1983) studied the effect of agitation on the final cell population and multiplicative increase over the seeding density of a human fibroblast line and found a sharp drop in relative growth extent at higher agitation intensities. Extensive data on cell damage and growth retardation at higher agitation intensities have been reported for both bioreactor and spinner-flask cultures. Generally, the approach used is to measure either the reduction of the extent of cell growth based on the maximum number of cells or the number of doublings, the apparent growth rate based on the increase of the number of attached viable cells, or a calculated "death rate."

Croughan et al. (1987) and Cherry and Papoutsakis (1988) correlated the data of Sinskey et al. (1981), Hu (1983), and their own data using the size of smallest turbulent eddies or the η/d ratio. For predicted eddy sizes below 100 μm, Croughan found that growth of FS-4 human fibroblasts was significantly reduced. They calculated η based on an ε calculation that assumes the agitation energy is uniformly distributed in the entire reactor-liquid volume. As mentioned in Section III,B, Cherry and Papoutsakis (1988) found that the eddy size should be calculated using the turbulent power dissipation rate in the volume surrounding the impeller. Using bovine embryonic kidney (BEK) cells, Cherry and Papoutsakis (1988) found that the cell growth rate decreased linearly with η/d starting at a ratio value of 1, with little growth observed at ratio values below 0.5. Using the total reactor volume for power dissipation, the growth rate reduction begins at an η/d of 1.8. This is similar to the results of Croughan et al. (1988). To further study the hydrodynamic effects on cells, Cherry and Papoutsakis *directly* measured the death rate of BEK cells by agitating the cells in a medium that did not support cell growth. The death-rate data gave results similar to the studies based on growth-rate measurements with a linearly decreasing death rate for η/d decreasing between 1.0 and 0.6. Croughan et al. (1989) refined the model by assuming that cell death is proportional to the Kolmogorov-eddy "concentration," which assumes that cell damage occurs when the eddy size is smaller than a critical eddy size. Similarly, the expressions for TCS can be used to correlate cell damage data. Unfortunately, the experimental data are not accurate enough to discriminate between these model expressions based on quantitative differences. In addition, the TCS and Kolmogrov-eddy "concentration" correlations primarily reflect the effect of changing one parameter, namely agitation intensity. TCS and η/d are both functions of ε, so the effect of TCS on the growth and death rates cannot be distinguished from the effect of η/d on the growth and death rates. Microcarrier bead concentration was varied to distinguish the bead–bead collisions from bead–eddy interactions. Based on experiments carried out at one agitation rate, Croughan et al. (1988) showed that bead–bead interactions are the predominant mechanism of cell damage for microcarrier concentrations above 4–5 g/L. The determination of the contribution of each damage mechanism to the overall cell damage is necessary for quantifying how variables such as viscosity, agitation intensity, and microcarrier concentration alter specific growth and death rates. Viscosity and bead diameter were also altered to determine the dependence of cell damage on these variables (Cherry and Papoutsakis, 1989; Croughan et al., 1989). Results show that the effect of viscosity on specific growth and death rate depends on the level of agitation, with increasing agitation amplifying the dependence of cell damage on viscosity. The effect of increased medium viscosity in reducing the specific death rate is amplified as the agitation rate is increased, with no effect on specific death rate when the agitation is below a critical level. The data show that there is a strong cross-parametric dependence of the death rate on the viscosity and the agitation intensity that can be characterized by the agitation input per unit fluid volume ε. If the data are to be modeled with a correlation-type expression, we have $q = K' \mu_f \beta \varepsilon \gamma$, with μ_f representing the medium viscosity (g/cm s). The data of Lakhotia and Papoutsakis (1992) show that β varies as ε varies, showing an inconsistency with all the aforementioned modeling efforts.

To improve the cell-damage correlations, a model based on the turbulent energy content of the eddies in the dissipation spectrum of turbulence has been developed accounting for cell death due to both bead–bead and bead–eddy interactions. In this model, the properties from a spectrum of eddies instead of the Kolmogorov-scale eddy size are used. This includes energy dissipation of the viscous as well as the inertial subrange. An expression that describes the entire universal-equilibrium range of the turbulent spectrum was used (the Pao–Corrsin model) given by (Pao, 1965; Hinze, 1975)

$$E(k) = A\varepsilon^{2/3}k^{-5/3}\exp\left[-1.5 A\nu\varepsilon^{-1/3}k^{4/3}\right] \qquad (9)$$

with $E(k)$ the spectrum function of turbulent kinetic energy (cm^3/s^2), A a constant experimentally determined to be 1.7, and k the wavenumber (cm^{-1}) of the turbulent spectrum. The specific death rate was taken to be proportional to the energy of eddies (in the viscous dissipation range) that cause cell damage by the bead–bead and bead–eddy interactions, and therefore proportional to the integral of $E(k)$

$$q = B \int_{k_c}^{\infty} 1.7\varepsilon^{2/3} k^{-5/3} \exp(-2.55\nu\varepsilon^{-1/3} k^{4/3}) \, dk = BI \qquad (10)$$

where I is a definite integral of turbulent kinetic energy spectrum (cm^2/s^2), q is the specific death rate (h^{-1}), and B is a proportionality constant (in s/cm^2) accounting for the dependence of the death rate on biological (e.g., cell fragility) and physicochemical parameters (e.g., bead concentration). The lower limit k_c is the wavenumber on (cm^{-1}) the order of $1/d$ with I representing the definite integral of the equation. The results of Lakhotia show that there is always a small specific death rate (q^*) present when using maintenance medium, even at the lowest agitation rates used. Additionally, death occurs only after a critical agitation rate (E_0) is surpassed. The model takes the form

$$q = BI \qquad I > E_0$$

$$q = q^* \qquad I < E_0 \qquad (11)$$

with E_0 [the minimum turbulent kinetic energy needed to injure cells (cm^3/s^2)] dependent on the cell and the attachment quality of the cell to the microcarrier along with other physiological variables that effect the resistance of the cell to hydrodynamic forces. The values of q^*, E_0, and B are determined experimentally. The model helps explain and predict the varying functional dependence of the specific death rates on the medium viscosity at varying agitation intensities. Their results (Lakhotia and Papoutsakis, 1992) suggest that increased viscosity decreases the death rates by a magnitude that depends on the agitation intensity. The protective effect of increased viscosity on the specific death rate is amplified as the agitation rate is increased. This emphasizes a cross-parametric effect of the viscosity and the agitation intensity on the death rates in maintenance-medium cultures. The values of the parameters k_c, E_0, and B were calculated to be 65 cm^{-1}, 0.044 cm^2 s^{-2}, and 4.0×10^{-3} for these cultures. For k_c, 65 cm^{-1} corresponds to an eddy size of 154 μm. Using this model, cell damage in maintenance medium cultures would be caused by the energy contained in eddies of size smaller or equal to 154 μm (compared to an average bead diameter of 185 μm).

Thus far, cell injury in microcarrier bioreactors has been assessed by studying cell death in media that do not allow cell proliferation and by growth rate reduction in regular growth media. In all of the aforementioned damage mechanisms, local shear and normal forces will injure the cells, but the cells will resist the shear because of its attachment to the microcarrier. However, the cell's membrane and cytoskeleton integrity can be partially affected, possibly damaging the protein bridges through which it attaches to the substratum. After repeated exposure to damaging conditions, either the protein bridges through which the cell attaches to the substratum are severely damaged and the cell detaches from the bead, and/or the cell membrane and components are severely damaged, detrimentally affecting cell integrity and proliferation. We will briefly discuss these possibilities in the next section.

B. Microcarrier and Cell Aggregation

Microcarrier and cell aggregation are two other phenomena that can affect the growth of cells on microcarrier beads. Consequently, cells in certain parts of the aggregates may be starved of oxygen and other nutrients. On the other hand, cells that are part of an aggregate may create a potentially beneficial microenvironment because of the release of many growth factors and other glycoproteins, and/or also as a result of cell–cell and cell–extracellular matrix interactions.

The phenomena of bead bridging has been commonly observed in microcarrier cultures (Mered et al., 1980; Varani et al., 1983; Scattergood et al., 1980; Cherry and Papoutsakis, 1988, 1990). Bridging occurs when two beads collide and one or more cells at the point of impact stick to the other bead. It has been suggested that the formation of a bridge requires the impact of a bead with a fairly high cell coverage to a bead with low or zero coverage (Cherry and Papoutsakis, 1988). A ring of three or four crescent-shaped cells then forms between the beads, leaving a bare circle 20–50 μm in diameter where the beads are in actual contact with each other. In some cases, the two beads are not in actual contact and are connected through a double layer of cells or a large cellular clump. Clump formation is more prevalent at lower levels of agitation, with clump size increasing as agitation speed decreases. It has become evident that bridging increases linearly with time, but it is not known if this is the general kinetic form of bridging. Bridging was found to decrease with increased agitation presumably because higher levels of agitation reduce the probability of aggregate formation and increase the probability of breaking the formed bridge. At higher levels of agitation, the initial collision between two beads may be more energetic, so there is less likelihood of a cell adhering to the colliding bead, and the bridges that do form may be broken apart more quickly. The clumps tend to be more compact rather than elongated or branched in structure as is found with random attachment. This suggests a greater removal of a single-bridged bead connection compared to a multiply connected one. The formation of large clumps (6–12 beads) is fluid-mechanically equivalent to having microcarriers with effective diameters 2 or more times larger than the diameter of the individual beads. One would expect from the eddy/bead size ratio that the cells growing on the outer surface of these large clumps would be subject to hydrodynamic damage. However, Cherry and Papoutsakis (1988) found no obvious visual evidence of this with bovine embryonic kidney cells, although the measured net growth rate was lower at minimal agitation levels where clump formation was significant. Dissolved oxygen and pH levels were controlled, and the clumps were only up to about 10 beads in size, so mass transfer problems should not have been the cause of this decreased growth rate. The actual mechanism may be the death of cells on the clump exterior, or death of bridge cells when fluid forces or collisions manage to break apart a bridge. Using transformed Chinese hamster ovary (CHO) cells, Borys (1990) provided visual evidence that when bead aggregates are formed, the cells tend to disappear from the externally exposed surface of the microcarriers and tend to grow in the bridging area between beads, thus forming large cellular masses as time progresses. Photographic evidence suggests that the cellular masses grow more elongated and larger with time, until one of the beads is removed and the cellular mass becomes more spherical and attached to only one bead. Eventually, some cellular masses detach completely from the beads and cells grow in this aggregate form with no attachment to solid support. All available evidence from our laboratory shows very high viability for the cells in these large cellular masses, which is somewhat surprising considering their size. Visual evidence suggests (Borys, 1990) that the cells grow preferentially as part of these cellular masses rather than on the microcarriers. We theorize that this is due to two reasons. First, cells are less susceptible to fluid (local shear) forces because these masses are elastic and the cell aggregate can deform under a stress without transmitting substantial stresses to the individual cells. Second, cells in these cellular masses create a potentially beneficial microenvironment due to release of autocrine growth factors. As long as there are no mass transfer problems either for the cell nutrients or for the protein products, these cellular masses are apparently beneficial for bioprocessing since the cells in these aggregates can condition their local microenvironment more effectively (growth factors) and since these aggregates can be retained in bioreactors for prolonged protein expression more easily than single cells and without the need to add more microcarriers.

In terms of modeling, the presence of bead aggregates complicates the situation because of the larger variation of effective microcarrier size. This will alter the characteristics of the eddy–bead, bead–bead, and bead–internals interactions, and thus cells will experience an even larger variation of forces depending on whether they are part of a cell aggregate, bead aggregate, or attached on a single microcarrier.

V. FLUID-MECHANICAL EFFECTS ON FREELY SUSPENDED CELLS

A. Biological Aspects

Shear effects in laminar flows on various blood and tumor cells have been widely studied in the biomedical engineering literature [see, e.g., Chittur et al. (1988), Hellums and Hardwick (1981), Martin et al. (1979), McIntire et al. (1987), McIntire and Martin (1981), O'Rear et al. (1982), Petersen et al. (1988)]. This subject is discussed in some detail in other chapters of this volume, so only a brief and partial discussion will be included here for the sake of completion. The objective of these studies is to assess the effect on blood cells of various types of blood flow in the human body, in circulatory assist devices, and artificial organs. Such blood flows are complex, but are simpler than the flows in agitated bioreactors. For this reason, flow effects have been investigated using devices that produce well defined flows and measurable shear stresses on the entire sample volume. These include stagnation-point flows, laminar flows through cylindrical capillaries, cylindrical, cone-and-plate, cone-and-cone Couette flows in viscometers, and plane Couette flows (Hellums and Hardwick, 1981; Martin et al., 1979; McIntire and Martin, 1981). In all cases, only the fluid shear stresses (bulk stresses) are well defined and measurable. The membrane stresses experienced by the rotating cell depend on the type of flow (Martin et al., 1979). Even though they are difficult to calculate precisely, they are reproducible and result in reproducible biological effects. Two key issues that were settled a number of years ago in the biomedical-engineering literature are that normal stresses are of secondary importance, and that mechanical-stress effects on a given cell in steady flows are dependent on the bulk shear stress rather than the shear rate.

Reports on shear effects on cells for biotechnological applications are less numerous. Mechanical-stress effects on insect cells in agitated and aerated suspensions, in a bubble-column bioreactor, and in a viscometer have been reported (Tramper and Vlak, 1986, 1988; Tramper et al., 1986,

1988). Handa-Corrigan and co-workers (Handa et al., 1987; Handa-Corrigan, et al., 1989) have studied the damage mechanisms of various suspended cells due to gas sparging in bubble-column reactors. Shear damage of hybridoma cells in viscometric flows has been examined recently by various investigators (Abu-Reesh and Kargi, 1989; Petersen et al., 1988; Schuerch et al., 1988; Smith et al., 1987a, 1987b; Ramirez and Mutharasan, 1990). These studies confirmed that, as in the blood-cell studies, cell damage increases with the time of exposure to shear and the level of shear stress. Flow effects in various channel flows on mouse myeloma cells, HeLa cells, and mouse L929 cells, have also been reported (Augenstein et al., 1971; McQueen and Bailey, 1989; McQueen et al., 1987). Damage of suspended animal or protozoa cells in agitated reactors has been reported from specific or indirect studies [Midler and Finn (1966) (for protozoa cells), Backer et al (1988), de St. Groth (1983), Dodge and Hu (1986), Gardner et al. (1990), Kunas and Papoutsakis (1989, 1990a, 1990b), Oh et al. (1989)].

A brief review on the effect of fluid shear on the structure and function of freely suspended blood cells and on hybridoma and other cells of biotechnological interest will be given below. All studies on freely suspended blood cells were conducted in devices that produce well-defined shear. It has been shown that shear forces that will not lyse normal erythrocytes produce subhemolytic effects on normal erythrocytes including changes in morphology, deformability, metabolism, and lifespan (McIntire and Martin, 1981; O'Rear et al., 1979, 1982; Frangos et al., 1985). Sublytic shear stresses on platelet cells may result in alterations of morphology, aggregation properties, release reactions that include the liberation of acid phosphatase, serotonin, norepinephrine, platelet factor 3 and ADP, and impaired serotonin-uptake function (Hellums and Hardwick, 1981; McIntire and Martin, 1981). The release products of sheared platelets modulate the shear-induced aggregation of and β-glucuronidase release by polymorphonuclear leukocytes (PMNLs) through a complex mechanism apparently involving the C-12 or C-5 lipoxygenase activity (Rhee and McIntire, 1986b; Rhee et al., 1986). It was found that high levels of fluid mechanical stress (600 dyn/cm^2) cause lysis of leukocytes or major alterations in cell morphology, including large vacuoles, fewer granules, and condensed chromatin (Dewitz et al., 1979). Shearing at 300 dyn/cm^2 (sublytic stresses) on leukocytes cause slight changes in cell morphology, alter the cell-membrane permeability, and result in increased participation in microaggregate formation, and impaired chemotaxis and random migration. These sublytic stresses also cause lysosomal enzyme release and degranulation, decreased chemilluminescence or phagocytosis, and increased hexosemonophosphate shunt activity (Dewitz et al., 1979; Martin et al., 1979). Micropipette studies indicate that when a stress is suddenly applied to a leukocyte (all types), it responds with an initial deformation immediately, and then continues to deform more slowly. After release of stress, the cell will return to its original spherical shape. It has been shown that this viscoelastic behavior is associated with the cytoplasm of the cell (Schmid-Schonbein et al., 1981). Sublytic shear stimulates the arachidonic acid metabolism in both platelets and PMNL cells (McIntire et al., 1987; Rajagopalan et al., 1988). When T and B lymphocytes and monocytes were exposed to defined laminar shear of 100 and 200 dyn/cm^2 for 10 min, the controlled exposure to these sublytic shear stresses resulted in alterations that affected the proliferative response of the T-lymphocyte population in a cell-density-dependent fashion (Chittur et al. 1988).

Physiological or biochemical studies involving cells of biotechnological interest are fewer. Al-Rubai et al. (1990) demonstrated that intense hydrodynamic stress in a bioreactor inhibits DNA synthesis and alters the metabolism of hybridoma cells. Ramirez and Mutharasan (1990) have recently reported that higher plasma membrane fluidity of hybridoma cells, as measured by steady-state fluorescence anisotropy, correlates with increasing shear sensitivity in viscometric Couette flows. *Membrane fluidity* is a term used to represent the degree of packing and the motions of the various components of a biological membrane. They used benzyl alcohol, cholesterol enrichment, and temperature changes to alter the membrane fluidity.

One of the earliest and the most detailed phenomenological studies on the effects of laminar viscometric shear on hybridoma cells was published by Petersen et al. (1988). Samples of the hybridoma cultures were subjected to well-defined laminar shear in a specially designed Couette viscometer. Exposure of the samples to increasing levels of shear stress (0–50 dyn/cm^2 for 10 min) or time of exposure to shear (50 dyn/cm^2 for 0–10 min) resulted in higher levels of cellular damage and death. Cell death in the viscometer was shown to exhibit trends similar to cell death caused by excessive agitation in spinner flasks, suggesting that viscometric shear can be used to model some of the fluid mechanical aspects of damage to cells caused by agitation. Cells cultured with low levels of fluid stresses (T-flask and slowly stirred spinner cultures) were more sensitive to shear than cells from rapidly agitated cultures (see Fig. 2). This shows that cells respond and adapt to some extent to the fluid environment they are exposed to. The issue of cell adaptation to high levels of shear was further pursued by Petersen (1989), who demonstrated this through subculturing the cells in progressively more intensely agitated spinner cultures. Petersen et al. (1988) also showed that cells from either the lag or stationary phases of batch cultures were more sensitive to mechanical damage than exponentially growing cells (see Fig. 2). Accumu-

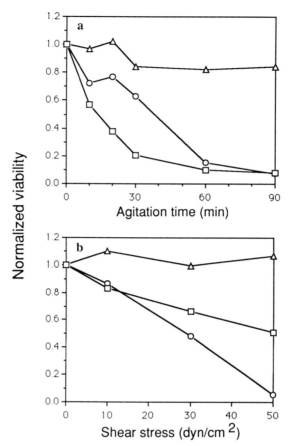

FIGURE 2   Comparison of the effects of (a) 400-rpm agitation in a spinner flask and (b) increasing levels of well-defined shear stress (10-min exposure) in the viscometer on CRL8018 hybridoma culture viability. Spinner-flask cultures were seeded with cells from routine T-flask cultures that were 3 days old. The agitation rate was 100 rpm. Cell samples were taken from the spinner-flask cultures at the times shown, and viability was measured before and after shearing in the viscometer (a). The agitation rate of the spinner flask cultures was then increased to 400 rpm and the culture viability was measured during the period of high agitation (b). Cultures were compared at times [O] 24 h; [△] 48 h; and [□] 72 h after subculture. [Taken from Petersen et al. (1988).]

lation of ammonia and changes in pH of the batch culture can contribute to this increase in shear sensitivity. However, the effects of ammonia and low pH are detrimental to the cells only on prolonged exposure. In a second paper Petersen et al. (1990) examined the possible reasons that may account for their observation that cells are more shear-sensitive in the lag and stationary phase of growth. They employed fed-batch and continuous cultures to show that the shear sensitivity is independent of the growth rate or the metabolic state of the cells, as long as the cells are actively growing. This suggests that the reasons for increased fragility in the lag and stationary phases is related to the fact that the cells are not actively growing due to adaptation to the new culture conditions or the exhaustion of nutrients and accumulation of inhibitory metabolites, respectively. The variability of cell sensitivity with the stage of the batch culture was also later confirmed by Lee et al. (1988) and by Ramirez and Mutharasan [1990].

Papoutsakis et al. (1991) pursued the issue of the possible factors that affect the shear fragility of cells by examining the possible effects of energy metabolism and of the cytoskeletal integrity. They employed several effectors (drugs) to specifically probe the involvement of the cell's cytoskeletal structure and energy metabolism in the ability of cells to resist shear injury. Cell injury was quantitated by the fractional normalized cell viability and the release of lactate dehydrogenase after exposing the cells for a short time period (10 min) to well-defined, laminar shear in a rotational Couette viscometer. Treatment of their hybridoma cells with either cytochalasin E or B, which disrupt the microfilament (actin) network, results in a marked increase in shear sensitivity. On the contrary, treatment with colchicine, which disrupts the microtubule network, did not affect the cell's shear fragility. When glycolysis was inhibited by treatment with deoxy-D-glucose, or when respiration was separately inhibited with KCN treatment, small effects were observed on the cell's shear sensitivity. A combined inhibition of glycolysis and respiration resulted in larger increases in shear injury. These results were further strengthened by additional studies using the more potent glycolysis inhibitor iodoacetate (results to be published). In view of the fact that the dynamic integrity of the actin network is energy-dependent, these results show that the cytoskeleton is apparently a key determinant of the cell's ability to resist fluid forces. Therefore, when the energy metabolism is either inhibited or downgraded, this affects the cytoskeletal integrity and thus increases the shear fragility. This explains the increased fragility during the lag and stationary phases or in the presence of inhibitory metabolites or low pH.

B. Damage of Freely Suspended Cells in Bubble Columns and Airlift and Agitated Bioreactors

The damage mechanisms of freely suspended cells in bioreactors due to mechanical stresses are only partially understood. For bubble-column (and, thus, possibly, airlift) bioreactors, a reasonable qualitative picture is emerging as a result of the work by Tramper et al. (Tramper and Vlak, 1986, 1988; Tramper et al., 1986, 1988) and Handa-Corrigan et al. (Handa et al., 1987; Handa-Corrigan et al., 1989). Handa-Corrigan and co-workers used hybridoma, myeloma, and baby hamster kidney cells in bubble columns to study the damaging effects of bubbles. The critical "damage-causing" area was deemed to be the bubble disengagement portion at the gas–liquid surface. They have more recently theorized that the damaging effects of bubbles and gas–liquid interfaces on cells are due to the rapid oscillations of cells caused by bursting bubbles, and also to the shear forces in draining films of unstable foams. They varied parameters such as bubble-column height, superficial gas velocity, bubble size, and concentration of foam stabilizers to visualize the possible mechanisms of cell damage by bubbles. Handa-Corrigan and co-workers (Handa, 1986; Handa-Corrigan et al., 1989) also found that small bubbles are considerably more damaging to cells, but only a small variation in bubble size was investigated (all were larger than 1.6 mm). Tramper et al. (1986) pointed out that the forces at the location where bubbles disengage from the sparger are potentially damaging to the cells, in addition to the forces at the free surface of the bubble column. In contrast to the conclusions of Handa-Corrigan et al., Tramper, and co-workers have shown that cell damage depends very little on bubble size (at least for bubbles larger than 2 mm). This apparent inconsistency is probably due to the fact that other parameters affect cell damage, and that bubble size by itself is not a good correlator of cell injury. Tramper et al. (1988) proposed a simple model to correlate cell death to bubble-column parameters. They correlated cell death to the air flow, the geometry of the bubble column, the size of the sparged air bubbles, and a hypothetical "killing volume" around the bubble in which all viable cells are killed. They showed that the shear forces associated with the rise of bubbles from a sparger through the culture medium did not create sufficient shear forces to damage the cells, similar to the conclusion by Handa-Corrigan et al. (1989). They also concluded that the "killing volume" is independent of airflow and the height of the bubble column, and is proportional only to the size of the bubble. The death-rate correlation was compared to the specific surface of the air bubble and led them to conclude that the height/diameter ratio was the key parameter to adjust to minimize cell damage while supplying sufficient oxygen to the cells.

C. Freely Suspended Cells in Agitated Bioreactors: Interaction with Bubbles is Again the Source of Cell Damage

Cell damage in agitated bioreactors with or without aeration had not been systematically examined until very recently. Cell damage begins at agitation rates anywhere between 150 and 350 rpm in most agitated bioreactors and spinner flasks. These data are consistent with the expectation that different reactor-vessel designs cause different mechanical stresses to the cells, and that different cell types exhibit different responses to such forces. In the case of freely suspended cells, the cell–cell or cell–solid-surface interactions are not plausible damage mechanisms because of the small size and inertia of the cells. On the other hand, interactions of cells with Kolmogorov-size eddies is a plausible mechanism, at least for protozoa cells (approx. 80 μm in diameter) (Midler and Finn, 1966), as discussed by Croughan et al. (1987) and Cherry and Papoutsakis (1988), assuming, as we pointed out in Section III, C, that no bubble entrainment occurred under the conditions that Midler and Finn (1966) carried out their experiments with the protozoa cells. For the aforementioned bioreactors and agitation intensities between 150 and 350 rpm, the Kolmogorov-eddy size can be calculated to be at least 6–10 times larger than the typical sizes (9–15 μm in diameter) of animal cells. It is indeed unlikely that cell damage in this case is caused by this mechanism. A different mechanism must therefore be explored as the cause of cell damage under these bioreactor conditions. This was recently elucidated by Kunas and Papoutsakis (1990b).

Kunas and Papoutsakis (1990b) employed two identical 2-liters agitated biorectors with round bottoms. The bioreactors were operated in parallel with the one serving as the control for the effects observed under various conditions in the other. On a close visual and photographic examination, they observed that cell damage is initiated at agitation levels (160–200 rpm) whereby air-bubble entrainment and breakup at the bottom of the formed vortex are initiated, despite the absence of any foaming. A vortex is formed at these agitation rates since the bioreactor is only partially baffled by dissolved oxygen, pH, and temperature probes. Under these conditions, few but relatively large (1–3 mm in diameter), easily deformable bubbles appear to be the source of cell damage. This was confirmed as follows. Once the vortex and the associated air entrainment were either reduced, by increasing the liquid volume in the reactor, or completely eliminated, by filling the reactor with liquid completely and using membrane oxygenation, the agitation could be increased up to 800 rpm before severe cell damage was observed. At 800 rpm, the Kolmogorov-eddy size is comparable to the cell size (see Section III, C). At high values ($\geq$ 300 rpm), and in the absence of a gas phase, air drawn in the bioreactor leads to the formation of an enormous number ($\sim$ 5000 bubbles per milliliter) of very small (50–300-μm) bubbles. These bubbles are rigid and nondeforming or coalescing. Under these conditions, the fluid suspension in the bioreactor has the appearance of a whitish emulsion, which is characteristic of bacterial fermentations. Kunas and Papoutsakis (1990b) found that these very rapidly moving bubbles are not detrimental to the cells at agitation rates below 600–700 rpm (see Fig. 3). These experiments show that cell damage in agitated bioreactors is due to two distinct fluid-mechanical mechanisms depending on the agitation rate. The first is present at relatively low agitation rates only when there is a gas phase, and is associated with vortex formation accompanied by bubble entrainment and breakup. This situation is typical of all experiments that have been reported in the literature, and also of large-scale operations. The second mechanism prevails in the absence of a vortex and bubble entrainment and only at very high agitation intensities [in the Kunas and Papoutsakis (1990b) case above 600 rpm]; in this case, cell damage is caused by stresses in the bulk turbulent liquid, and correlates with Kolmogorov-eddy sizes similar to or smaller than the cell size. These results also show that bubble size is not a relevant quantity to correlate cell damage. In fact, these experiments show that the presence of many, fast moving bubbles are not necessarily detrimental to the cells, and may suggest that high agitation rates with direct sparging (of proper quality, which at the present is not known) may be perfectly compatible with the growth of suspended animal cells. These conclusions are consistent with the experiments of Oh et al. (1989) and Smith (1990). Smith (1990) used baffles in his 2-liter-capacity bioreactor in order to uncouple the effects of turbulent fluid forces and gas-entrainment effects on the cells. A decrease of gas entrainment in the reactor from 14% to 2% allowed suitable growth conditions for the hybridoma cells used.

In summary, it is now clear that in most cases cell damage in agitated bioreactors is solely the result of air entrainment and bubble breakup. This is not inconsistent with the fact that cell damage in viscometric laminar-shear flows shows similar trends with cell damage in agitated reactors as was earlier demonstrated by Petersen et al. (1988). Indeed, the actual stresses that injure or kill cells are likely to be shear stresses associated with bubble breakup and related interfacial phenomena. So, in both viscometric flows and during bubble breakup, shear stresses acting on the cells appear to be the source of cell injury. In this light, the mechanisms that cause cell damage in agitated bioreactors are apparently *qualitatively* similar to those in bubble-column (and airlift reactors). Although there is little understanding of the forces released and shear stresses created during breakup, the available information and a possible scenario that may account for cell injury under such conditions are discussed in the next section.

D. Bubble Breakup, Thin Films, and Rheological Properties of Interfaces

As previously discussed, the key mechanism of cell damage in agitated and bubble-column bioreactors are the mechanical stresses generated near

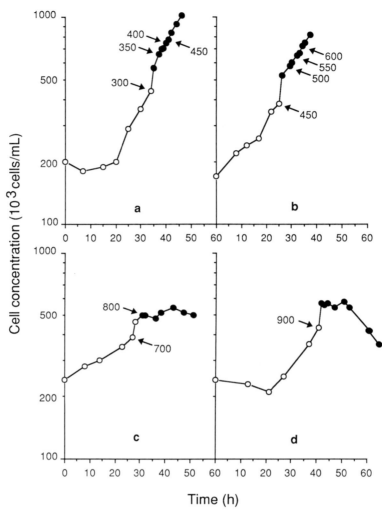

FIGURE 3   Batch growth curves for hybridoma cultures grown in a 2-L agitated bioreactor without a gas headspace. Oxygenation and pH control was accomplished via silicone tubing. Cultures were stirred at an agitation rate of 60 rpm until a cell concentration of $4-5 \times 10^5$ cells/mL was obtained. The agitation was increased to the given values at the points indicated by the arrows. During the course of each run, large amounts of bubbles ranging in size from 50 to 300 μm were entrained into the culture medium, although a vortex was absent. The filled symbols represent the points of exponential growth cell growth used to calculate the apparent growth rates of (a) 0.0511 ± 0.0042 h^{-1}, (b) 0.0417 ± 0.0035 h^{-1}, (c) 0.0015 ± 0.0047 h^{-1}, and (d) 0.0187 ± 0.0084 h^{-1}. The error estimates represent the 95% confidence limit for a regression parameter. [Taken from Kunas and Papoutsakis (1990b).]

the bubble interface during breakup. Cell damage *may* also occur as a result of the formation and coalescence of bubbles. Understanding the mechanisms of these phenomena and the related rheological properties of the thin-film bubble interfaces is crucial for understanding, quantitating, and preventing cell damage in bioreactors. Bubble coalescence and breakup have been reported in the literature, but relatively little of that information is useful in the context of cell damage. In this context, one must understand the factors that affect the frequency and location of bubble breakup, the magnitude of the shear stresses in the neighborhood of the breaking bubble, and the concentration of the collected cells near the bubble interface at the time of the breakup. These factors are important since it has been shown by viscometric and bioreactor studies that the cell death rate is proportional to the frequency and magnitude of the shear stresses experienced by the cells.

Bubbles formed in the bulk of the liquid, whether in a bubble-column or agitated bioreactor, eventually move to the liquid surface. Bubbles released from an air sparger in an intensely agitated bioreactor will flow through the bulk turbulent liquid before reaching the liquid surface. It has been suggested that bubbles rising in the bulk turbulent liquid of an agitated reactor may break up when the exciting frequency of eddy shedding matches the natural frequency of the bubble (Elzinga and Banchero, 1961; Hu and Kintner, 1955). Sevik and Park (1973) hypothesized that the magnitude of bubble response to external pressure increases as one of the natural oscillation frequencies of the bubble is reached. They related the natural frequency of a bubble undergoing small-amplitude oscillations to the characteristic frequency of turbulent flow in order to obtain an expression for the lowest Weber number where breakup will occur. Analogous to cell damage caused by eddies of scale lengths approaching the size of the cell and microcarrier beads, it has been postulated that the energy associated with eddies of scale lengths less than the diameter of the bubble are responsible for bubble breakup, whereas eddies larger than bubbles merely transport the bubbles.

In bubble columns and agitated bioreactors, the bubble-bursting phenomenon at the liquid surface is important. When a bubble reaches the surface, a hemispherical film cap (Azbel et al., 1979; MacIntyre, 1972, Prins and van't Riet, 1987) is formed with the typical film-cap thickness of $1-10\ \mu\mathrm{m}$. The film of the cap drains as a result of gravity and film-curvature-generated suction. When the film cap drains to a critical thickness (typically less than $0.1\ \mu\mathrm{m}$), the probability of generating a hole in the film cap and bubble rupture increases. Azbel et al. (1979) suggested that the bubble-rupture event is due to the growth of film-cap oscillations from various physical fluctuations. Briefly, when the natural frequencies in the bubble cavity and film cap approach each other, the coupling of the film cap to the bulk liquid becomes capable of causing bubble rupture. Once a hole in the film cap has been formed, the decreasing pressure within the bubble causes a flow (typically at velocities of 1–50 m/s) of the surrounding liquid into the bubble "crater." If the bubbles are small enough, the potential energy at the bottom of the bubble crater is high enough to lead to the ejection of an upward liquid jet from that area. The jet arises above the liquid surface and disintegrates into droplets before disappearing into the liquid. The shear stresses generated near the surface of the collapsing bubble are apparently damaging the cells that have been collected near the bubble. The situation is more complex when bubbles collide and deform on the surface before breakup, but for our purposes we will assume that the factors that affect the breakup process and its associated shear stresses can be sufficiently analyzed by considering the breakup of a single bubble.

The highest shear stresses generated near the bubble during breakup are apparently due to the fast velocity of the collapsing cap film down the bubble cavity (MacIntyre,1972). Since animal cells are probably too large ($9–20$ μm) to be incorporated in the bubble interfacial film, the cells experience stresses generated near the collapsing thin film. These stresses will be proportional to the bulk-liquid viscosity and the rate of strain. Detailed calculations for the latter can be carried out based on boundary-layer theory (MacIntyre, 1972). However, in view of the many uncertainties and assumptions that are necessary for such calculations, we will take the rate of strain to be proportional to the film velocity, since the bulk liquid velocity away from the collapsing bubble surface is relatively small. Theoretical and experimental analyses show that the maximum velocity of the collapsing-bubble thin film is (MacIntyre, 1972)

$$v_{\text{col}} = O\left[\left(\frac{2\sigma}{\rho\delta}\right)^{0.5}\right] \tag{12}$$

where O stands for order of, σ is the surface tension (dyn/cm), ρ is the density, v_{col} is the maximum velocity of a collapsing bubble thin film (cm/s), and δ is the thickness of the collapsing film (cm). The elasticity of the bubble surface and the dynamic adsorption of surface-active molecules are likely to change both σ and δ as the bubble collapses, and so a precise calculation of the collapsing velocity is not possible (MacIntyre, 1972). Since film drainage, which leads to film thinning and rupture, is affected by both the curvature (bubble size and shape) and rheological properties of the bubble surface, it is clear that the rheological properties of the bubble surface (elasticity, surface viscosity, etc.) affect both the likelihood of breakup and the severity of the shear stresses generated near the breaking bubble. For example, film thinning and drainage is more rapid in a pure liquid. Several impurities, monomolecular films and dissolved material (such as proteins and chemical additives) generally stabilize the bubble resulting in slower liquid drainage and a lower frequency of bubble rupture. Rigid bubble surfaces will thin more slowly, and once ruptured, will flow more slowly, thus generating smaller shear stresses. The frequency and severity of the bubble breakup mechanism at and away from the liquid surface and the associated stresses are affected by the bubble rigidity (a function of bubble size and surface active agents), surface tension, and rheological properties at the interface. These parameters are affected by the medium composition, sparger design, and the fluid-mechanical characteristics of mixing in the particular reactor employed. The bulk medium viscosity may also play an important role in transmission of forces.

Bubble coalescence becomes an important process to consider when bubbles are larger than about 100 μm, since attractive forces (termed Bjerknes forces) are present between pulsating bubbles. The attractive forces can be explained by noting that a bubble tends to move against a pressure gradient with an acceleration proportional to its volume. The bubbles will drift in the direction that the pressure gradient has when the bubble volume is the greatest. A quantitative description has been given by Prosperetti (1982) for variation of the pressure field only slightly over a distance comparable to the bubble size. The time necessary for the coalescence of two bubbles is given by balancing secondary Bjerknes forces with the Stokes drag forces acting on a bubble. An equation for an estimation of this time is given by the following correlation if the two bubbles considered for coalescence are the same size:

$$t_{\text{coalescence}} = \frac{8\pi^2 \mu_L R_b}{\rho V_0^2 \phi_\omega^2} x(0) \tag{13}$$

where R_b is the bubble radius (cm), V_b is the bubble volume, ϕ is the gas-void fraction, π is the surface pressure (n/m^2), ω is the frequency (or oscillation in liquid), and x represents the initial separation $|x_1 - x_2|$ between the two bubbles. With the volume term in the denominator, the time necessary for coalescence decreases with increasing R_b as R_b^{-5}. An idea of the orders of magnitude of $t_{\text{coalescence}}$ for pure water is given by Prosperetti (1982) for water with an initial separation of 2 mm (this would relate to about 100 bubbles per milliliter). With $\omega = 2\pi \times 20$ kHz and $\phi = 0.1$, $t_{\text{coalescence}}$ of 2.3×10^8 s, 2300 s, and 0.023 s are obtained for R_b of 1, 10, and 100 μm, respectively.

Coalescence of two or more bubbles may occur if (1) bubbles are rising in line (the upper bubble shelters the lower bubble), (2) there is forced contact of a bubble growing from an orifice with the previous departing bubble, and (3) the rise of a bubble toward a free surface where the surface can be seen as a second, infinite bubble (de Vries, 1972; Chesters and Hofman, 1982). It has been suggested that two processes are in competition in the determination as to whether two bubbles coalesce. When coalescence occurs, the liquid between the bubbles is squeezed out, accompanied by a flattening and dimpling of the bubble surfaces that causes an increase in surface area. When the residual film reaches thicknesses of order 1 μm, van der Waals pressures become dominant and a hole is rapidly formed. Surface tension expands this hole and the bubbles become one (Chesters, 1975). The kinetic energy of the system decreases with an increase of the free energy (from increased surface area), and the bubbles decelerate and bounce apart if the van der Waals forces are not sufficient for hole formation. This brief discussion serves as only a perspective of the bubble coalescence phenomenon since a detailed analysis would become very involved if a serious attempt at completeness is made. The work of Oolman and Blanch (1986), Prince (1989), and Prince and Blanch (1990) gives detailed studies on bubble coalescence phenomena in bioreactors.

The term "bubble rigidity or stiffness" is used to characterize bubbles that do not coalesce or break easily (Prins and van't Riet, 1987; MacIntyre, 1972; Phillips, 1977). It is a useful concept especially in the context of cell damage due to bubble breakup and/or coalescence. The reasons why rigid small bubbles do not coalesce has been extensively addressed by Sebba (1987) and Prosperetti (1982), to name only a few. Bubble rigidity depends on the size as well as on the properties of the interfacial film (Joly, 1972a, 1972b; MacIntyre, 1972). Rigidity can be viewed as being inversely proportional to some power of the bubble radius R_b, depending on the fluid-mechanical situation (MacIntyre, 1972). Thus, small bubbles are more rigid than large bubbles. It has not been well established which surface rheological property is best suited for characterizing bubble rigidity, and it is likely that different properties will be more useful for different uses of the rigidity concept. All of them are in some way related to one another either through basic fundamental relationships or on the basis of experimental data. Joly (1972b) suggests that the shear (as opposed to the dilatational) surface viscosity η_s (g/cm s) and the surface elastic (or shear) modulus G are very useful for characterizing bubble rigidity (and the associated bubble and foam stability) in the present context. Higher η_s and G values characterize more rigid bubbles. Prins and van't Riet (1987) suggest the surface dilatational modulus, $E = d\sigma/d(\ln A)$, to characterize bubble rigidity (A is the bubble surface area); however, E may be suitable only for characterizing expanding or contracting bubbles, and not for characterizing bubble rigidity to breaking. The surface dilatational modulus is a measurement for the resistance against compressional or dilatational deformation. The higher the surface dilatational elasticity, the more rapidly the restoration of the uniformity of the surface tension of the surface (Lucassen and van den Tempel, 1972). Theoretical considerations and experimental data show that η_s and G are related. In fact, both increase as σ decreases (or as π, the surface pressure, increases) (Joly, 1972a, 1972b; MacRitchie, 1978; Phillips, 1977). In this sense, either of the two properties appears to be suitable as a convenient measure of bubble rigidity, in addition to the bubble size.

In addition to parameters affecting bubble breakup and coalescence, the collection of cells near the bubble interface becomes an important facet to consider. This process is affected by the bubble interfacial properties, which are dependent on the dynamic accumulation of surfactants and the composition of the culture medium. A review of chemical additives effective in protecting cells from bubble breakup damage will further emphasize the importance of interfacial properties.

E. Chemicals That Protect Cells against Fluid-Mechanical Damage and the General Nature of Their Effect

The search for chemical additives to protect cells from fluid-mechanical damage started over 30 years ago with the pioneering work of Earle et al. (1954), McLimans et al. (1957), Swim and Parker (1960), Runyan and Gayer (1963), and Kilburn and Webb (1968). Among the additives that have been used, serum and the pluronic family of nonionic surfactants are the best documented and most widely studied. Several other additives have also been employed, including other polyalcohols, derivatized celluloses, cell-derived fractions, and proteins.

As we discussed in Sections V,B and V,C, all evidence presented thus far shows that damage of suspended cells in agitated and/or aerated bioreactors is due to the interactions of cells with bubbles and rearranging gas–liquid interfaces. Thus, all additives that have been found to protect suspension cells from fluid-mechanical damage either decrease the fragility of the cells (by a nutritional or other biological mechanism) or affect the forces on the cells due to their interactions with gas–liquid interfaces (by a physicochemical mechanism).

Bryant (1966) studied the use of chemically defined media by using a protein-free medium that successfully allowed various cells to grow in static cultures, but he soon realized that the same cells cultured in the serum free medium but grown in shaker cultures would lyse within 2 or 3 days after inoculation. Earlier, derivatized celluloses had been used with suspension cell cultures (Earle et al., 1954; McLimans et al., 1957; Kuchler et al., 1960). Kuchler et al. (1960) had grown mammalian cells in shaker cultures in a serum-free medium that contained methylcellulose [Dow's Methocel, 15 cP (measured at 2% w/v (weight/volume ratio) in water, 20°C, with an Ubelohde viscometer)]. They hypothesized that serum provided protection to agitated cells by providing nutrients needed for growth, buffering capacity, and protective action against hydrodynamic stresses associated with mixing (Kuchler et al., 1960), and suggested that serum proteins and methylcelluloses may behave similarly in protecting cells from shear damage since methylcelluloses probably bind to the surface of cells just as do mucopolysaccharides. Serum has been shown to allow better cell growth in agitated and/or aerated cultures in dosage-dependent fashion (Handa-Corrigan et al., 1989; Kilburn and Webb, 1968; Mizrahi and Moore, 1970). However, until recently, it has not been clear whether this was due to faster cell growth stimulated by higher serum concentrations, or due to protection from fluid-mechanical damage by physicochemical or biological mechanisms. It is certainly clear that low serum or serum-free cultures are more susceptible to fluid-mechanical damage (Kunas and Papoutsakis, 1990a).

Two problems need to be resolved in order to study the protective effects of additives. The first is the ability to assert that an additive's effect constitutes protection from fluid-mechanical damage. The second is the ability to assess quantitatively and reproducibly the protective effect of various additives. Both of these difficulties derive from the fact that there is substantial variability in the bioreactor and cellular factors that affect cell damage as discussed in Sections V,A and V,C. Kunas and Papoutsakis (1989, 1990a, 1990b) resolved these problems, to a large extent, by carrying out their experiments in two identical, well-controlled bioreactors run in parallel. In this system, one reactor is serving as the control for the other. They found that fetal bovine serum (FBS) at concentrations greater than 5% could protect the cells from detrimental hydrodynamic stresses. They noted that 10% FBS could protect cells even after less than one hour of exposure of cells to serum. Viscometric studies using FBS in addition to the bioreactor studies suggested that the protective effect of FBS in bioreactors is both physical and metabolic in nature (Michaels et al., 1991). Under the viscometric conditions, the effect of an additive is assessed for protection against laminar, well-defined shear in a Couette viscometer. These studies showed that FBS protects cells against shear damage in the viscometer after prolonged, but not after short, exposure. These results suggest that the protective effect of serum is both biological and fluid-mechanical in nature, and that the biological protection requires prolonged exposure to FBS. The protective effect of serum has been also demonstrated by the experiments of Lee et al. (1988). It is difficult, however, to quantitatively assess the extent to which the FBS (or any other protectant's) protection is due to physical or biological mechanisms, and even more difficult to explain the nature of each mechanism.

When Bryant (1966) used methylcellulose as a protective agent in large stationary and shaker flask cultures for the growth of human skin epithelial, mouse fibroblast, and monkey kidney cells, he found that the use of Methocel decreased glucose utilization while maximizing cell concentration. His studies showed that in the absence of serum, methylcellulose was required for growth in shaker flasks, especially for the relatively fragile monkey kidney cells. Several viscosity grades of Methocel (10, 25, 1500, 4000 cP) were tested, and it was found that they all provided the same protection. He concluded the large proteins present in serum provided a protective effect that was physical rather than nutritional in shaker flasks, and furthermore suggested that large molecules of nutritionally inert polymers (such as methylcellulose) should also exert a physical protective effect on suspended cells. Telling and Elsworth (1965) included sodium carboxymethylcellulose (CMC) and tryptose phosphate broth (TPB) to help protect baby hamster kidney cells grown in a 30 liter bioreactor at agitation rates of 460 rpm. However, the protective effect of the two additives was not clearly documented since they did not compare growth of these cells to a control that did not include CMC and TPB. More recently, Goldblum et al. (1990) used various Methocels in viscometric studies to show that the Methocels could increase the resistance of suspended insect cells (relatively fragile Sf9 and TN-368 cells) to lysis by a factor of from 58 to 76 when subjected to 50 dyn/cm^2 for 5 mins. The higher-molecular-weight (MW) Methocels and the higher concentrations used offered the best protection. In addition, the Methocels providing the highest viscosity of culture medium tested (4–25 times the unsupplemented medium viscosity) also provided better protection than did the low-viscosity Methocel additives. All media contained 10% FBS in addition to the additive. When they used high concentrations (4.5% w/v) of dextran (MW = 476,000) as an additive, they once again found a protection from cell damage due to viscometric flows. The viscosity of the medium with dextran was increased to 6.6-fold over the unsupplemented medium. The fact that the protective effect provided by the Methocels and dextran was found to be medium-viscosity dependent is undesirable (increased power requirements for agitation and decreased mass transfer) and in contrast to Bryant's (1966, 1969) studies. Nevertheless, these studies show that both the Methocels and dextran increase the shear robustness of the insect cells used in the studies of Goldblum et al. (1990).

In a further attempt to provide protection to the cells and also to decrease the use of serum in culture media, Mizrahi and Moore (1970) used pluronics, hydroxyethyl starch, polyvinylpyrrolidone, dextrans, and modified gelatin in addition to CMC. Pluronics are block copolymers of polyoxyethylene and polyoxypropylene. The use of Pluronic F68 in suspension culture systems has been documented since 1960 (Swim and Parker, 1960). It has been shown by many that the addition of the nonionic block copolymer to culture media for both mammalian and insect cells can provide cell protection for sparged systems (Kilburn and Webb, 1968; Radlett et al., 1971; Handa et al., 1987; Handa-Corrigan et al., 1989; Maiorella et al., 1988; Murhammer and Goochee, 1988, 1990). The protection mechanism has not been investigated by these studies, although several hypotheses have been given as to what might be happening. First, in many of the studies where Pluronics have been used, it is not clear that the additive was protecting cell from shear damage, because cells were grown either in shake flasks or low or unreported-rpm spinner flasks. In other words, the enhancement of cell growth by the addition of Pluronics could have been irrelevant to its shear-protection capabilities as was clearly demonstrated by the static-culture studies of Bentley et al. (1989). Specifically, Bentley et al. (1989) have shown that Pluronics have a concentration-dependent positive or negative effect on cell growth *in static cultures*, thus establishing that at certain concentrations these polyols affect cell growth independent of agitation or aeration. Handa et al. (1987) and Handa-Corrigan et al. (1987, 1989) speculated that Pluronic F68 provides protection because it acts as a foam stabilizer preventing cells from being exposed to damaging forces in draining foam films or forces from bubble rupture. Murhammer and Goochee (1990) proposed that stable foams do not provide the only protection to cells, but that the cell-protecting pluronic polyols may imbed into and help stabilize the cell's plasma membrane. They tested several pluronic polyols (including Pluronics, reverse Pluronics, and diblock copolymers) to determine how their protecting capabilities were affected by the relative sizes of the molecules, the relative positions of the hydrophilic and hydrophobic blocks, and the molecular weights of the polyoxypropylene and polyoxyethylene blocks. They showed that the protective capabilities correlated with an empirical measure of the emulsifying ability of the surfactant molecule. This empirical measure is given by a hydrophilic–lipophilic balance (HLB) value. The surfactants with a low HLB are more oil-soluble whereas those with a higher HLB are more water-soluble. In their studies, they found that the pluronic polyols with a low HLB lysed cells, while those with high HLB values protected the cells. Cultured insect cells were grown in spinner flasks (50 rpm), airlift, and sparged agitated bioreactors at 200 rpm for their experiments. Plurafac linear alcohol ethoxylates, diblock copolymers with a hydrocarbon block and polyoxyethylene block, lysed the cells. The two Plurafacs they used had high HLB values similar to those of the pluronics that protected cells. They suggested that the saturated hydrocarbon chain in Plurafacs interacts with the cell's plasma membrane in a negative manner compared to the hydrophobic polyoxypropylene portion of the effective Pluronics. Although these polyglycols are not metabolized by the cells (Mizrahi, 1984), by their nature as surface-active agents they do interact with the cell plasma membrane, and thus Murhammer and Goochee (1989, 1990) have suggested that their protective effect is to some or large extent due to this interaction. Interestingly, however, the correlation of the high HLB of these polyglycols with increased protection is also consistent with the mechanism suggested by Handa-Corrigan et al. (1989) and Kilburn and Webb (1968). As mentioned already, Bentley et al. (1989) have shown that Pluronic surfactants have a concentration-dependent effect on cell growth independent of agitation or aeration. However, this does not imply that their protective effect against shear damage is of biological nature. Murhammer and Goochee (1990) made an analogy between what happens when the Pluronics lyse the cells with the solubilization of membranes by surfactants used to intentionally lyse the cells (Helenius and Simons, 1975). After the surfactants (detergents) adsorb onto and penetrate the cell membrane, they cause a change in molecular organization in a manner that alters permeability, leading to leakage of the cell to cause lysis.

Smith (1990) showed that Pluronic F68 prevented vortexing damage in the bioreactor during exponential growth of hybridoma cells. Ramirez and Mutharasan (1990) have measured the plasma membrane fluidity (PMF) using measurement of fluorescence anisotropy (r_s) (see Section V.B) of their hybridoma cells grown in the presence of F68. They found that F68 increases the r_s by an average of 0.01 unit. Since an r_s increase implies a PMF decrease, they suggested that F68 interacts with the plasma membrane and decreases its PMF thus making the cells more resistant to shear damage. This is consistent with the Murhammer and Goochee (1989; 1990) hypothesis. Unfortunately, this small r_s increase is not sufficient to prove an increase of the cell resistance to shear damage. According to their data an r_s increase of 0.04–0.05 unit is necessary for this effect. In addition, the simple and direct-evidence experiment of studying the effect of F68 on cell resistance to their viscometric shear was not reported. Such studies were however carried out in our laboratory (Petersen, 1989; Michaels et al., 1991). The results show that F68 does not affect the resistance of our hybridoma cells (CRL8018) to viscometric shear. On the other hand, Goldblum found that Sf9 insect cells grown in the presence of

0.1% w/v F68 were 15.5 times more resistant to laminar shear in a cone-and-plate viscometer, and growth with 0.2% w/v F68 makes the cells about 42 times more shear resistant than cells grown without F68. The TN-368 insect cells were 6.3 times more shear resistant. In their viscometric studies they used insect cells from the late exponential phase of static T-flask cultures. These contrasting results indicate that the protection mechanism of F68 may well be cell-type-dependent. We will further discuss the results of our studies shortly after a discussion of the protective effect of other polyalcohols.

Murhammer and Goochee (1990) also found that poly(oxyethylene) glycol (more commonly known as polyethylene glycol (PEG), a potent fusogen at high concentrations) protected the S9f cells in agitated and sparged cultures but not in airlift cultures. Handa-Corrigan had previously used PEG to successfully protect cells from bubble damage in bubble columns (Handa, 1986).

In addition to the studies to examine the protective effect of serum against shear damage (Kunas and Papoutsakis, 1989, 1990a) in our laboratory, we have used a similar methodology of bioreactor experiments to examine the protective effects of polyethylene glycols (PEG) of various molecular weights and concentrations, and of polyvinyl alcohol (PVA, MW = 10,000). The effects of these were also compared to the cell-protecting capabilities of Pluronic F68 (Michaels et al., 1991; Michaels and Papoutsakis, 1991). The results from an experiment are shown in Figure 4. PEGs of molecular weights above 1400 and PVA have a profound protective effect under high agitation intensities, but do not affect cell growth under static or mild-agitation conditions. Their protective effect was found to be stronger than the effect of Pluronic F68. In contrast, under high-agitation conditions, 1–3% w/v dextran has a detrimental effect on cell growth, but no effect under static or mild-agitation growth conditions (Papoutsakis and Kunas, 1989). We examined if the protective effect of the various additives required prolonged exposure, or was fast-acting after a short (approximately an hour, which is as fast as we can measure an effect on cell growth and viability in the bioreactor) exposure to the additive (Michaels et al., 1991). We thought that a difference between the prolonged-exposure versus short-exposure effects might indicate a biological mechanism of action. We found that PEG, PVA, and F68 could protect cells from fluid-mechanical damage even after a short exposure. The interpretation that the effects of PEG, PVA, and F68 are of nonbiological nature was further strengthened by parallel viscometric studies (Petersen, 1989; Michaels et al., 1991), according to the methodology reported in our earlier paper (Petersen et al., 1988). Under the viscometric conditions, the effect of an additive is assessed for protection against laminar, well-

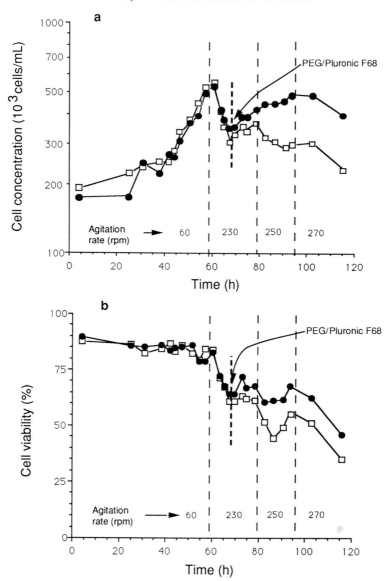

FIGURE 4   Comparison of the short-term exposure protective capabilities of PEG and Pluronic F68 in identical, surface-aerated bioreactors run in parallel. Cells were cultured at low agitation (60 rpm) until midexponential growth and then the agitation rate was increased to the values shown on the batch growth curves of the graph. Cell damage was seen at 230 rpm by a reduction in viable cell concentration. 0.1% w/v [●] F68 or [□] PEG (MW = 8000) was added as indicated by the arrow. (a) Viable cell concentration; (b) cell viabilities.

defined shear in a Couette viscometer. These studies showed that unlike serum, PEG and F68 do not protect cells against shear damage in the viscometer, either after prolonged or short exposure. In fact, PEG was found to be mildly detrimental to cell growth. We also interpreted the different (positive versus negative or no) effect of the other additives in the bioreactor and in the viscometer to imply that their effect is purely fluid mechanical (i.e., due to changes in the interactions between bubbles and cells). If their effect was biological, cells would have been protected in both shear environments. Similarly, if the additives protect cells from shear stresses through the often hypothesized "coating" of the cells, cells would have been protected by the viscometric shear as well. The protective effect in the bioreactor can then be explained by the effect of these additives on the surface tension and rheological properties of the gas-liquid interface, since these properties affect bubble breakup and thus the associated shear stresses (see Section II).

F. Interfacial Properties and Bubble Entrainment and Breakup

1. Correlations with Surface Tension and Fluid Viscosity

Since it is assumed that the surfactants PEG, PVA, and F68 provide protection by changing bubble stability, bubble rupture, and other associated interfacial phenomena, it is important to analyze the factors that affect the processes that lead to cell damage due to cell–bubble interactions. In order to provide a better physicochemical characterization of the effect of the additives, attempts have been made to correlate viscosity and static surface tension of media with cell protecting capability (Handa-Corrigan et al., 1987; Mizrahi, 1975; Smith, 1990; Michaels and Papoutsakis, 1991). These two parameters are generally chosen since they can be easily measured. Viscosity affects the structure of turbulence and the transmission of mechanical forces to cells (Cherry and Papoutsakis, 1986; Kunas and Papoutsakis, 1990b), and has been established to be an important parameter in the damage of cells in microcarrier bioreactors (Cherry and Papoutsakis, 1990). Surface tension affects bubble entrainment, coalescence and breakup, and has been suggested as an important parameter that affects cell damage due to bubble breakup (Handa-Corrigan et al., 1987; 1989). Michaels and Papoutsakis (1991) found that 0.1% w/v of both PVA and F68 lower the surface tension of the serum-free medium (SFM) they used by 10–12%. However, their measurements show that the addition of PEG (an effective cell protectant) to the SFM does not lower the static surface tension. It was thought that the addition of PEG would lower the static surface tension as compared to SFM without additives, since it is a surfactant. Their surface tension measurements indicate that the presence of the defined proteins (biosurfactants) in the SFM substantially lower the surface tension of the SFM compared to the basal medium and that the addition of PEG cannot lower the surface tension any further. Similarly, PVA and F68 concentrations above 0.1% w/v do not lower the surface tension of the medium much further. On the basis of the static surface tension measurements, a correlation between the protective effect of an additive and a lower surface tension cannot be established. However, it should be noted that in bioreactor systems there is a constant rearrangement of gas–liquid interfaces, leading to constantly changing interfacial surface tensions with subsequent changes in adsorbed surfactants and/or proteins. Analysis of dynamic properties is essential in describing the behavior of fluid interfaces when interfacial motion is involved and includes cell-bubble attachment characteristics. Measurements of the interfacial shear and surface-dilatational viscosities would be more appropriate measurements in an attempt to correlate cell protection with interfacial properties.

2. Characterization of the Fluid Environment of the Bioreactor

Several empirical equations characterizing the fluid environment of mechanically agitated reactors based on experimental data have been presented in the mixing literature. These correlations may be used to help understand how the hydrodynamic environment of the cell-culture system may be affected by additives and other reactor parameters. We will briefly discuss some of these correlations and their significance in an attempt to clarify the processes involved. As an example, we will refer to correlations applicable to the 2 liter, surface aerated bioreactor used in our laboratory, and include some correlations applicable to other systems (with sparging).

Four important reactor parameters to consider in the surface aerated reactors used in our experiments are: (1) the critical agitation rate for surface aerated bubble entrainment, (2) the critical agitation rate for dispersion of these bubbles, (3) the average diameter of entrained bubbles, and (4) the depth of the vortex in the reactor under high-agitation conditions. We will examine how these parameters are affected by changes in the interfacial tension and medium viscosity.

For surface aeration, bubbles are generated at the surface because of turbulence in the liquid phase. They will remain in the bulk by liquid flow generated by the impeller. The extent of entrainment is a function of the turbulence at the liquid surface and the downward volumetric flow rate with key parameters being the impeller characteristics (design, diameter, and location) along with the surface tension of the liquid medium. A correlation for determining the critical impeller speed for bubble entrainment for surface aeration was given by Joshi et al. (1982) by making predictions based on the liquid flow generated by the impeller. They noted that surface aeration will occur when the downward liquid velocity exceeds the terminal rise velocity of the bubbles. Their prediction for critical agitation rate for gas entrainment was therefore given by combining correlations for bubble diameter, bubble terminal rise velocity, and liquid circulation velocity given by others. Bhavaraju et al. (1978) predicted the average entrained bubble diameter using six-bladed rushton turbines as

$$d_B = 1.21 \frac{\sigma^{0.6}}{(P/V)^{0.4}(\rho_L)^{0.2}} \left(\frac{\mu_L}{\mu_G}\right)^{0.1} \tag{14}$$

where d_B is the average diameter (cm) of an entrained bubble, P the power consumption of the agitator (kW) [or agitation power input rate (g cm^2/s^3)], V the volume of liquid (cm^3) in the reactor, σ the surface tension (N/m), and μ the viscosity of the gas or liquid (Pa/s). Van Krevelen and Hoftizer (1950) give the bubble terminal rise velocity, v_b (cm/s) as

$$v_b = 0.71(gd_B)^{1/2} \tag{15}$$

where g is acceleration due to gravity (cm^2/s). It is then assumed that the surface aeration begins when the liquid circulation velocity is equal to the bubble terminal rise velocity (Joshi et al., 1982). Uhl and Gray (1966) suggested the following equation for liquid circulation velocity v_c (cm/s) with the correlation

$$v_c = 0.53 \left(\frac{d_i}{W}\right) nd_i \left(\frac{d_i}{T}\right)^{7/6} \tag{16}$$

where W is the impeller width (cm), d_i the impeller diameter (cm), T the tank diameter (cm), and n the agitation rate (rev/s). By setting the bubble terminal rise velocity equal to the liquid circulation velocity and substituting for the bubble diameter, a prediction of the minimum impeller speed for entrainment is given by

$$N_e = \frac{1.65}{N_P^{0.125}} \frac{T^{1.1}}{d_i^{1.98}} \left(\frac{\sigma g}{\rho_L}\right)^{0.19} \left(\frac{\mu_L}{\mu_G}\right)^{0.031} \left(\frac{W}{d_i}\right)^{0.625} \tag{17}$$

where N_e is the minimum agitation rate for surface aerated bubble entrainment (rev/s). Therefore, this critical agitation rate is a function of the power number $N_p(Pg/\rho_L n^3 d_i^5)$, the W/d_i ratio, the characteristic bubble

TABLE 2

Prediction of (a) Critical Agitation Rates for Bubble Entrainment and Dispersion
in a Surface Aerated 2-liter Bioreactor and (b) Relative Rates
of Bubble Entrainment in a Surface Aerated Reactor as a Function of
Surface Tension

(a)

Critical Agitation Rates for Bubble Entrainment and Dispersion

Surface tension (dyn / cm)	N_e (rpm)	N_d (rpm)
70	209	224
60	203	217
50	196	210
40	188	201

(b)

Relative Rate of Entrainment as a Function of Surface Tension Calculated

Surface tension (dyn / cm)	Entrainment rate (relative to 70 dyn / cm)
70	1
65	0.95
60	0.89
55	0.84
50	0.78
45	0.73
40	0.67

[a]Based on Eq. (17) given by Joshi et al. (1982) and Tanaka and Izumi (1987).
[b]The rates of bubble entrainment given are relative to the entrainment rate predicted for a surface tension of 70 dyn/cm. Correlated by Matsumara et al. (1977).

rise velocity ($\sigma g/\rho_L$), and the impeller and tank diameters. By using only one reactor configuration for comparison experiments (constant W, d_i, T, V), one can evaluate the effect that changes in surface tension and viscosity of the medium should have. Once the N_e is surpassed in the turbulent regime, entrained bubbles undergo breakup and coalescence to form a more uniform bubble dispersion. A correlation for the agitation rate for a uniform dispersion (N_d) [or minimum agitation rate for dispersion of entrained bubbles (rev/s)] was given for a four-bladed 50°-pitched impeller by Tanaka and Izumi (1987) as 1.07 times the N_e. Table 2(a) shows the N_e and N_d values for various surface tensions for the laboratory-size reactor systems used in our laboratory.

The average diameter of entrained bubbles varies with surface tension to the 0.6 power as predicted in Eq. (14) by Bhavaraju et al. (1978). The estimated average bubble diameters and critical agitation rates for entrainment are given in Figure 5 using our reactor configuration and for culture volumes of 1–2 L. The power number is estimated from Nagata (1975). These predictions are consistent with the measurements taken by Kunas and Papoutsakis (1990b) with the same reactor set up for estimated bubble diameters when comparing entrainment at 200–250 rpm (0.5–3 mm) with those at 800 rpm (50–300 μm). A more important change that surfactants can bring about in the mechanically agitated bioreactor may be in the *rate* of gas entrainment. Correlations for this have been given by Matsumara et al. (1977). These correlations predict that the bubble entrainment is proportional to $(\sigma)^{0.72}$, with the impeller diameter/tank diameter ratio having the most substantial influence on the rate of gas entrainment. Changes in relative rates of entrainment are shown in Table 2(b) for a variation in σ from 70 to 45 dyn/cm.

In summary, surface active additives should promote gas entrainment and dispersion at lower agitation rates, but conversely would lead to a lower relative rate of entrainment and smaller bubble size for entrained bubbles. According to Tramper et al. (1988), the smaller entrained bubble volume would imply a smaller "killing volume," and, thus, reduced cell death.

Many workers (Van de Vusse, 1955; Nagata, 1955; Brennan, 1976) have reported that vortex depth is a function of the impeller Froude number $(n^2 d_i/g)$ and a proportionality factor based on reactor geometry for unbaffled reactors. Our reactors are partially baffled, but a vortex about the impeller shaft does form beginning at an agitation rate of 190 rpm (1.2 working volume). The vortex geometry is more importantly a function of the reactor geometry and the liquid viscosity and should remain unaffected by the changes in surface tension.

Equations for similar types of correlations used with direct sparging can also be applied. For example, Van Dierendonck et al. (1971) have given the critical impeller speed below which the impeller speed has no influence on gas phase holdup. Gas-phase holdup is determined usually by measuring the liquid levels in the vessel with and without aeration.

$$N_0 = \left(\frac{\sigma g}{\rho} \right)^{0.25} \left(\frac{A}{d_i} + B \frac{T}{d_i^2} \right) \qquad (18)$$

where N_0 represents the just-mentioned critical impeller speed and A and B are constants. Keeping reactor configurations constant, N_0 is propor-

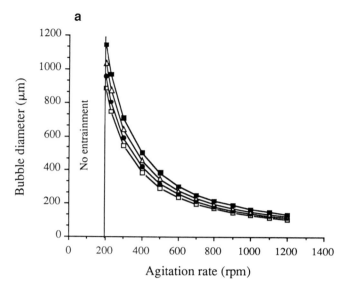

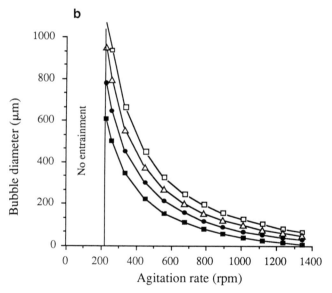

FIGURE 5   Prediction of the average diameter of an entrained bubble as a function of agitation rate in a surface aerated 2-L bioreactor for various (a) working volumes—[□] 1.0 L, [●] 1.2 L, [△] 1.5 L, [■] 1.9 L and (b) surface tensions—[□] 70 dyn/cm, [△] 60 dyn/cm, [●] 50 dyn/m, [■] 40 dyn/cm. The average bubble diameters are predicted from Eq. (14) of Bhavaraju et al. (1978) using a power number of 2.0. The surface tension in (a) is taken as 60 dyn/cm. The reactor volume in (b) is taken as 1.2 L.

tional to the surface tension to the fourth power. Above this speed, gas-phase holdup was dependent only on the impeller speed and was independent of the gas flow rate.

Another important parameter to consider is the critical film thickness for bubble rupture to occur. A relation for this critical film thickness (Azbel et al., 1979) is given by

$$\delta_{crit} = \frac{0.57\sigma}{\rho_L c_s^2 \left[(\beta + 1)(\beta - 2\beta + 9)^{0.5} - (\beta^2 + 3) \right]} \tag{19}$$

where β represents $6\sigma/\rho_G - \rho_L)gR_b$; R_b represents the bubble radius and c_s the speed of sound (cm/s). This relation correlates interfacial surface tension to the critical film cap thickness. Since film drainage, which leads to film thinning and rupture, is affected by both bubble size and rheological properties of the bubble surface, it is clear that surface elasticity and surface viscosity affect the likelihood of breakup. Note that lower surface tensions cause both smaller bubbles and a thinner critical film-cap thickness before rupture. In addition, film-cap thinning is less rapid in a liquid containing surfactants as compared to pure liquids. A positive feature of the additives used as cell protectants is that they help stabilize the frequency of the bubble rupture process by causing slower liquid drainage and by lowering the critical film cap thickness and thus possibly decreasing cell–bubble attatchment..

3. Discussion of the Correlations Presented

There is evidence that shear stresses generated near the bioreactor surface damage cells collected near the bubble (Handa, 1986; Handa-Corrigan et al., 1989; Tramper et al., 1986). Therefore, a review of some of the possible effects on bubble dynamics caused by the addition of surfactants to the medium indicates the importance of bubble rheological properties and may be important in understanding of the shear protection mechanisms by additives such as PEG, PVA, and Pluronics. It becomes evident that the interfacial surface tension controls directly the size of entrained bubbles, and also influences the number of entrained bubbles. The positive aspect of the use of surfactants is that they result in smaller bubbles (smaller "killing volume"), a slower entrainment rate, and a thinner critical film-cap thickness as shown based on correlations on bubble dynamics given in the literature. It has been shown in studies on the effect of surfactant concentration on the critical film thickness of microscopic thin films that as the surfactant concentration is increased, the critical film thickness decreases to a certain point and gradually reaches a constant value (Manev et al., 1974). It has been hypothesized that the alteration of the critical film-cap thickness can be due to both surface shear viscosity and surface dilatational viscosity (Ivanov and Dimitrov, 1974; Benjamins et al., 1975). For most models of bubble formation, coalescence, and rupture presented in the literature, the interfacial surface tension is usually determined under static conditions, even though bubble phenomena are dynamic (Clift et al., 1978) and the interfacial surface tension is a function of the amount of surfactant (and protein) in the solution and the way that it is adsorbed. This should emphasize the difficulties involved in predicting bubble dynamics, and the review given in this chapter is only meant to elucidate how the additives used as shear protectants may be protecting the cells from damage related to bubble entrainment and breakup. It does not seem likely that the changes in σ and μ_L imparted by the additives changed the rate of gas entrainment or the average bubble size enough to account for the differences seen in cell damage or protection with the different additives.

Another important factor to consider is the collection of cells near the bubble interface when a bubble ruptures. Studies have shown that the additives that provide shear protection may reduce the concentration of cells attached to or in the vicinity of bursting bubbles. For example, Handa (1986) visualized that hybridoma cells cultured in sparged bubble columns in the presence of Pluronic F68 did not interact with the gas/liquid surface and that the cells were not carried to the vicinity of the bursting bubbles at the top of the foam layer. Handa also noted that bubbles tended to slip off each other in the presence of Pluronic F68. Cell-bubble attachment experiments comparing different additives may provide information as to which additives may provide the best protection.

VI. SUMMARY

We have reviewed the effects of hydrodynamic forces on mammalian cells cultured in agitated, airlift, and bubble column bioreactors in an attempt to understand the mechanisms through which these forces affect cell growth and death. An understanding of these effects is important for the design and scale-up of bioreactors. On identifying and classifying the various interactions between cells grown in free suspension and on microcarrier beads with fluid forces or solid objects, we discussed the importance and extent to which each may account for cell damage. Models that were based on experimental evidence for cell damage in microcarrier bioreactors have been reviewed. We have listed some of the various biological responses elicited by freely suspended cells when subjected to hydrodynamic forces. Studies indicate that responses more complex than growth and death need to be considered when attempting to understand cell damage in bioreactors. These await further investigation, especially with cells of biotechnological interest. A review of chemicals that protect cells against fluid-mechanical damage and the nature of their effect was given, along with characteristics of bubble breakup, thin films, and rheological properties that play important roles in cell protection from fluid-mechanical forces.

REFERENCES

Abu-Reesh, I., and Kargi, F. (1989). Biological responses of hybridoma cells to defined hydrodynamic shear stress. *J. Biotechnol.* **9**, 167–178.

Aebersold, P., Carter, C. S., Hyatt, C., Johnson, S., Ottaway, K., Leitman, S. F., and Rosenberg, S. A. (1988). A simplified automated procedure for generation of human lymphokine-activated killer cells for use in clinical trials. *J. Immunol. Methods* **112**, 1–7.

Al-Rubai, M., Oh, S. K. W., Musaheb, R., and Emery, A. N. (1990). Modified cellular metabolism in hybridomas subjected to hydrodynamic and other stresses. *Biotechnol. Lett.* **12**, 323–328.

Augenstein, D. C., Sinskey, A. J., and Wang, D. I. C. (1971). Effect of shear on the death of two strains of mammalian tissue cells. *Biotechnol. Bioengr.* **13**, 409–418.

Aunins, J. G., Woodson, B. A., Hale, T. K., and Wang, D. I. C. (1989). Effects of paddle impeller geometry on power input and mass transfer in small scale animal cell culture vessels. *Biotechnol. Bioeng.* **34**, 1127–1132.

Azbel, D. S., Lee, S. L., and Lee, T. S. (1979). Acoustic resonance theory for the rupture of film cap of a gas bubble at a horizontal gas-liquid interface. In *Two-Phase Momentum, Heat and Mass Transfer in Chemical, Process, and Energy Engineering Systems*, Vol. 1, pp. 159–169, Afgan, N. H., Durst, F., and Tsiklauri, G. V. (eds.). Hemisphere Publishing Corp., New York.

Backer, M. P., Metzger, L. S., Slaber, P. L., Nevitt, K. L., and Boder, G. B. (1988). Large-scale production of monoclonal antibodies in suspension culture. *Biotechnol. Bioeng.* **32**, 993–1000.

Batchelor, G. K. (1980). Mass transfer from small particles suspended in turbulent fluid. *J. Fluid Mech.* **98**, 609–623.

Ben-Ze'ev, A., Farmer, S. R., and Penman, S. (1980). Protein synthesis requires cell-surface contact while nuclear events respond to cell shape in anchorage-dependent fibroblasts. *Cell* **21**, 365–372.

Benjamins, J., de Feijter, J. A., Evans, M. T. A., Graham, D. E., and Phillips, M. C. (1975). Dynamic and static properties of proteins adsorbed at the air/water interface. *J. Chem. Soc. Faraday Disc.* **1975**, 218–229.

Bentley, P. K., Gates, R. M. C., Lowe, K. C., de Pomerai, D. I., and Walker, J. A. L. (1989). In vitro cellular responses to a non-ionic surfactant, Pluronic F-68. *Biotechnol. Lett.* **11**, 111–114.

Bhavaraju, S. M., Russel, T. W., and Blanch, H. W. (1978). The design of gas sparged devices for viscous liquid systems. *AIChE J.* **24**, 454–467.

Bird, R. B., Stewart, W. E., and Lightfoot, E. N. (1960). *Transport Phenomena*. Wiley, New York.

Borys, M. C. (1990). Effect of microcarrier culture conditions on recombinant protein production by CHO cells. M.S. thesis, Northwestern University, Evanston, IL.

Brennan, D. J. (1976). Vortex geometry in unbafled vessels with impeller agitation. *Trans. Inst. Chem. Eng.* **54**, 209–217.

Bryant, J. (1966). Mammalian cells in chemically defined media in suspension cultures. *Ann. N.Y. Acad. Sci.* **139**, 143–161.

Bryant, J. C. (1969). Methylcellulose effect on cell proliferation and glucose utilization in chemically defined medium in large stationary cultures. *Biotechnol. Bioengr.* **11**, 155–179.

Cherry, R. S., and Kwon, K. Y. (1990. Transient shear stresses on a suspension cell in turbulence. *Biotechnol. Bioeng.* **36**, 563–571.

Cherry, R. S., and Papoutsakis, E. T. (1986). Hydrodynamic effects on cells in agitated tissue culture reactors. *Bioproc. Eng.* **1**, 29–41.

Cherry, R. S., and Papoutsakis, E. T. (1988). Physical mechanisms of cell damage in microcarrier cell culture bioreactors. *Biotechnol. Bioeng.* **32**, 1001–1014.

Cherry, R. S., and Papoutsakis, E. T. (1990). Understanding and controling injury of animal cells in bioreactors. In *Animal Cell Biotechnology*, Vol. 4, pp. 71–121, Spier, R. E., and Griffiths, J. B. (eds.). Academic Press, Boston.

Chesters, A. K. (1975). The applicability of dynamic similarity criteria to isothermal liquid-gas two-phase flows without mass transfer. *Int. J. Multiphase Flow* **2**, 191–212.

Chesters, A. K., and Hofman, G. (1982). Bubble coalescence in pure liquids. *Appl. Sci. Res.* **38**, 353–361.

Chittur, K. K., McIntire, L. V., and Rich, R. R. (1988). Shear stress effects on human T cell function. *Biotechnol. Prog.* **4**, 89–96.

Clift, R., Grace, J. R., and Weber, M. E. (1978). *Bubbles, Drops and Particles*, pp. 339–347. Academic Press, New York.

Croughan, M. S., Hamel, J. F., and Wang, D. I. C. (1987). Hydrodynamic effects on animal cells grown in microcarrier cultures. *Biotechnol. Bioeng.* **29**, 130–141.

Croughan, M. S., Hamel, J. F. P., and Wang, D. I. C. (1988). Effects of microcarrier concentration in animal cell culture. *Biotechnol. Bioengr.* **32**, 975–982.

Croughan, M. S., Sayre, E. S., and Wang, D. I. C. (1989). Viscous reduction of turbulent damage in microcarrier cell cultures. *Biotechnol. Bioengr.* **33**, 862–872.

Culliton, B. J. (1989). Designing cells to deliver drugs. *Science* **246**, 746.

de St. Groth, F. (1983). Automated production of monoclonal antibodies in a cytostat. *J. Immunol. Methods* **57**, 121–136.

de Vries, A. J. (1972). Morphology, coalescence, and size distribution of foam bubbles. In *Adsorptive Bubble Separation Techniques*, pp. 7–31, Lemlich, R. (ed.). Academic Press, New York.

Dewitz, T. S., McIntire, L. V., Martin, R. R., and Sybers, H. D. (1979). Enzyme release and morphological changes in leukocytes induced by mechanical trauma. *Blood Cells* **5** 499–510.

Dodge, T. C., and Hu, W. S. (1986). Growth of hybridoma cells under different agitation conditions, *Biotechnol. Lett.* **8**, 683–686.

Earle, W. R., Schilling, R. I., Bryant, J. C., and Evans, V. J. (1954). The growth of pure strain L cells in fluid suspension cultures. *J. Natl. Cancer Inst.* **14**, 1159–1171.

Elzinga, E. R., and Banchero, J. T. (1961). Some observations on the mechanics of drops in liquid-liquid systems. *AIChE J.* **7**, 394–399.

Erickson, L. E., Patel, S. A., Glasgow, L. A., and Lee, C. H. (1983). Effects of viscosity and small bubble segregation on mass transfer in airlift fermenters. *Process Biochem.* **1**, 16–37.

Folkman, J., and Greenspan, H. P. (1975). Influence of geometry on control of cell growth. *Biochim. Biophys. Acta* 417, 211–236.

Folkman, J., and Moscona, A. (1978). Role of cell shape in growth control. *Nature* 273, 345–349.

Frangos, J. A., Eskin, S. G., McIntire, L. V., and Ives., C. L. (1985). Flow effects of prostacyclin production by cultured human endothelial cells. *Science* 227, 1477–1479.

Gardner, A. R., Gainer, J. L., and Kirwan, D. J. (1990). Effects of stirring and sparging on cultured hybridoma cells. *Biotechnol. Bioeng.* 35, 940–947.

Glasgow, L. A., Erikson, L. E., and Patel, S. A. (1984). Wall pressure fluctuations and bubble size distributions at several positions in an airlift fermentor. *Chem. Eng. Commun.* 29, 311–336.

Goldblum, S., Bae, Y. K., Hink, W. H., and Chalmers, J. (1990). Protective effect of methylcellulose and other polymers on insect cells subjected to laminar shear stress. *Biotechnol. Prog.* 6, 383–390.

Golde, D. W., and Gasson, J. C. (1988). Hormones that stimulate the growth of blood cells. Each hemopoieten regulates the production of a specific set of blood cells. Now made by recombinant-DNA methods, these hormones promise to transform the practice of medicine. *Sci. Am.* 259, 62–70.

Greenaway, P. J., and Farrar, G. H. (1990). The growth and production of human immunodeficiency virus. In *Animal Cell Biotechnology*, Vol. 4, pp. 379–411, (Griffiths, J. B., and Spier, R. E. eds.). Academic Press, Boston.

Handa, A. (1986). Gas-liquid interfacial effects on the growth of hybridomas and other suspended mammalian cells. Ph.D. thesis, University of Birmingham, Birmingham, England.

Handa, A., Emery, A. N., and Spier, R. E. (1987). On the evaluation of gas-liquid interfacial effects on the hybridoma viability in bubble column bioreactors. *Devel. Biol. Standard.* 66, 241–253.

Handa-Corrigan, A., Emery, A. N., and Spier, R. E. (1989). Effect of gas-liquid interfaces on the growth of suspended mammalian cells: mechanisms of cell damage by bubbles. *Enzm. Microb. Technol.* 11, 230–235.

Hansen, J. T., and Sladek, J. R., Jr. (1989). Fetal research. *Science* 246, 775–779.

Helenius, A., and Simons, K. (1975). Solubilization of membranes by detergents. *Biochim. Biophys. Acta* 415, 29–79.

Hellums, J. D., and Hardwick, R. A. (1981). Response of platelets to shear stress—a review. In *The Rheology of Blood, Blood Vessels and Associated Tissues*, pp. 160–183, Gross, D. R. and Hwang, N. D. C. (eds.). Sijthoff and Noordhoff, Rockville, MD.

Hinze, J. O. (1975). *Turbulence*, McGraw Hill, New York, NY.

Hu, W.-S. (1983). Quantiative and mechanistic analysis of mammalian cell cultivation on microcarriers. Ph.D. Thesis, Massachusetts Institute of Technology, Cambridge, MA.

Hu, S., and Kintner, R. C. (1955). The fall of single liquid drops through water. *AIChE J.* 1, 42–50.

Hulscher, M., and Onken, U. (1988). Influence of bovine serum albumin on the growth of hybridoma cells in airlift loop reactors using serum-free medium. *Biotechnol. Lett.* 10, 689–694.

Ivanov, I. V., and Dimitrov, D. S. (1974). Hydrodynamics of thin liquid films: Effect of surface viscosity on thinning and rupture of foam films. *Colloid Polym. Sci.* 252, 982–990.

Jadus, M. R., Thurman, G. B., Mrowca-Bastin, A., and Yannelli, J. R. (1988). The generation of human lymphokine-activated killer cells in various serum-free media. *J. Immunol. Methods* 109, 169–174.

Joly, M. (1972a). Rheological properties of monomolecular films. Part I: Basic concepts and experimental methods. In *Surface and Colloid Science*, Vol. 5, pp. 1–77, Matijevic, E. (ed.). Wiley-Interscience, New York.

Joly, M. (1972b). Rheological properties of monomolecular films. Part II: Experimental results. Theoretical interpretation. Applications. In *Surface and Colloid Science*, Vol. 5, pp. 79–193, Matijevic, E. (ed.). Wiley-Interscience, New York.

Joshi, J. B., Pandit, A. B., and Sharma, M. M. (1982). Mechanically agitated gas-liquid reactors. *Chem. Eng. Sci.* 37, 813–844.

Kawase, Y., and Moo-Young, M. (1990). Mathematical models for design of bioreactors: Applications of Kolmogoroff's theory of isotropic turbulence. *Chem. Eng. J.* 43, B19–B41.

Kilburn, D. G., and Webb, F. C. (1968). The cultivation of animal cells at controlled dissolved oxygen partial pressure. *Biotechnol. Bioeng.* 10, 801–814.

Kuboi, R., Komosawa, I., and Otake, T. (1974). Fluid and particle motion in turbulent dispersion. *Chem. Engr. Sci.* 29, 641–657.

Kuchler, R. J., Marlowe, M. L., and Merchant, D. J. (1960). The mechanism of cell binding and cell sheet formation in L-strain fibroblasts. *Exp. Cell Res.* 10, 428–437.

Kunas, K. T. (1990). Growth and injury of freely suspended animal cells in an agitated and surface-aerated bioreactor. Ph.D. thesis, Rice University, Houston, TX.

Kunas, K. T., and Papoutsakis, E. T. (1989). Increasing serum concentrations decrease cell death and allow growth of hybridoma cells at higher agitation rates. *Biotechnol. Lett.* 11, 525–530.

Kunas, K. T., and Papoutsakis, E. T. (1990a). The protective effect of serum against hydrodynamic damage of hybridoma cells in agitated and surface-aerated bioreactors. *J. Biotechnol.* 15, 57–70.

Kunas, K. T., and Papoutsakis, E. T. (1990b). Damage mechanisms of suspended animal cells in agitated bioreactors with and without bubble entrainment. *Biotechnol. Bioeng.* 36, 476–483.

Lakhotia, S., and Papoutsakis, E. T. (1992). Agitation induced injury in microcarrier cultures. The protective effect of viscosity is agitation intensity dependent: Experiments and modeling. *Biotechnol. Bioengr.* 39, 95–107.

Lee, G. M., Huard, T. K., Kaminski, M. S., and Palsson, B. O. (1988). Effect of mechanical agitation on hybridoma cell growth, *Biotechnol. Lett.* 10, 625–628.

Lucassen, J., and van den Tempel, M. (1972). Dynamic measurement of dilational properties of a liquid interface. *Chem. Eng. Sci.* 27, 1283–1291.

MacIntyre, F. (1972). Flow patterns in breaking bubbles. *J. Geophys. Res.* 27, 5211–5228.

MacRitchie, F. (1978). Proteins at interfaces. *Adv. Protein Chem.* 32, 283–326.

Maiorella, B., Inlow, D., Shauger, A., and Harano, D. (1988). Large-scale insect cell-culture for recombinant protein production. *Bio/Technology* 6, 1406–1410.

Manev, E., Scheludko, A., and Exerowa, D. (1974). Effect of surfactant concentration on the critical thicknesses of films. *Colloid Polym. Sci.* 252, 586–593.

Martin, R. S., Dewitz, T. S., and McIntire, L. V. (1979). Alterations in leucocyte structure and function due to mechanical trauma. In *Quantitative Cardiovascular Studies*, pp. 419–454, Hwang, N. H. C. ed.). University Park Press, Baltimore.

Matsumura, M., Masunya, H., and Kobayashi, J. (1977). Correlation for flow rate of gas entrained from free liquid surface of an aerated stirred tank. *J. Ferm. Tech.* 55, 388–400.

McIntire, L. V., Frangos, J. A., Rhee, B. G., Eskin, S. G., and Hall, E. R. (1987). The effect of fluid mechanical stress on cellular arachidonic acid metabolism. *Ann. N.Y. Acad. Sci.* 516, 513–524.

McIntire, L. V., and Martin, R. R. (1981). Mechanical trauma induced PMN leukocyte dysfunction. In *The Rheology of Blood, Blood Vessels and Associated Tissues*, pp. 214–235, Gross, D. R., and Hwang, N. H. C. (eds.). Sijthoff and Noordhoff, Rockville, MD.

McLimans, W. F., Giardinello, F. E., Davis, E. V., Kucera, C. J., and Rale, G. W. (1957). Submerged culture of mammalian cells: The five liter fermentor. *J. Bacteriol.* **74**, 768–774.

McQueen, A., and Bailey, J. E. (1989). Influence of serum level, cell line, flow type and viscosity on flow-induced lysis of suspended mammalian cells. *Biotechnol. Lett.* **11**, 531–536.

McQueen, A., Meilhoc, E., and Bailey, J. E. (1987). Flow effects on the viability and lysis of suspended mammalian cells. *Biotechnol. Lett.* **9**, 831–836.

Mered, B., Albrecht, P., and Hopps, H. E. (1980). Cell growth optimization in microcarrier culture. *In Vitro Cell Devel. Biol.* **16**, 859.

Michaels, J. D., Petersen, J. F., McIntire, L. V., and Papoutsakis, E. T. (1991). Protection mechanisms of freely suspended cells (CRL 8018) from fluid-mechanical injury. Viscometric and bioreactor studies using serum, Pluronic F68, and polyethylene glycol. *Biotechnol. Bioeng.* **38**, 169–180.

Michaels, J. D., and Papoutsakis, E. T. (1991). Polyvinyl alcohol and polyethylene glycol as protectants against fluid-mechanical injury of freely suspended animal cells (CRL 8018). *J. Biotechnol.* **19**, 241–258.

Michaels, J. D., Kunas, K. T., and Papoutsakis, E. T. (1992). Fluid-mechanical damage of freely-suspended animal cells in agitated bioreactors: Effects of dextran, derivatized celluloses, and polyvinyl alcohol. *Chem. Engr. Comm.* (in press).

Midler, M., Jr., and Finn, R. K. (1966). A model system for evaluating shear in the design of stirred fermentors. *Biotechnol. Bioeng.* **8**, 71–84.

Mizrahi, A. (1975). Pluoronic polyols in human lymphocyte cell line cultures. *J. Clin. Microbiol.* **2**, 11–13.

Mizrahi, A. (1977). Primatone RL in mammalian cell culture media. *Biotechnol. Bioeng.* **19**, 1557–1561.

Mizrahi, A. (1984). Oxygen in human lymphoblastoid cell line cultures and effect of polymers in agitated and aerated cultures. *Devel. Biol. Standard.* **55**, 93–102.

Mizrahi, A., and Moore, G. E. (1970). Partial substitution of serum in hematopoietic cell line media by synthetic polymers. *Appl. Microbiol.* **19**, 906–910.

Murhammer, D. W., and Goochee, C. F. (1988). Scaleup of insect cell cultures: Protective effects of Pluronic F-68. *Bio/Technology* **6**, 1411–1418.

Murhammer, D. W., and Goochee, C. F. (1990). Structural features of nonionic polygonal polymer molecules responsible for the protective effect in sparged animal cell bioreactors. *Biotechnol. Prog.* **6**, 142–148.

Muul, L. M., Director, E. P., Hyatt, C. L., and Rosenberg, S. A. (1986). Large scale production of human lymphokine activated killer cells for use in adoptive immunotherapy. *J. Immunol. Methods* **88**, 265–275.

Muul, L. M., Nason-Burchenal, K., Hyatt, C., Schwarz, S., Slavin, D., Director, E. P., and Rosenberg, S. A. (1987). Studies of serum-free culture medium in the generation of lymphokine activated killer cells. *J. Immunol. Methods* **105**, 183–192.

Nagata, S. (1975). *Mixing: Principles and Applications*, Halsted Press, Tokyo.

Oh, S. K. W., Nienow, A. W., Al-Rubeai, M., and Emery, A. N. (1989). The effects of agitation with and without continuous sparging on the growth and antibody production of hybridoma cells. *J. Biotechnol.* **12**, 45–62.

Oolman, T. and Blanch, H. W. (1986). Bubble coalescence in stagnant liquids. *Chem. Engr. Comm.* **43**, 237–261.

O'Rear, E. A., McIntire, L. V., Shah, E. C., and Lynch, E. C. (1979). Use of rheological techniques to evaluate erythrocyte membrane alterations. *J. Rheology* **23**, 721–733.

O'Rear, E. A., Udden, M. M., McIntire, L. V., Lynch, E. C. (1982). Reduced erythrocyte deformability associated with calcium accumulation. *Biochim. Biophys. Acta* **691**, 274–280.

Pao, Y. H. (1965). Structure of turbulent velocity and scalar fields at large wavenumbers. *Phys. Fluids* **8**, 1063–1075.

Papoutsakis, E. T., and Kunas, K. T. (1989). Hydrodynamic effects on cultured hybridoma cells CRL 8018 in an agitated bioreactor. In *Advances in Animal Cell Biology and Technology for Bioprocesses*, pp. 203–208, (Proc. 9th Meeting Eur. Soc. Animal Cell Technol. Knokke, Belgium, 1988) Spier, R. E. (ed.). Butterworths, London.

Papoutsakis, E. T., Petersen, J. F., and McIntire, L. V. (1991). Cytoskeletal microfilament network and energy metabolism effect ability of animal cells to resist shear injury, In *Production of Biologicals from Animal Cells in Culture* (Proceedings of the 10th ESACT Meeting, Avignon, France, May, 1990). Spier, R. E., Griffiths, J. B., and Meigner, B., (eds.), pp. 229–234, Butterworths, England.

Petersen, J. F. (1989). Shear stress effects on cultured hybridoma cells in a rotational couette viscometer. Ph.D. thesis, Rice University, Houston, TX.

Petersen, J. F., McIntire, L. V., and Papoutsakis, E. T. (1988). Shear sensitivity of cultured hybridoma cells (CRL-8018) depends on mode of growth, culture age and metabolite concentration, *J. Biotechnol.* **7**, 229–246.

Petersen, J. F., McIntire, L. V., Papoutsakis, E. T. (1990). Shear sensitivity of freely suspended animal cells in batch, fedbatch, and continuous cultures. *Biotechnol. Prog.* **6**, 114–120.

Phillips, M. C. (1977). The conformation and properties of proteins at liquid interfaces. *Chem. Ind.* (March 5) 170–176.

Placek, J., Tavlarides, L. L., Smith, G. W., and Fort, I. (1986). Turbulent flow in stirred tanks. Part II: A two-scale model of turbulence. *AIChE J.* **32**, 1771–1786.

Prince, M. (1989). Bubble coalescence and break-up in air-sparged biochemical reactors. Ph.D. Thesis, Univ. of California, Berkeley, CA.

Prince, M. and Blanch, H. W. (1990). Transition electrolyte concentrations for bubble coalescence. *AIChE J.* **36**, 1425–1429.

Prins, A., and van't Riet, K. (1987). Proteins and surface effects in fermentation: Foam, antifoam and mass transfer. *Trends Biotechnol.* **3**, 296–301.

Prokop, A., and Rosenberg, M. Z. (1989). Bioreactor for mammalian cell culture. In *Advances in Biochemical Engineering/Biotechnology*, Vol. 39, pp. 30–67, Fiechter, A. (ed.). Springer-Verlag, New York.

Prosperetti, A. (1982). Bubble dynamics: A review and some recent results. *Appl. Sci. Res.* **38**, 145–164.

Radlett, P. J., Telling, R. C., Stone, C. J., and Whiteside, J. P. (1971). Improvement in the growth of BHK-21 cells in submerged culture. *Appl. Microbiol.* **22**, 534–537.

Rajagopalan, S., McIntire, L. V., Hall, E. R., and Wu, K. K. (1988). The stimulation of arachidonic acid metabolism in human platelets by hydrodynamic stresses. *Biochim. Biophys. Acta* **958**, 108–115.

Ramirez, O. T., and Mutharasan, R. (1990). The role of the plasma membrane fluidity on the shear sensitivity of hybridomas grown under hydrodynamic stress. *Biotechnol. Bioeng.* **36**, 911–920.

Reiter, T., Penman, S., and Capco, D. G. (1985). Shape-dependent regulation of cytoskeletal protein synthesis in anchorage-dependent and anchorage-independent cells. *J. Cell Sci.* 76, 17–33.

Rhee, B. G., and McIntire, L. V. (1986). Effect of shear stress on platelet-PMN leukocyte interactions. *Chem. Eng. Commun.* 47, 147–161.

Rhee, B. G., Hall, E. R., and McIntire, L. V. (1986). Platelet modulation of polymorphonuclear leukocyte shear induced aggregation. *Blood* 267, 240–246.

Rosenberg, S. A. (1990). Adoptive immunotherapy for cancer—also called cell-transfer therapy, it is one of a new class of approaches being developed to strengthen the innate ability of the immune system to fight cancer. *Sci. Am.* 262, 62–69.

Rosenberg, S. A., Lotze, M. T., Muul, L. M., Leitman, S., Chang, A. E., Ettinghausen, S. E., Matory, Y. L., Skibber, J. M., Shiloni, E., Vetto, J. T., Seipp, C. A., Simpson, C., and Reichert, C. (1985). Observations on the administration of autologous lumphokine-activated killer cells and recombinant interleukin-2 to patients with metastatic cancer. *New Engl. J. Med.* 313, 1485–1492.

Rosenberg, S. A., Packard, B. S., Aebersold, P. M., Soloman, D., Topalian, S. L., Toy, S. T., Simon, P., Lotze, M. T., Yang, J. L., Seipp, C. A., Simpson, C., Carter, C., Block, S., Schwartzentruber, D., Wei, J. P., and White, D. (1988). Use of tumor-infiltrating lymphocytes in the immunotherapy of patients with metastatic melanoma. *New Engl. J. Med.* 319, 1676–1680.

Runyan, W. S., and Geyer, R. P. (1963). Growth of L cell suspension in Warburg apparatus. *Proc. Soc. Exp. Biol. Med.* 112, 1027–1030.

Scattergood, E. M., Schlaback, A. J., McAleer, W. J., and Hilleman, M. R. (1980). Scale-up of chick cell growth on microcarriers in fermentors for vaccine production. *Ann. N.Y. Acad. Sci.* 413, 332–341.

Schmid-Schonbein, G. W., Shih, Y. Y., and Schien, K. (1980). Morphometry of human leukocytes. *Blood* 56, 866–875.

Schuerch, U., Kramer, H., Einsele, A., Widmer, F., and Eppenberger, H. M. (1988). Experimental evaluation of laminar shear stress on the behaviour of hybridoma mass cell cultures, producing monoclonal antibodies against mitochondrial creatine kinase. *J. Biotechnol.* 7, 179–184.

Sebba, F. (1987). *Foams and Biliquid Foams-Aphrons.* Wiley, Chichester (UK).

Sevik, M., and Park, S. H. (1973). The splitting of drops and bubbles by turbulent fluid flow. *J. Fluids. Eng.* 95, 53–60.

Shinnar, R., and Church, J. M. (1960). Predicting particle size in agitated dispersions. *Ind. Eng. Chem.* 52, 253–256.

Sinskey, A. J., Fleischaker, R. J., Tyo, M. A., Giard, D. J., and Wang, D. I. C. (1981). Production of cell derived products: virus and interferon. *Ann. N.Y. Acad. Sci.* 369, 47–59.

Smith, C. G., Greenfield, P. F., and Randerson, D. H. (1987a). A technique for determining shear sensitivity of mammalian cells in suspension culture. *Biotechnol. Techn.* 1, 39–44.

Smith, C. G., Greenfield, P. F., and Randerson, D. H. (1987b). Shear sensitivity of three hybridoma cell lines in suspension culture. In *Modern Approaches to Animal Cell Technology*, pp. 316–327. Spier, R. E., and Griffith, J. B. (eds.). Butterworths, Boston, MA.

Smith, C. G. (1990). Mechanical shear effect on hybridoma cells in suspension culture. Ph.D. Thesis. University of Queensland, Australia.

Swim, H. E., and Parker, R. F. (1960). Effect of Pluronic F68 on growth of fibroblasts in suspension on rotaryshakers. *Proc. Soc. Exp. Biol. Med.* 103, 252–254.

Tanaka, M., and Izumi, T. (1987). Gas entrainment in stirred-tank reactors. *Chem. Eng. Res. Des.* **65**, 195–198.

Telling, R. C., and Elsworth, R. (1965). Submerged culture of hamster kidney cells in a stainless steel vessel. *Biotechnol. Bioeng.* **7**, 417–434.

Tramper, J., and Vlak, J. M. (1986). Some engineering and economic aspects of continuous cultivation of insect cells for the production of baculoviruses. *Ann. N.Y. Acad. Sci.* **469**, 279–288.

Tramper, J., and Vlak, J. M. (1988). Bioreactor design for growth of shear-sensitive mammalian and insect cells. In *Advances in Biotechnological Processes* Vol. 7, pp. 199–208, Mizrahi, A. (ed.). Alan R. Liss, New York.

Tramper, J., Smit, D., Straatman, J., and Vlak, J. M. (1988). Bubble-column design for growth of fragile insect cells. *Bioproc. Eng.* **3**, 37–41.

Tramper, J., Willimas, J. B., and Joustra, D. (1986). Shear sensitivity of insect cells in suspension; *Enz. Microb. Technol.* **8**, 33–36.

Uhl, V. W., and Gray, J. B. (1966). *Mixing*, Vol. 1. Academic Press, New York.

Van de Vusse, J. G. (1955). Mixing by agitation of miscible liquids. *Chem. Eng. Sci.* **4**, 178–220.

Van Dierendonck, L. L., Fortain, J. M. H., and Vanderboss, D. (1971). The specific contact area in gas-liquid reactors. In *Chemical Reaction Engineering, Proceedings of the Fourth European Symposium*, pp. 205–215. Pergamon Press, Oxford, UK.

Varani, J., Dame, M., Veals, T. F., and Wass, J. A. (1983). Growth of three established cell lines on glass microcarriers. *Biotechnol. Bioeng.* **25**, 1359.

Wittelsberger, S. C., Kleene, K., and Penman, S. (1981). Progressive loss of shape-responsive metabolic controls in cells with increasingly transformed phenotype. *Cell* **24**, 859–866.

Yssel, H., de Vries, J. E., Koken, M., van Blitterswijk, W., and Spits, H. (1984). Serum-free medium for generation and propagation of functional human cytotoxic and helper T cell clones. *J. Immunol. Methods* **72**, 219–227.Eleftherios T. Papoutsakis and James

CHAPTER 11

∎ ∎ ∎ ∎ ∎

Gravity
and the
Mammalian Cell*

∎ ∎ ∎ ∎ ∎

Paul Todd

I. INTRODUCTION

This chapter is a limited treatment of the influence of inertial acceleration on events critical to the mammalian cell. Most of the experimental
examples considered are drawn from experience with cells exposed to
$(1 \times g)$ or less. Accelerations well above $20 \times g$ are a routine experience
for cells subjected to centrifugation in research, but, apart from the
resulting intense cell–cell interactions, the effect of this process per se is
not studied extensively beyond verifying that viable cells are recovered.
The more fundamental questions concerning the role of inertial accelerations around $1g$ are emphasized in this chapter. This subject is presented in
three major steps. First, the physical processes that need to be considered
are introduced along with some statements about their action in the
mammalian cell environment, especially considering small size, high viscosity, strong and weak intermolecular forces, and the role of the cytoskeleton
in intracellular transport. Second, a selection of experimental observations

*The research for this chapter, performed by the National Institute of Standards and
Technology, an agency of the U.S. government, is not subject to U.S. copyright.

is presented with emphasis on two categories of experiments on mammalian cells in vitro, namely, "low gravity" experiments in orbital spaceflight and modifications of the gravity vector in laboratory experiments, including clinorotation. Third, a list of hypotheses is reviewed with, in some cases, statements concerning conditions under which they might be valid and experiments with which they might be tested.

A. Gravity-Dependent and Interacting Processes

1. Physical Processes in Cells

Physical phenomena that could influence cell behavior include sedimentation, droplet sedimentation, isothermal settling, convection, streaming potential, sedimentation potential, hydrostatic pressure, potential energy, and interactions among physical transport processes. Thermal motion and fluid viscosity play a significant (but not always dominant) role in all transport processes at the cellular level (Purcell, 1977). The sedimentation of intracellular organelles tends to be counteracted by the cytoskeleton. Intracellular convective transport occurs in large cells. In the low-gravity environment extracellular solutes must be transported by diffusion in the absence of convection, and flocculation and coalescence are reduced by the lack of motion of aggregates. Research in gravitational cell biology depends on the evaluation of the full variety of physical phenomena affected by gravity and the roles played by these phenomena in extracellular, intercellular, and intracellular processes.

The study of the behavior of particles in fluids begins with consideration of three gravity-dependent processes: particle sedimentation, zone sedimentation, and convection. Electrostatic, diffusive, and inelastic processes are major nongravitational processes acting at the same time as gravity on all objects in and outside the cell.

2. Brief Descriptions of the Relevant Physical Processes

a. Sedimentation   Stokes sedimentation of particles will occur inside and outside the cell, and the cell itself will sediment or float if freely suspended in a fluid of different density. If we think of the cell as a suspension of Stokes particles in a Newtonian fluid, which it is not (see the following paragraph), then we might treat a large organelle as a Stokes particle with finite dimensions and a drag force, a buoyant force, and a gravitational force, and balance these forces and calculate its velocity using the well-known sedimentation equation for spheres (Todd, 1977, 1989a).

If we survey the physical properties of certain organelles, we find that all of them could sediment within the cell on the basis of their diameters if they were considered to be suspended in an unconfined Newtonian fluid (Table 1). But nearly all organelles are attached to something in the cell, and their motility within the cell is due to the action of the cytoskeleton. This role of the cytoskeleton is a major scientific issue as is the dynamics of the cytoskeleton in intracellular graviception. The cytoskeleton is actually part of the metabolic machinery of the cell. The actomyosin fibers of the cytoskeleton have ATPase functions, and the ATPase functions can translate chemical energy into mechanical energy. Nevertheless, as Table 1 implies, some organelles are known to sediment, especially amyloplasts in plant root-cap cells and otoconia (otoliths) in the organ of balance of mammals and most other animals; the latter are suspended in extracellular fluid. Additionally, the influence of Brownian movement on organelle-sized particles is substantial, and diffusive processes combine with the gravity vector to produce such phenomena as droplet sedimentation and isothermal settling.

b. Diffusion   Diffusion, or Brownian motion is not caused or modified by gravitational acceleration. Three broad categories of diffusing substances are of interest in mammalian cell science: small molecules (sucrose) with $D = 5 \times 10^{-6}$ cm^2/s, large molecules (albumin) with $D = 6 \times 10^{-7}$ cm^2/s, and whole cells or membranous organelles with $D < 10^{-10}$ cm^2/s. Diffusion and sedimentation velocities are sometimes similar, and their sum results in gradual settling; and under certain combinations of diffusivity, viscosity, and concentration gradients, the collective behavior of dissolved molecules and/or particles results in droplet (or zone) sedimentation.

c. Droplet sedimentation   Droplet sedimentation is due to a diffusion-driven local unstable density gradient. The diffusion coefficients of small molecules are in the range 10^{-6}–10^{-5} cm^2/s, of macromolecules 10^{-7}–10^{-6}, and of whole cells and particles 10^{-12}–10^{-9}. If a small zone, or droplet, contains particles or macromolecules whose diffusivity is much less than that of the solutes outside, then rapid diffusion of solutes in and slow diffusion of particles out of the droplet (with conservation of mass) leads to a locally increased density of the droplet, whose motion, at least temporarily, follows that of a Stokes particle (Mason, 1976), causing the formation of "streamers." Under conditions of droplet sedimentation, particles still react individually to other forces unless the ionic environment also permits aggregation (Boltz and Todd, 1979; Omenyi et al., 1981; Todd and Hjertén, 1985; Todd, 1985).

In the case of erythrocytes there is sufficient electrostatic repulsion among cells to permit the maintenance of stable dispersions up to at least

TABLE 1

Physical Properties of Organelles Used to Calculate Stokes Sedimentation Velocities

Organelle	Volume (μm^3)	ρ (g/cm^3)	$\rho - \rho_0$ (g/cm^3)	v (cm/s)	t (s)	x (μm)	Feature
Mitochondrion	2–100	1.1	0.01–0.02	0.1–4×10^{-8}	10^3	0.1	Convoluted structure
Nucleolus	10–20	1.4	0.3	2×10^{-7}	10^4	20	Suspended by chromatin
Chromosome	5–50	1.35	0.3	2×10^{-7}	10^3	2	Suspended by microtubules
Amyloplast	100	1.5	0.4	1×10^{-6}	$< 10^3$		Sedimenting particle
Otolith	1000	2.0	0.8	$> 1 \times 10^{-5}$	1	0.1	Known g-sensor
Dictyosome	100	1.2	0.15	$> 3 \times 10^{-7}$	10^3	2	Internal membrane

3×10^8 cells/ml (Snyder et al., 1985; Omenyi et al., 1981) up to at least 0.15-*M* ionic strength. Droplet sedimentation occurs in solutions of macromolecules as well as in suspensions of particles, and the region around a secreting mammalian cell may be considered a zone in which concentration (and hence potentially droplet sedimentation) of macromolecules occurs. Mammalian cells in vivo are normally surrounded by other cells or a very thin (a few nanometers) unstirred layer in contact with a flowing body of fluid, but cells studied in vitro may have thicker unstirred layers and more possibilities for the creation of sedimenting microscopic zones within a few hundred nanometers of the cell surface. In such cases droplet sedimentation could give way to steady convective flow.

d. Isothermal settling Isothermal settling produces vertical concentration gradients in which Stokes sedimentation and Brownian motion are at equilibrium. If the temperature T does not change over the height h of an ensemble of particles, then the mean kinetic energy, which is proportional to kT, of all particles is the same at all heights. The potential energy of a particle of mass m is usually expressed as mgh, but if the particles are subject to buoyant forces in the fluid, the potential energy becomes $V(\rho - \rho_0)gh$, if the particle volume is V. From the Boltzmann distribution rule the concentration of particles at height h is

$$c(h) = c(0)\exp\left[\frac{-V(\rho - \rho_0)gh}{kT}\right] \tag{1}$$

The value of kT is 0.04 eV/molecule or 3 kJ/mol at 310 K. This means concentration is an exponential function of height under isothermal conditions and that large, dense particles with potential energy $\gg kT$ (from mammalian cells to marbles) will be concentrated at $h = 0$ and that small particles (from molecules to microbodies) will have $c(h) \approx$ constant. However, submicrometer organic particles, such as certain organelles, have values of V and ρ that lead to measurable exponential distributions of $c(h)$.

Stokes' sedimentation in the context of intracellular settling was considered in the papers of Pollard (1965, 1971) and others, including Tobias et al. (1973) that address the role of Brownian movement. Brownian movement in a sense counterbalances sedimentation in the case of very small particles, less than about 0.5 μm in diameter and below a density difference of about 0.03 g/cm^3.

If we consider, for example, a collection of 0.2-μm particles, suspended in a 20-μm vessel, possibly a cell, then the particles will be distributed isothermally and adiabatically in the vertical direction with many more at the bottom than at the top, and if we reduce inertial acceleration to $(0.01)g$, those same $0.2\text{-}\mu\text{m}$ particles will be almost uniformly distributed through the same small volume.

   e. *Natural convection*   Convection may be due to a density gradient caused by a temperature gradient (thermal convection) or a solute concentration gradient (solutal convection). This motion can be spatially patterned (Bénard cells) and may be important or unimportant in intracellular processes. Thermal convection could require higher thermal gradients than are possible in most living cells, but, in addition to thermal convection, solutal convection can occur when concentration gradients lead to dense solutions being found above or beside less-dense solutions, even under isothermal conditions. Owing to the lack of good quantification of natural convection at small dimensions and poorly understood hydrodynamic properties of the cell, we do not know whether convection inside a single cell is possible. However, Kessler (1978) conducted a dimensionless analysis of fluid motion in plant cells by estimating the Peclet number Pe using the ratio of characteristic time for diffusive transport L^2/D to the characteristic time for streaming transport l/q, where L is a characteristic length of the cell, D is diffusion coefficient, l is one-half the cell circumference, and q is streaming velocity. Or

$$\text{Pe} = \frac{L^2/D}{l/q} \qquad (2)$$

For approximately round cells, $L = 2l$, and $\text{Pe} = 4lq/D$, and it was found that streaming transport slightly dominates diffusion.

   It is apparent that convective forces due to buoyancy play a role in early postnucleation events during the growth of submicrometer crystals of proteins from solution (Kam et al., 1978)—a process that resembles the self-assembly of cytoskeletal structures in cells.

   f. *Electrokinetic phenomena*   Electrophoresis and sedimentation potential are electrokinetic phenomena. The former is gravity-independent, while the latter is caused by inertial acceleration. Electrophoresis is the motion of particles (molecules, small particles, and whole biological cells) in an electric field and is one of several electrokinetic transport processes. The velocity of a particle or molecule per unit applied electric field is its electrophoretic mobility, μ, a characteristic of individual particles.

   The surface charge of suspended particles prevents their coagulation and leads to stability of lyophobic colloids including, in some cases, suspensions of subcellular particles and cells. The surface charge also leads to motion when such particles are suspended in an electric field. The particle surface has an electrokinetic (zeta) potential ζ proportional to σ_e, its surface charge density—a few millivolts at the hydrodynamic surface of stable, nonconducting particles, including biological cells and organelles, in aqueous suspension (O'Brien and White, 1978).

If a charged particle sediments in an electrolyte solution a potential will be created, and this potential will impart motion to other charges in the environment, including dissolved ions. While the ζ potential of a stationary particle is only "felt" by charges up to 7 Å or so away (the Debye length), this electric field is swept through a greater distance as the particle sediments. If a particle is caused to move by the acceleration due to gravity (upward or downward), the strength (V/cm) of the electric field generated is directly proportional to the inertial acceleration, or g. This potential could be as great as 20 mV, which is comparable to the negative potential at the surface of a typical cell, organelle, or colloidal particle. A streaming potential can also be developed by passing fluid over a charged surface.

g. Combined effect   A combination of fields is customarily present, and most objects are acted on by a combination of forces. To deal with this fact, all types of flow (mass, charge, magnetic flux, etc.) are assumed to be interdependent, and transport relationships are described by a flow-and-field matrix. Thus, in the generalized Onsager relationships, more than one type of field can cause more than one type of flow, so, for example, electric potentials can move charged masses and inertial potentials can move charges associated with mass. In most cases, the cross-term coefficients (the effect of gravity on a current and the effect of the electric field on sedimentation, respectively) are considered to be negligible; however, at subcellular dimensions, it may not be possible to ignore cross terms in small-particle transport (Tobias et al., 1973).

h. Phase separation   Aqueous liquid-phase separation can occur. This happens when, for example, two polymers A and B are dissolved in aqueous solution at concentrations that cause phase separation, and an upper phase forms that is rich in A and poor in B, and a lower phase forms that is rich in B and poor in A. In laboratory applications, typically A is polyethylene glycol (PEG), considered a relatively hydrophobic solute, and B is dextran or a similar polysaccharide. Solute B can also be a salt at high concentration. The phase separation process is normally driven by buoyancy in the presence of gravity or a centrifugal force. In low gravity "top" and "bottom" lose their meaning, and A-rich and B-rich phases are defined (Van Alstine et al., 1987).

As an example, one two-phase aqueous system consists of top and bottom phases with densities of 1.0164 and 1.1059 g/cm^3, respectively, and the corresponding viscosities are 0.0569 and 4.60 P. Because of the high concentration of macromolecular solutes in the cell (Fulton, 1982), boundaries between aqueous phases should not be unexpected. In macroscopic experiments, organelles and liposomes are known to partition between phases (Albertsson, 1986), and phase separation can also be driven by an applied electric field (Rao et al., 1990).

i. Interfacial tension   Interfacial tension is a property of cell membranes that establishes their stability with respect to internal and external aqueous phases. Wetting layers form in cells, presumably, as they do in laboratory vessels that contain fluid; indeed, intrusion (wetting) layers in vessels can be about the same thickness as membranes in cells, and, at least macroscopically, the process is affected by gravity (Kayser et al., 1986). The surface tension of the mammalian plasma membrane is surprisingly low, owing to the surfactant effect of the transmembrane proteins—about 0.02–0.1 dyn/cm—so low that some large cells could sag under their own weight, based on approximate calculations.

j. Hydrostatic pressure   Hydrostatic pressure is a gravity-dependent component of the cellular environment. Most cells are found in a submerged environment subjected to experimentally determined hydrostatic pressure. Under certain experimental conditions, such as low gravity, the hydrostatic pressure can be nearly zero. Intracellular processes that involve a volume change, such as secretion, cytokinesis or fission, and multimolecular chemical reactions that involve changes in partial molal volume should be affected by changes in hydrostatic pressure.

k. Transmembrane phenomena   Mammalian transmembrane channel proteins effect communication between the cell and the outside world, including its inertial accelerations. As pointed out in another chapter in this book, the physical laws that govern the properties and functions of transmembrane channel proteins have been learned over the past decade.

Like most transmembrane proteins, channel proteins are subjected to many tugs at different parts of their structure, and these tugs are generally of the order of 1 kT in energy. For example, the cytoskeleton, through normal contractile processes, tugs at the amino terminus; effector molecules that cause "patching" on the cell surface tug at the extracellular moiety; while thermal and nonthermal motions of the lipid bilayer tug at the structure of the channel itself, admitting ions into the cell at random. These processes, typically studied by the patch-clamp technique, have indicated that the living cell is capable of detecting (and responding to?) single events of the order of kT (Sachs, 1988; Morris and Sigurdson, 1989). These characteristics are significant because they represent the cell's most sensitive responses to its environment, and the sensitivity is of the order of the energy of cell-sized objects under the influence of gravity.

B. Cell Biology and Biotechnology

Bioprocessing exposes mammalian cells to forces that do not normally occur during their in vivo lifetime. The presence of buoyant bubbles of gases, ample space into which to sediment or float, shear stresses, stirred and unstirred free fluids around the cell, the absence of interacting cell types, and other factors conspire to create conditions under which gravitational acceleration can influence biological processes not normally affected by gravity.

Cultivation techniques subject the mammalian cell to vigorous tumbling (as in fluidized-bed culture systems, including microcarrier methods) moving monolayers (roller bottles and tubes) or sessile monolayers. None of these conditions approximate the relationship of the mammalian cell to the gravity vector in vivo. There are numerous reasons to predict that this modification of lifestyle should have little or no effect at the intracellular level. However, cultivation of cells in a sessile monolayer has the unusual effect of a never-reorienting gravity vector in the inertial frame of the cell.

C. Physics of Cell Clinorotation

If a cell is frequently reoriented by rotation, there will be too little time for statolith motion that is significant to the cell. This hypothesis underlies the design of plant gravitropism experiments on clinostats. Mammalian tissue cells are not known to be equipped with organelles that cause them to be responsive to the gravity vector. Cells may be rotated while suspended or attached.

1. Clinorotation of Suspended Cells

Clinorotation of suspended particles is physically different from the rotation of a solid body, such as attached cells or whole organisms (plants). The concept is illustrated in Figure 1, which shows that the total vertical velocity vector oscillates, so that circular motion results with a radius vector that can be derived from equations of motion in which the sample zone is treated as a solid particle. In actuality, the sample zone is more like

FIGURE 1   Trajectory followed by a high-density zone in a rotating tube. In the presence of gravity the center of the circle followed by the zone is below the center of rotation (adapted from Hjertén, 1962; Todd, 1990).

a sedimenting droplet, which can be treated as a particle with density ρ_D (see Section I,A,2,c, on droplet sedimentation, above). Due to gravity and centrifugal acceleration, the center of the circle (coordinates k, l) is not the center of the rotating vessel, and the vertical circle, in x and y, described by the sample zone is

$$(x - k)^2 + (y - l)^2 = r^2 \exp(2\gamma t) \qquad (3)$$

in which a zone is stabilized in suspension when $k^2 + l^2$ and γ, the inverse time constant for centrifugal motion are minimized. Hjertén solved the equations of motion for these values and found that neither has a minimum value as a function of ω or of any other controllable variable, so reasonable values must be determined from the solutions

$$\gamma = \frac{V_D(\rho_D - \rho_0)}{6\pi\eta R/\omega^2 + 32\pi R^5\rho\rho_0/27\eta} \qquad (4)$$

$$\sqrt{k^2 + l^2} = \frac{V_D(\rho_D - \rho_0)g}{\omega\sqrt{(4\pi R^3/3)^2\omega^2\rho_0^2 + (6\pi\eta R)^2}} \qquad (5)$$

It is possible to define conditions of zone density ρ_D and radius R, solution density ρ_0, viscosity, and angular velocity ω, such that, in a tube of radius R, the zone will be maintained in suspension. For example, if the Hjertén number

$$Hj \equiv \gamma\tau \qquad (6)$$

[recently defined (Todd, 1990)] is < 1.0, then acceptable conditions for stability of the sample zone exist. In a typical simulation experiment τ, the residence time, is between 10^4 and 10^6 s, and γ is between 10^{-5} and 10^{-3} s^{-1}. On the other hand, if g is set to zero in Eq. (5), the sample zone is not displaced vertically from its original position except by centrifugal

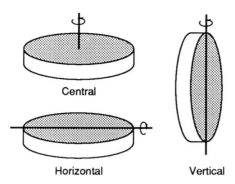

Central

Horizontal   Vertical

FIGURE 2   Diagram showing definitions of axes and orientations of rotated cylindrical culture vessel.

acceleration, so, in the absence of g the tube need not rotate, and $\omega = 0$ in Eq. (4), giving $\gamma = 0$ and no centrifugal motion. Thus, low gravity represents an improvement over the rotating tube in terms of vertical sample zone stability (Gordon and Shen-Miller, 1971).

2. Clinorotation of Attached Cells

When a monolayer cell culture in a cylindrical chamber is rotated about one of its three main axes (horizontal, vertical, or central, as defined in Fig. 2), the physical variables affected are g vector, centrifugal force, and hydrostatic pressure, with their attendant effects on concentration and density gradients in the culture medium. Shear stress and convective mixing could be modified in a rotating culture relative to a static culture (Todd, 1992).

a. Inertial acceleration   The constant acceleration to which the entire experiment is subjected is $1g$. For clinorotation to eliminate the directional bias caused by this acceleration (nonvectorial gravity) requires a rotation frequency f that eliminates the response of a cell's sensor, if any, such as a statolith as in plant root statocytes or *Chara* rhizoids (Sievers and Hensel, in press). Such a rotation would limit an object's fall distance x to x_r, where x_r is the minimum fall distance that causes a cellular response. Thus, the object's sedimentation velocity v cannot exceed $2fx$ ($v < 2fx$). A "critical" rotation frequency for an object obeying Stokes's law for spheres and whose terminal sedimentation velocity is $2fx$ is thus given by

$$f = \frac{(\rho - \rho_0)a^2 g}{9\eta x} \tag{7}$$

where a is the radius of the sphere, ρ_0 is the medium density, and η is viscosity. If the gravity sensor is a sedimenting intracellular object ("statolith"), then there are constraints on $\rho - \rho_0$, a, and x; namely, $\rho - \rho_0 = 0.0002 - 0.15$ g/cm^3, $a = 0.01 - 1.0$ μm, and $x = 0.1 - 10$ μm. Using these constraints leads to a minimum value of $f = 10^{-9}$ Hz for the smallest, least dense, most distantly sedimenting statolith and a maximum value of $f = 0.4$ Hz. Only the upper limits can be valid on thermodynamic grounds (Pollard, 1965; Tobias et al., 1973; Todd, 1989a). If a statolith hypothesis were to apply to a mammalian cell, then the properties of the intracellular particle would be $a > 1.0$ μm, density > 1.20 g/cm^3, and fall distance ≈ 0.1 μm, if the cytoplasmic viscosity $= 10$ cP. These properties are based on Eq. (7) and the fact that $0.01 < f < 2$ Hz in most experimental studies.

b. Centrifugal force Centrifugal force depends on the distance between the cell and the center of rotation. If cells that are observed after rotation about a horizontal axis occupy a band 4 mm wide along the axis of rotation, then, in the standard relations the centripetal acceleration a' is given by $a' = \omega^2 r$, $r = 2$ mm. At this radius $a' = (0.01)g$ at 60 rpm, and $a' = (2 \times 10^{-4})g$ at 1 rpm. Centrifugal-force control cultures can be created by rotating them about their central axis as defined in Figure 2 (Gruener and Hoeger, 1990), and these would experience up to $(0.04)g$ lateral acceleration. The steady-state velocity of a "statolith," of the type mentioned in the previous section in a 60-rpm acceleration field, would be 2×10^{-5} cm/s.

c. Hydrostatic pressure Hydrostatic pressure changes are experienced by a cell positioned at the center of an interior surface of a flat cylindrical chamber and at the axis of rotation. It fluctuates from $h\rho g$ when the cell is on the lower wall of the disk and subsequently through variable depths of medium defined by trigonometric relationships (Todd, 1991). In a culture chamber of typical size the maximum hydrostatic pressure encountered is about 100 Pa (1 mm H$_2$O corresponds to roughly 10 Pa). The maximum hydrostatic pressure encountered by nonrotated controls would be similar.

The effect of pressure modulation on differentiated function in vitro has been investigated by Levesque and Nerem (1989). Amplitudes of approximately 30 dyn/cm^2 (3 Pa) at a frequency of 1 Hz were found to alter orientation, morphology, proliferation, and migration of cultured bovine endothelial cells. Additional details can be found in Chapter 6 of this book.

Modified hydrostatic pressure might be expected to affect certain cellular processes that involve volume changes. Results of some experiments on single cells in space suggest the following. In the case of secretion, Hymer et al. (1988) reported preliminary evidence that rat anterior pituitary cells secrete significantly less growth hormone when maintained in vitro in spaceflight ($P = 0$ Pa vs. $P = 500$ Pa on earth)—this finding has been confirmed in the same cell type only in in vivo experiments, where systemic effects of reduced gravity on the whole animal could also play a role (Grindeland et al., 1987). *Paramecium aurelia* displayed a lack (and possibly reversal) of exocytosis in the form of unreleased trychocysts buried beneath the plasma membrane (Richoilley et al., 1988). On the ground this organism also lives under about 500-Pa hydrostatic pressure. Organelle anomalies were noticeable in electron micrographs of lymphocytes subjected to low gravity in vitro (Cogoli et al., 1988).

d. Shear stress   Shear stress has not been measured in small rotating culture systems during steady rotation. The lowest possible shear stress at the cell layer is zero if one assumes that a stagnant fluid layer exists that is greater than the thickness of a cell attached to the tumbled surface. This condition is expected in constant-velocity bulk fluid rotation in the absence of density gradients, such as those imposed by air bubbles. The maximum possible shear stress, which could occur only during onset and termination of rotation, assuming the medium is a Newtonian fluid (in which viscosity is independent of shear stress), can be estimated from

$$\tau_{yz} = -\eta \left(\frac{dv_z}{dy} \right) \tag{8}$$

where

$$\frac{dv_z}{dy} = \frac{(\omega h - 0)}{2h} = \frac{\omega}{2} \; s^{-1} \tag{9}$$

so that the velocity gradient is simply equal to one-half the angular rotation frequency. The viscosity in mammalian-cell culture medium at 20°C is typically 0.01 dyn · s/cm^2 (0.001 Pa · s), and the angular frequency at 1 rpm is 0.105 rad/s. Thus the maximum possible shear stress is 0.0012 dyn/cm^2 (0.00012 Pa). Consequently, the shear stress resulting from the rotation appears to be entirely negligible, as the minimum shear stress at which effects on secretion in cultured cells in monolayer has been reported is around 2 dyn/cm^2 (0.2 Pa) (Stathopoulos and Hellums, 1985), and studies using up to 90 dyn/cm² (9 Pa) (Cherry and Papoutsakis, 1986a, 1986b, 1987; Frangos et al., 1985; Levesque and Nerem, 1985) have demonstrated several-fold increases in secretion rates. These effects occur at shear stresses some three orders of magnitude higher than expected in most clinorotation experiments. Nevertheless, increased shear stress at the surface of cell monolayers can modify transmembrane potential (de Souza et al., 1986), cytoskeletal rigidity (Sato et al., 1985), and receptor-mediated binding (Sprague et al., 1987).

e. Convection Extracellular convective mixing of cell products and metabolites is modified by clinorotation. Differentiating or secretory cells in vitro are known to secrete macromolecules that concentrate near the secreting cell either as secretion granules (such as presynaptic acetylcholine vesicles in the case of nervous system cells, typical radius = 200 nm, typical density = 1.06) or single molecules (such as acetylcholine receptor clusters, typical molecular weight = 60,000–300,000, typical diffusion coefficient = 10^{-6} cm²/s). In a static, nonrotating monolayer culture with the cells at the bottom, accumulation of secreted materials may exceed diffusion. With tumbling, these products, if they form a dense zone over the cell, will be convected away when the vessel is reoriented. In the case of a horizontal culture rotating about its central vertical axis, only centrifugal acceleration will cause convection; in the case of a vertical culture that is static or rotating about its vertical axis, steady downward convection of products will lead to a lowered equilibrium concentration of them in the vicinity of the secreting cell; and in the case of a culture rotating about its horizontal axis (tumbling), convection will sweep away products in one direction and then the opposite direction with each half rotation of the vessel.

The purely diffusive mixing that occurs in static cultures is slow. For example, the time required for a protein molecule, with typical molecular weight (10^4–10^5) and diffusion coefficient $D = 10^{-6}$ cm²/s, to diffuse between cells 300 μm apart or to any location 300 μm away, can be estimated from Einstein's relationship

$$t = \frac{\langle x \rangle^2}{2D} \tag{10}$$

where $\langle x \rangle^2$ = root-mean-square diffusion distance and D is diffusion coefficient. Using Eq. (4) with $\langle x \rangle^2 = 0.1$ mm² and $D = 10^{-6}$ yields about 12 min for the mean transit time for this process.

On the other hand, when the density gradient becomes inverted (opposite to the gravitational acceleration vector) and the Rayleigh–Taylor conditions

$$\frac{d\rho}{dz} > \frac{67.94\mu D}{gr^4} \tag{11}$$

are met, the stratified fluid system will become unstable and zone or droplet sedimentation will occur. Here, $\mu = \eta/\rho$, the kinematic viscosity. Under certain combinations of D, η, and dc/dz, the collective behavior of dissolved molecules and/or particles results in droplet (or zone) sedimentation, and Stokes sedimentation will occur depending on the volume and density of the zone (Mason, 1976; Tobias et al., 1973). Under convecting conditions the time required for a solitary droplet to travel the same 300-μm distance can be estimated using

$$t = \frac{x}{v} = \frac{9x\eta}{2R^2(\rho_D - \rho_0)g} \tag{12}$$

If, for example, a cell secretes 10^{-11} g of protein (0.01 of its own mass) into a hemispherical volume of radius $R = 20$ μm, so that $\rho_D - \rho_0$ becomes 6.25×10^{-4} g/cm^3, then t determined from Eq. (12) is 617 s or 10 min, comparable to the time required for a molecule to travel the same distance by diffusion. The secretory rate of the cell types in these cultures (in grams per hour per cell) is thus of some importance (Hymer et al., 1988).

In each of the three rotating cases it is possible to estimate a value for the Peclet number Pe [see Eq. (2)], the ratio of characteristic time for diffusive transport to that for convective transport. In a static horizontal culture Pe ≈ 0, because $v \approx 0$. In the above conditions Pe $= 12$ min/10 min ≈ 1.2, indicating that convection dominates *very* slightly. Pe will be sensitive to molecular weight and secretion rate.

The characteristic times are of the order of 10 min for both transport processes, while the rotation time is of the order of $1.6 - 60$ s, generally shorter than either characteristic time. Thus, at 1 rpm, a new dense volume over the secreting cell will be swept away approximately once per 10 revolutions, assuming that cell secretion rate (grams of protein per cell per hour) is rapid compared to the diffusive removal of product. The clinorotation effect should be sensitive to changes in revolutions per minute (rpm) in the 0.01–0.1-Hz range if secretion product removal explains the effect of tumbling.

Chemical differentiation processes that occur in the embryo, such as induction of the neural protein En-2, which requires the presence of dorsal mesoderm cells (Hemmati-Brivanlou et al., 1990) are presumably the same in vivo and in vitro. However, transport processes on which effector–receptor interactions depend are very different in vivo and in vitro and are presumably modified in rotating cultures. Tactic motions of animal cells, which usually require some kind of chemical gradient, will almost certainly have their gradients modified by clinorotation.

3. Conclusions Concerning Cell Clinorotation

Not all of the variables mentioned above are indicators of the cells' intrinsic responsiveness to inertial acceleration, so transport modeling of attached-cell clinorotation is crucial to the development of an understanding of gravitational effects when they occur in single cells.

II. EXPERIMENTAL OBSERVATIONS ON CELLS UNDER MODIFIED INERTIAL ACCELERATION

A. Tabulation of Experiments Exposing Mammalian Cells to Altered Inertial Acceleration

Several single-cell systems have been investigated under increased and decreased inertial acceleration, relative to $1g$, using centrifuges, clinostats, and orbital spaceflight. Table 2 is a tabulation of several such experiments involving mammalian cells and the observed results. The table cannot be considered a complete listing, and it does not include interpretations. Further details can be found in the references cited for each entry. A few selected cases are presented in detail in Section II,B.

B. Selected Experimental Observations

1. Cultured Mammalian Fibroblasts: Proliferation and Metabolism

a. Human fibroblasts in low gravity   Human diploid fibroblasts in orbital flight exhibited little response to the altered g vector. Early work in the U.S. Space Program indicated little or no effect of microgravity on the growth of human fibroblasts in vitro (Montgomery et al., 1977). Cultured human WI-38 fibroblasts were grown during the 59-day mission of Skylab. The population doubling time in flight, 22.3 ± 3.1 h, did not differ significantly from that at $1g$, 20.4 ± 4.8. The speed of cell migration on the culture vessel surface was the same, and no ultrastructural or karyotypic differences could be observed by the investigators. Cells that had

TABLE 2

Effects of Modified Inertial Acceleration on Various Mammalian Cells

Cell type	Modified acceleration	End point	Result	References
Human diploid fibroblasts	Spaceflight $10\,g$	Growth rate metabolism	No change 25% less glucose consumed	Montgomery et al., 1977
Chinese hamster	Vertical substratum	Direction of division	No effect of orientation	Todd, 1977
Human lymphocytes	Spaceflight $10\,g$, $1\,g$	Blastogenesis	95% reduced DNA synthesis	Cogoli et al., 1988
Human lymphocytes	Clinostat	Blastogenesis	50% reduced DNA synthesis	Cogoli et al., 1988
Mouse hybridoma	Spaceflight $(1.4)\,g$	Ultrastructural RNA synthesis	Pending	d'Augères et al., 1988
Human leukocytes	Spaceflight	Interferon production	Fivefold increase	Talas et al., 1983
Human leukocytes	Spaceflight storage	> 10 storage variables	Three variables changed	Surgenor et al., 1990
Human leukocytes	Spaceflight irradiation	Cytogenetic damage	Increased single break on 1 flight	Bender et al., 1968
Human fetal kidney	Spaceflight electrophoresis	Plasminogen activators	No obvious differences	Lewis et al., 1987
Human fetal kidney	Spaceflight electrophoresis	Morphology	No obvious differences	Todd et al., 1985
Human fetal kidney	Spaceflight incubation	Attachment to microcarriers	Normal attachment	Tschopp et al., 1984
Rat pituitary	Spaceflight electrophoresis	Electrophoretic mobility	Normal mobility	Hymer et al., 1987
Rat pituitary	Spaceflight incubation	Hormone secretion	50–90% less GH,[a] Same PRL[b]	Hymer and Grindeland, 1989
Human RBC	Spaceflight storage	> 10 metabolism, morphological variables	Small change in pH, p_{CO_2}	Surgenor et al., 1990
Human RBC	Spaceflight stopped flow	Aggregation in low shear	Less aggregation	Dintenfass et al., 1985
Human platelets	Spaceflight storage	> 10 metabolism, morphological variables	Decreased rate of deterioration	Surgenor et al., 1990
Neuromuscular junction	Clinostat rotation	Junctional complex formation	Reduced receptor formation	Gruener and Hoeger, 1990
Mouse hybridoma	Rocket flight	Electrofusion	Less size dependence	Schnettler et al., 1990

[a] Growth hormone.
[b] Prolactin.

rounded for mitosis did not seem to require the gravitational force to reattach to the surface on which they were growing. A small, but statistically significant, reduction (about 25%) in glucose consumption was noted.

b. Oriented cell cultures   Cultured fibroblasts on a vertical surface were studied in certain laboratory experiments. In these lab experiments the orientation of cell divisions was monitored in horizontally attached and vertically attached cultured mammalian cells in monolayer. Results indicated that the cell division process in cultured mammalian cells is rather insensitive to the influence of gravity (Todd, 1977). Chinese hamster fibroblasts with distinct polarity were cultured attached to the surface of plastic T flasks filled with medium and incubated for several days in the vertical or horizontal position. The angle subtended by the plane of cell division and the long axis of the flask was measured on a few hundred cells, and the data collected consisted of the ratio of the fraction of cells V in each angular interval on vertical flasks divided by the corresponding fraction of cells H on horizontal flasks. An example of the resulting histogram of V/H ratios is given in Figure 3, where it is seen that there

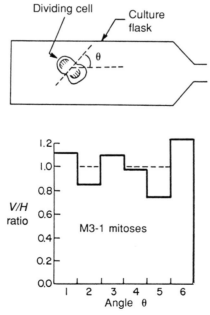

FIGURE 3   Histogram showing the ratios of mitoses in vertical to those in horizontal culture flasks at each interval of the mitosis orientation angle, defined in the upper drawing (Todd, 1977).

was no evidence for a preferred orientation of cell division in vertical cultures.

2. Lymphocytes and Leukocyte Culture Systems

a. Blastogenic response The blastogenic response of stimulated human lymphocytes is profoundly affected by spaceflight conditions. One of the most notable effects of gravity at the cell (or cell culture) level was reported by Cogoli et al. (1984, 1988), who observed a 95% reduction in the incorporation of radioactive thymidine into DNA of human lymphocytes stimulated by concanavalin A. This effect was observed in identical experiments on three spaceflights, and was insignificant on a $1g$ centrifuge in orbital spaceflight aboard Spacelab mission D-1 (STS 61-A, Oct.–Nov. 1985). A synopsis of these results is provided in Figure 4, which implies a steadily increasing rate of blastogenesis with increasing inertial acceleration. However, a confirmed hypothesis explaining this conspicuous effect of reduced inertial loading is still lacking (see discussion in Section III, below).

b. Mouse hybridoma cell line A continuously proliferating mouse hybridoma cell line, AM2, produces a specific antibody in vitro during continuous multiplication. The cells have a polarized structure with an

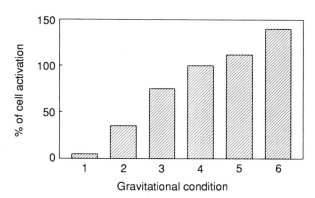

FIGURE 4 Lymphocyte activation by concanavalin A in 72 h based on percent labeled cells in 72 h of exposure to ^{3}H-thymidine. Percent activation is percent of labeling index compared to ground controls. Cells were isolated from human peripheral blood by centrifugation on Ficoll. (1) Orbital spaceflight, (2) clinostat, (3) orbital spaceflight with reference $1g$ centrifuge, (4) synchronous ground control at $1g$, (5) synchronous ground control centrifuged at $(1.4)g$ total acceleration (vector sum of $\omega^2 r + g$), (6) centrifuged at $10g$. [Constructed from data of Cogoli et al., 1988; Lorenzi et al., 1988.]

eccentric nucleus. After 168 h of low-gravity exposure in spaceflight, during which some cells were fixed with glutaraldehyde in culture, fixed cells were prepared for electron microscopy and uridine radioautography, and live cells were collected and recultured for viability and growth tests. Low-gravity cultivation was achieved on the German Spacelab Mission "D-1" in October 1985. To date, three aspects of the experiment have been investigated: ultrastructure through serial sectioning of cells fixed during flight, RNA synthesis as determined by uridine radioautography, and the viability of returned cells on the basis of trypan blue staining and their ability to resume growth (d'Augères et al., 1988). Viability data indicate that the conditions chosen for cultivation were suboptimal. Ground control cultures grown in flight containers under flight conditions (except low gravity) grew more slowly and lost about 5% of their viability. With this baseline, cells cultivated at $1g$ and low gravity grew at the same rate, whereas cultures incubated at $(1.4)g$ grew much more slowly (165-h doubling time vs. 50-h doubling time). Further reports on this set of experiments are expected as data analysis (especially ultrastructure work) proceeds.

c. Interferon release   In a Soviet-Hungarian program, a fivefold increased interferon secretion by human cells has been reported (Talas et al., 1983).

d. Radiation effect   The radiation responses of leukocytes in two nearly identical experiments utilizing human leukocyte cultures were examined in orbital spacecraft. Leukocytes from two normal subjects were stimulated to divide with phytohemagglutinin and exposed to ^{32}P β-irradiation for 20 min during the flight of Gemini III and for 70 min (at a lower dose rate) on Gemini XI. Total low-gravity periods were 4.7 and 71 h, respectively, and the dose range studied was 4–283 cGy (centigrays; 1 gray unit of absorbed radiation dose = 1 J/kg of energy). Single- and multiple-break chromosome aberrations were scored in colchicine-arrested metaphases, and no differences were observed in the frequency of multiple-break aberrations (rings and dicentrics), but the frequency of single-break aberrations (chromosome deletions) increased approximately twofold (statistically significant) in cells that were irradiated on Gemini III (Bender et al., 1967) compared to synchronous ground controls. However, this difference was not reproduced in a similar experiment on the 3-day Gemini XI mission or in launch simulation experiments (Bender et al., 1968). The only explanation offered for the differences in results between these two experiments is random sampling error, but this explanation is not considered satisfactory, and the experiments have not been repeated (Shank, 1974).

3. Functional Mammalian Cell Cultures

a. Kidney cell electrophoresis in low gravity   Human embryonic kidney (HEK) cells were subjected to electrophoresis in the low-gravity environment of a space shuttle flight. Early-passage cultures of human embryonic kidney cells contain a small fraction of cells that produce plasminogen activators, and those that produce urokinase consistently appear in a high-electrophoretic-mobility fraction (Todd et al., 1986). Such cell populations were separated at high concentrations during electrophoresis in low gravity aboard shuttle flight STS-8, and their capacities to produce plasminogen activators was evaluated (Barlow et al., 1988). The journey to low gravity and back did not appear to abolish or impair differentiated functions of electrophoretically separated cells or their progeny as judged by morphology (Todd et al., 1985), production of immunologically identifiable tissue plasminogen activator (Lewis et al., 1987), or production of urokinase (59 ± 20 vs. 59 ± 9 unweighted mean units per milliliter after flight and in control cultures, respectively) (Todd et al., 1985).

b. Kidney cell attachment to microcarriers in low gravity   There is interest in maintaining cells in space for microgravity bioprocessing purposes (Morrison, 1977; Cogoli and Tschopp, 1982; Morrison, 1988). With this purpose in mind, Tschopp et al. (1984) investigated the ability of anchorage-dependent cells to form attachments during spaceflight and found that cultured human kidney cells attach normally to dextran (Cytodex) microcarrier beads in culture in microgravity. The number of cells attached per bead was slightly, but not significantly, elevated in comparison with ground controls. This finding may not necessarily shed light on cell–cell or cell–vessel wall attachments during low gravity—interactions that are especially important in lymphocyte activation.

c. Pituitary cell electrophoresis in low gravity   It is known that preparative electrophoresis separates growth-hormone-producing rat pituitary cells from other cells of the anterior pituitary (Plank et al., 1983). Rat anterior pituitary cells, which are known to include a high-mobility fraction rich in growth hormone production (Plank et al., 1983), were separated according to electrophoretic mobility, and separate fractions rich in growth hormone and prolactin production were characterized (Hymer et al., 1985, 1986, 1987, 1988). Results of separation experiments were in agreement with experiments at $1g$ using free-flow electrophoresis but not with those using density gradient electrophoresis, which separates cells by a combination of electrophoresis and sedimentation since growth-hormone producing cells are more dense than other cells (Hymer et al., 1987).

d. Hormone secretion   Pituitary cell secretion anomalies have been observed in several cellular systems subjected to prolonged low gravity. Hymer and Grindeland found anomalies in growth hormone production in spaceflight experiments using fresh suspensions of dispersed cells from rat pituitaries (Hymer et al., 1985, 1988; Hymer and Grindeland, 1989). The anomalies consisted of a 2- to 20-fold reduction in growth hormone secreted by somatotrophic cells and a corresponding amount of hormone retention, while mammotrophic cells (from male rat pituitaries) in the same suspensions released and retained normal amounts of prolactin. The same secretion anomalies were later observed in intact rats during their whole-body responses to spaceflight conditions (Grindeland et al., 1987; Hymer and Grindeland, 1989).

4. Hematologic Cell Systems

a. Erythrocytes in stored whole blood   Living cells of whole blood stored for 6 days in orbital spaceflight under typical blood-banking conditions have been subjected to extensive analysis (Surgenor et al., 1990; Meehan et al., 1989), and none of the following measured cellular variables changed significantly as a result of low-gravity exposure: percentage of echinocytes, glucose, ATP, intracellular and extracellular electrolytes, phospholipids, cholesterol, hemolysis, osmotic fragility, and cellular IgG (immunoglobulin G). Extracellular p_{CO_2} was slightly elevated and pH correspondingly depressed in orbited suspensions. It should be instructive to compare the resulting data to similar results obtained with vigorously metabolizing cells in low gravity.

b. Platelets   Human platelet storage at low gravity results in reduced rates of deterioration. Platelet concentrates of the type usually prepared for blood banking were prepared from 24 units of human blood and stored in specially designed "compressed bags" that were held between pairs of rigid meshes to minimize gas diffusion distance to the center of the bags. This technique resulted in superior preservation at $1g$ and avoided the need to agitate the platelet concentrates (Surgenor et al., 1990). The ground-control bags were held vertically, as were low-gravity samples prior to flight. Only two physiological variables remained unchanged following spaceflight: lactic acid content and ADP-induced platelet aggregation. All other variables—collagen-stimulated aggregation, serotonin uptake, mean platelet volume, thromboxane, and ultrastructure—were improved by factors as great as 8 in low gravity. Extracellular variables such as glucose, p_{CO_2}, p_{O_2}, and pH all indicated that platelets metabolized more aerobically in low gravity, consistent with improved oxygen availability owing to the uniform distribution of the suspended platelets (see Section III, below).

5. Clinorotation and Synapse Formation

Gravitational developmental biology experiments performed in the laboratory rotate the biological system on a clinostat at a frequency chosen to cancel the organism's gravitational response (Block et al., 1986). Gruener and Hoeger (1990) chose to study the formation of neuromuscular junctions in vitro using cultures of appropriately derived cells from embryos of the toad *Xenopus levis*. This model system is not mammalian, but it presumably resembles processes of cell differentiation that are common to higher vertebrates.

In mixed cultures of myoblasts and neuroblasts suspended from specific embryonic rudiments, these two cell types differentiate and form neuromuscular synapses, which can be quantitated by fluorescent staining. When such culture systems were rotated at 1 or 10 rpm, a 50% reduction in countable synaptic junctions was observed (Gruener and Hoeger, 1990).

III. HYPOTHESES OFFERED TO EXPLAIN CELLULAR EFFECTS OF GRAVITY

A. General Considerations

The physical processes affected by altered inertial acceleration in mammalian cell systems could be intracellular, extracellular, or intercellular. Very little transport modeling has been performed on these systems as a means of distinguishing among these possibilities. By evaluating the relative roles of diffusion, secretion rate, convection, and forced flow in transport (as implied in the previous section), it should be possible to distinguish among these possibilities. Some of the space experiments that have been performed in the area of applied cell biology (Taylor, 1977; Morrison, 1977) have been considered in terms of physical fundamentals (Tobias et al., 1973; Todd, 1977, 1989a, 1989b).

The spaceflight environment presents numerous technical difficulties that deter the performance of rigidly controlled scientific experiments (Schneider et al., 1988; Schopf et al., 1988). Before confidently interpreting results of space experiments on single cells, we should ask at least three technical questions (applicable to all scientific inquiry):

1. Has the experimental result been produced consistently on repetition of experiments?
2. Have proper "control" experiments been performed, and, if so, did they produce results that define the roles of spaceflight factors other than reduced gravity?
3. Have plausible hypotheses concerning several possible effects of microgravity-induced unloading been tested?

A. Cogoli, who proposed criteria 1 and 2, has identified three experiments in which these two issues have been addressed: enhanced growth of *Paramecium* (Richoilley et al., 1988), inhibition of lymphocyte blastogenesis in vitro (Cogoli et al., 1988; Tixador et al., 1978), and increased resistance of bacteria to an antibiotic (Lapchine et al., 1988).

Since all observations must ultimately be explained by physical processes, it is presumed that clusters of observations will be related to individual physical explanations. The intracellular motion of organelles plays an important role in the *essential* responses of plants and certain other eukaryotic organisms to gravity. Searches for intracellular gravitational effects in animal cells, however, have revealed little or no evidence for either *essential* or *fortuitous* responses at the subcellular level, apparently due to the dominant role of the cytoskeleton in organelle motion. The sedimentation of particles in cells may have been considered too simplistically, and it is necessary to consider additional phenomena such as isothermal settling, in which sedimentation is balanced with diffusion; the Dorn effect, in which an electric field results when a particle sediments; droplet sedimentation, which involves larger hydrodynamic units whose density depends on particle concentration; and convective transport within and near the cell (Tobias et al., 1973). Bodies that sediment, settle, convect, or deform, even in the presence of significant thermal fluctuations, may serve to activate or inactivate stretch-sensitive ion channels in cell membranes (Morris and Sigurdson, 1989). No hypotheses concerning direct intracellular responses of mammalian cells to modified inertial accelerations in the range $0-2g$ have yet been solidly tested. Hypotheses concerning modulations of the extracellular environment are offered in the paragraphs that follow.

B. Analysis of Specific Examples

1. Lymphocyte Blastogenesis in Space

The hypothesis that reduced loading prevents cell aggregation in a form required for the cell–cell interactions that lead to blastogenesis has been entertained, but it has been stated that cells do aggregate in space, and monocytes, required for lymphocyte activation, were present in all cultures (Cogoli et al., 1988). Experimental evidence indicates that suspended cultured human cells are able to attach to suspended microcarrier particles during orbital spaceflight (Tschopp et al., 1984). However, reduced cell aggregation has been demonstrated directly in microgravity experiments (Dintenfass et al., 1985), and deliberately maintaining cells in dilute suspension on the ground (under $1g$ conditions) and/or eliminating monocytes inhibits blastogenesis in vitro (Bauer and Hannig, 1986). Further low-gravity and laboratory experiments may aid in elucidating the relative roles of intracellular and intercellular processes in the blastogenic response. Meanwhile, calculations relevant to intercellular processes can be attempted.

How does buoyant suspension of a cell population differ from microgravity suspension? If cells, each of which has density ρ_i, are suspended in fluid of density ρ_0, the cell sedimentation velocity is as given by Eq. (13), but if $\rho_i = \rho_0$, then $v_i = 0$, and cells do not sediment. But if cells have a variety of densities around the mean $\bar{\rho} = \rho_0$, less dense cells will float and more dense cells will sediment. So the cell will sediment or float according to

$$v_i = \frac{2(\rho_i - \rho_0)a^2 g}{9\eta} \tag{13}$$

Consider two cells, one with $\rho_1 \geq \rho_0$ and one with $\rho_2 \leq \rho_0$. They will move apart at velocity $v_1 + v_2$, although the average population velocity

$$\bar{v} = \frac{\sum n_i v_i}{\sum n_i} \tag{14}$$

may be zero. In microgravity $v = 0$ because $g = 0$, and ρ_i values do not determine velocity. Some cell–cell contacts that occur among heterogeneous cells in in vitro suspensions [of lymphocytes; e.g., see Cogoli et al. (1988)], would not occur in microgravity. It can be calculated that, whether by diffusion or differential sedimentation at $(0.001)g$, two lymphocytes initially 1 cm apart would collide, on the average, after a few weeks.

Diffusion is the only mode of transport of molecules between noncolliding cells in zero gravity. How long does it take a protein molecule to travel from effector to receptor cell in space? Consider the simple case of two cells 3 mm apart, suspended in fluid with viscosity η. One cell secretes a single spherical molecule with molecular weight M; this molecule must

travel, by diffusion alone, 3 mm to the receptor cell. The time required for
the journey as estimated using Eq. (10) with $\langle x^2 \rangle = 0.1$ cm^2 and $D = 10^{-6}$ is about 2 h for the mean transit time for this process.

2. Attachment of Suspended Cells to Microcarrier Beads

Differential sedimentation is responsible for microcarrier spheres catching
up to cells or cells catching up to spheres in the presence of gravity. It is of
interest to calculate the rate of attachment to "Cytodex 3" microcarrier
beads (density = 1.08 g/cm^3) by cultured human kidney cells (density =
1.05 g/cm^3) using Eq. (13). The average radius of beads is 75 μm, and
that of cells is 7.5 μm. At 1g the sedimentation velocities are 700 and
4 μm/s, respectively. At $(0.0001)g$ the corresponding sedimentation velocities are 0.07 and 0.0004 μm/s. Thus, in a 1-cm vessel in about 40 h,
even in "zero-gravity" cells can contact beads by differential sedimentation
alone.

3. Isothermal Settling of Platelets

Human platelets stored in microgravity have a longer lifetime than do their
counterparts maintained on the ground (Surgenor, 1986). Interactions that
occur during settling are among the hypothetical causes of the short
lifespan of the thrombocyte in vitro. While a certain amount of flocculation occurs during platelet storage, it is nevertheless reasonable to ask
whether single-platelet suspensions actually settle. On the basis of experimental data on platelet sedimentation (Corash et al., 1984) and calculations
based on a Stokes sedimentation rate of 0.01 μm/s, which corresponds to
about one diameter settling distance every 2 min (Todd, 1989b), it can be
estimated that most of the platelets would sediment the 12.5 cm to the
bottom of the bag in about 240 h and would be distributed between 3 and
8 cm above the bottom of the bag after 200 h of settling during the storage
experiment. According to Eq. (1), Brownian movement will lead to a final
vertical distribution in which the concentration of platelets $c(h)$ is reduced
by $1/e$ every 9 μm from the bottom of the container. During the
low-gravity experiment cycle, however, platelets would still fill the bag up
to within 1.5 cm from the top.

Because ground-control platelets settled during the experiment, which
lasted longer than the usual storage time for platelets, the geometry and
dynamics of gas exchange differed significantly between samples stored on
the ground and in low gravity. Not surprisingly, platelets returned from
low-gravity storage after 6 days were in far superior condition relative to
their ground-stored counterparts. It thus appears that, with or without
flocculation, platelet settling is significant and cannot be dismissed as being
unrelated to their short (a few days) lifespan in vitro on the ground.

4. Neuromuscular Synapse Formation

The rotation of a monolayer cell culture in a cylindrical chamber is attended by effects on concentration and density gradients in the culture medium. Shear stress and convective mixing could be modified in a rotating culture relative to a static culture. Concentrations gradients that cells produce are presumably destroyed in a rotating culture.

Of the five phenomena characterized in Section I,C on physics of clinorotation, none stands out as potentially causative of the clinorotation effects found on neuromuscular synapse formation in vitro (Todd, 1992). A statolith hypothesis cannot be eliminated on the basis of the rotation frequencies at which effects were observed. The centrifugal acceleration applied to the cells is of the order of milli-g values. Hydrostatic pressure changes are small, but not negligible in terms of apparent effects of pulsatile pressure changes on cultured cells (Levesque and Nerem, 1989). The maximum possible shear stress in these experiments was several orders of magnitude below those at which effects on cultured cells have been claimed (Cherry and Papoutsakis, 1986a, 1987). The tumbling of cultures should lead to slightly enhanced mixing of the overlying medium relative to diffusive transport, depending on cellular secretion rate; plausibility arguments favor this hypothesis. It appears unlikely that significant electrokinetic streaming potential is generated by the tumbling of cultures. Genuine intracellular responses related to the cytoskeleton may be entertained (see Section III,C).

C. Examples of Hypotheses to Be Tested

The mammalian cell in vivo is subjected to chronic reorientation. This fact, arguing from evolutionary theory, suggests that the mammalian cell evolved to be insensitive to reorientation and possibly unadapted to sessile existence. Most of the preceding chapters in this book have dealt with mammalian cells in vitro. In vitro cultivation techniques subject the mammalian cell to vigorous tumbling (as in fluidized-bed culture systems, including microcarrier methods), reorienting monolayers (roller bottles and tubes), or sessile monolayers. None of these conditions approximate the relationship of the mammalian cell to the gravity vector in vivo. There are numerous reasons for predicting that this modification of lifestyle should have little or no effect at the intracellular level.

1. The Statolith Hypothesis

In the plant world, where sessile conditions are the rule, root gravitropism has become best understood in terms of the statolith concept. Although many arguments for and against the role of amyloplasts in the statocytes of the root caps of vascular plants have appeared, there is today very little doubt that this response to gravity involves the movement of these tiny objects caused by gravity.

Certain organelle systems mentioned in Table 1 are specific adaptations in the plant world. Animal cells, which have no cell wall, differ explicitly from plant cells in their lack of a need to synthesize a cell wall in a particular direction. If plant cells need to respond to gravity for this purpose only, then one would not expect the intracellular activities of animal cells to be very responsive to gravity. An analysis of the constituents of the mammalian cell should indicate whether there exist any organelles that can sediment under the influence of gravity. Earlier theoretical analysis indicated that the nucleolus might be a sufficiently large and dense structure to be influenced by gravity (Pollard, 1965). This would be the case if the nucleolus could be considered as a solid object suspended in a viscous liquid medium. However, although it is a densely packed structure, it is not isolated from the surrounding nucleoplasm as a solitary hydrodynamic unit. It is suspended in the nucleus by a number of threads, and its motion is constrained by the motion of the chromatin with which it is associated. Evidence has also been presented that the nucleolus is associated with specific nuclear membrane sites (Bourgeois et al., 1979) and that a cytoskeletal-type matrix exists in the nucleus (Abei et al., 1986) and forms a scaffold associated with the nuclear envelope (Bershadsky and Vasiliev, 1988).

Electron and visible-light micrographs of vertical sections of such cells contain both the position of the nucleolus and the direction of the gravity vector, so retrospective and prospective statistical studies that test Pollard's hypothesis are both possible. There is little or no evidence for the sedimentation of nucleoli to the lower face of nuclei in cultured human cells exposed to $1g$ unidirectionally for several days. On the average, the nucleolus is just about as close to the top of the nuclear membrane as it is to the lower side (Todd, 1977). More recently, a series of experiments was performed in which vertical sections of normal human fibroblasts and Chinese hamster cells grown on membranes were analyzed for the purpose of determining nuclear position (Cornforth et al., 1989; Carpenter et al., 1989). In published photographs of three cells, one nucleolus was attached to the lower nuclear membrane, one was attached to both membranes, and one was unattached in the plane of the section.

It appears that the nucleus is positioned in the cytoplasm under constraints imposed by the cytoskeleton. Interphase nuclei seem to be associated with intermediate (cytokeratin) filaments of the type associated with vesicular organelles (Bershadsky and Vasiliev, 1988). If cultured mammalian cells attached to coverslips are centrifuged at moderate speed (10–100 g), one finds that cells remain intact without significant displacement of their nuclei. If, on the other hand, one treats cultured cells attached to coverslips with cytochalasin B and then subjects the attached cells to a centrifugal field, the centrifugal acceleration is then adequate to enucleate the cells (Prescott et al., 1972). Interphase nuclei also seem to be associated with intermediate filaments (Bershadsky and Vasiliev, 1988). If one were to approximate the nucleus as a hydrodynamic unit equivalent to a sphere 12 μm in diameter with density $\rho = 1.14$ suspended in a fluid with viscosity 17 cP and density 1.03, then one would anticipate, from the Stokes equation, a sedimentation velocity of the cell nucleus equal to about 20 μm per hour. Clearly, all nuclei would sediment to the bottoms of their cells within a few minutes in earth's gravity. That this is not the case is observable in mammalian tissue sections in which the nuclei are positioned according to cell type and not according to the gravitational vector.

It is to be learned from this discussion that fibrous materials in the cell can influence the response of its organelles to gravity. Thorough experimental testing of the ability of the nucleus to sediment in sessile animal cells, nevertheless, has never been performed.

2. Convection and Macromolecular Assembly

A study of early lattice formation in nucleating protein crystals (Kam et al., 1978) indicates that critical assembly processes occur at the submicrometer level. During lattice formation, the Gibbs free energy of crystallization is released to the immediate environment as heat, and solute is depleted near the lattice-forming surface. Both events lead to a local density reduction with the potential for convection. The gravity-unloading of this process should, therefore, lead to higher quality crystal growth, which, evidently, it does (DeLucas et al., 1986; Bugg, 1986; Littke and John, 1984).

The growth of protein crystals in a convection-free environment may be considered a simplified model for self-assembly processes in cells. Some protein crystal growth processes are somewhat isotropic, and these result in ultra-high-quality crystals in low gravity (DeLucas et al., 1986). Other protein crystals grow in a highly anisotropic manner, and these grow to longer and more uniform crystals in low gravity (Littke and John, 1984; DeLucas et al., 1986). These latter might be considered a simplified model for the (metabolically supported) self-assembly of long-chain protein aggregates in cells, such as microfilaments, cytokeratin filaments, and microtubules. Similarly, the formation of such self-assembled structures as microtubules might be modified during gravity-unloading. Preliminary experiments by Moos et al. (1988) indicate significant differences between microtubules assembled during low-gravity and $2g$ phases of parabolic aircraft flight.

Such convective processes occur in free solution, but it remains to determine whether the cytosol is capable of supporting such convection. The notion that they might occur at the surface of free-living cells has been entertained (Albrecht-Buehler, 1990, 1991).

3. Hydrostatic Pressure

One might expect effects of reduced gravity on certain cellular processes that involve volume changes. Some, but certainly not all, experiments on single cells in space are suggestive. In the case of secretion, Hymer and co-workers (Hymer et al., 1988; Hymer and Grindeland, 1989) reported evidence that rat anterior pituitary cells release significantly less growth hormone in space (see Section II, above). *Paramecium tetraaurelia* displayed a lack (and possibly reversal) of exocytosis in the form of unreleased trychocysts buried beneath the plasma membrane (Richoilley et al., 1988). Ground-based hydrostatic pressure experiments should be performed on these systems.

REFERENCES

Abei, V., Cohn, J., Buhle, L. and Gerace, L. (1986). The nuclear lamina is a meshwork of intermediate-type filaments. *Nature* **323**, 560–564.

Albertsson, P.-Å. (1986). *Partition of Cell Particles and Macromolecules*, 3rd ed. Wiley, New York.

Albrecht-Buehler, G. (1990). In defense of "non-molecular" cell biology. *Int. Rev. Cytol.* **120**, 191–241.

Albrecht-Buehler, G. (1991). Possible mechanisms of indirect gravity sensing by cells. In *Gravity and the Cell*, Am. Soc. Grav. Space Biol. Bull. **4**, (2) 25–34.

Barlow, G. H., Lewis, M. L., and Morrison, D. R. (1988). Biochemical assays on plasminogen activators and hormones from kidney sources. In *Microgravity Science and Applications Flight Programs, January–March 1987. Selected Papers*, pp. 175–193. National Aeronautics and Space Administration (Report NASA TM-4069), Washington, DC.

Bauer, J., and Hannig, K. (1986). Free flow electrophoresis: An important step among physical cell separation procedures. In *Electrophoresis '86*, pp. 13–24, Dunn, M. J. (ed.). VCH Verlagsgesellschaft, Weinheim.

Bender, M. A., Gooch, P. C., and Kondo, S. (1967). The Gemini 3S-4 spaceflight-radiation interaction experiment. *Radiat. Res.* **31**, 91–111.

Bender, M. A., Gooch, P. C., and Kondo, S. (1968). The Gemini XI S-4 spaceflight-radiation interaction experiment. *Radiat. Res.* **34**, 228–238.

Bershadsky, A. D., and Vasiliev, J. M. (1988). *Cytoskeleton*, Plenum Press, New York.

Block, I., Briegleb, W., and Wohlfarth-Botterman, K. E. (1986). Gravisensitivity of the acellular slime mold *Physarum polycephalum* demonstrated on the fast-rotating clinostat. *Eur. J. Cell Biol.* **41**, 44–50.

Boltz, R. C., and Todd, P. (1979). Density gradient electrophoresis of cells in a vertical column. In *Electrokinetic Separation Methods*, pp. 229–250, Righetti, P. G., van Oss, C. J., and Vanderhoff, J. (eds.). Elsevier/North-Holland Biomedical Press, Amsterdam.

Bourgeois, C. A., Hemon, D., and Bouteille, M. (1979). Structural relationship between the nucleolus and the nuclear envelope. *J. Ultrastruct. Res.* 68, 328–340.

Bugg, C. E. (1986). The future of protein crystal growth. *J. Crystal Growth* 76, 535–544.

Carpenter, S., Cornforth, M. N., Harvey, W. F., Raju, M. R., Schillaci, M. E., Wilder, M. E., and Goodhead, D. T. (1989). Radiobiology of ultrasoft X rays IV. Flat and round-shaped Chinese hamster cells (CHO-10B, HS-23). *Radiat. Res.* 119, 523–533.

Cherry, R. S., and Papoutsakis, E. T. (1986a). Hydrodynamic effects on cells in agitated tissue culture reactors. *Bioproc. Eng.* 1, 29–41.

Cherry, R. S., and Papoutsakis, E. T. (1986b). Mechanism of cell damage in agitated microcarrier tissue culture reactors. *Proc. World Congr. Chem. Eng. III*, Tokyo (Sept. 21–25).

Cherry, R. S., and Papoutsakis, E. T. (1987). Effects of fluid forces and bead-bead bridging on cell death in microcarrier bioreactors. *AIChE Ann. Meeting Abstracts*.

Cogoli, A., and Tschopp, A. (1982). Biotechnology in space laboratories. *Advances in Biochemical Engineering*, Vol. 22, *Space and Terrestrial Biotechnology*, pp. 1–49, Riechter, A. (ed.). Springer-Verlag, New York.

Cogoli, A., Tschopp, A., and Fuchs-Bislin, P. (1984). Cell sensitivity to gravity. *Science* 225, 228–230.

Cogoli, A., Bechler, B., Müller, O., and Hunzinger, E. (1988). Effect of microgravity on lymphocyte activation. In *Biorack on Spacelab D1*, pp. 89–100, Longdon, N., and David, V. (eds.). European Space Agency (Report ESA SP-1091), Paris.

Corash, L., Costa, J. L., Shafer, B., Conlon, J. A., and Murphy, D. (1984). Heterogeneity of human whole blood platelet subpopulations. III. Density-dependent differences in subcellular constituent. *Blood* 64, 185–193.

Cornforth, M. N., Schillaci, M. E., Goodhead, D. T., Carpenter, S. G., Wilder, M. E., Sebring, R. J., and Raju, M. R. (1989). Radiobiology of ultrasoft X rays III. Normal human fibroblasts and the significance of terminal track structure in cell inactivation. *Radiat. Res.* 119, 511–522.

d'Augères, C. B., Arnoult, J., Bureau, J., Duie, P., Dupuy-Coin, A. M., Géraud, G., Laquerrière, F., Masson, C., Pestmal, M. and Bouteille, M. (1988). The effect of microgravity on mammalian cell polarization at the ultrastructural level. In *Biorack on Spacelab D1*, pp. 101–105, Longdon, N., and David, V. (eds.). European Space Agency (Report ESA SP-1091), Paris.

DeLucas, L. J., Suddath, F. L., Snyder, R., Naumann, R., Broom, M. B., Pusey, M., Yost, V., Herren, B., Carter, D., Nelson, B., Meehan, E. J., McPherson, A., and Bugg, C. E. (1986). Preliminary investigations of protein crystal growth using the space shuttle. *J. Crystal Growth* 76, 681–693.

de Souza, P. A., Levesque, M. J., and Nerrem, R. M. (1986). *Fed. Proc.* 45, 471.

Dintenfass, L., Osman, P., and Jedrzejczyk, H. (1985). First haemorheological experiment on NASA space shuttle "Discovery" STS 51-C: Aggregation of red cells. *Clin. Haemorheol.* 5, 917–936.

Frangos, J. A., McIntire, L. V., Eskin, S. G., and Ives, C. L. (1985). Flow effects on prostacyclin production by cultured human endothelial cells. *Science* 227, 1477–1479.

Fulton, A. B. (1982). How crowded is the cytoplasm? *Cell* 30, 345–347.

Gordon, S. A., and Shen-Miller, J. (1971). Simulated weightlessness studies by compensation. In *Gravity and the Organism*, pp. 415–426, Gordon, S. A., and Cohen, M. J. (eds.). University of Chicago Press, Chicago.

Grindeland, R. E., Hymer, W. C., Farrington, M., Fast, T., Hayes, C., Motter, K., Patil, L., and Vasques, M. (1987). Changes in pituitary growth hormone cells prepared from rats flown on Spacelab 3. *Am. J. Physiol.* **252**, 209–215.

Gruener, R., and Hoeger, G. (1990). Vector-free gravity disrupts synapse formation in cell culture. *Am J. Physiol.* **258** (*Cell Physiol.* 27), C489–C494.

Hemmati-Brivanlou, A., Stewart, R. M., and Harland, R. M. (1990). Region-specific neural induction of an *engrailed* protein by anterior notochord in *Xenopus*. *Science* **250**, 800–802.

Hjertén, S. (1962). *Free Zone Electrophoresis*. Almqvist and Wiksells Boktr. AB, Uppsala, Sweden.

Hymer, W. C., and Grindeland, R. E. (1989). The pituitary growth hormone cell in space. In *Cells in Space*, pp. 71–75, Sibonga, J. D., Mains, R. C., Fast, T. N., Callahan, P. X., and Winget, C. M. (eds.). National Aeronautics and Space Administration (NASA Conf. Publ. No. 10034), Moffett Field, California.

Hymer, W. C., Grindeland, R., Lanham, J. W., and Morrison, D. (1985). Continuous flow electrophoresis (CFE): Applications to growth hormone research and development. In *Pharm Tech Conference '85 Proceedings*, pp. 13–18, Aster Publishing Corp. Springfield, OR.

Hymer, W. C., Grindeland, R., and Lanham, J. W. (1986). Continuous flow electrophoresis (CFE) at unit and microgravity: Applications to growth hormone (GH) research and development. In *Microgravity Science and Applications*, pp. 197–211. National Academy Press, Washington, DC.

Hymer, W. C., Barlow, G. H., Cleveland, C., Farrington, M., Grindeland, R., Hatfield, J. M., Lanham, J. W., Lewis, M. L., Morrison, D. R., Rhodes, P. H., Richman, D., Rose, J., Snyder, R. S., Todd, P., and Wilfinger, W. (1987). Continuous flow electrophoretic separation of proteins and cells from mammalian tissues. *Cell Biophys.* **10**, 61–85.

Hymer, W. C., Grindeland, R., Hayes, C., Lanham, J. W., Cleveland, C., Todd, P., and Morrison, D. (1988). Heterogeneity in the growth hormone pituitary gland "system" of rats and humans: Implications to microgravity based research. In *Microgravity Science and Applications Flight Programs, January–March 1987, Selected Papers*, Vol. 1, pp. 47–88. National Aeronautics and Space Administration (Report NASA TM-4069), Washington, DC.

Kam, Z., Shore, H. B., and Feher, G. (1978). On the crystallization of proteins. *J. Mol. Biol.* **123**, 539–555.

Kayser, R. F., Moldover, M. R., and Schmidt, J. W. (1986). What controls the thickness of wetting layers. *J. Chem. Soc., Faraday Trans.* 2 **82**, 1701–1719 (1986).

Kessler, J. O. (1978). Convective control of long-range coherence in plant growth regulation. *Life Sci. Space Res.* **16**, 99–104.

Lapchine, L., Moatti, N., Richoilley, G., Templier, J., Gasset, G., and Tixador, R. (1988). The antibio experiment. In *Biorack on Spacelab D1*, pp. 45–51, Longdon, N., and David, V. (eds.). European Space Agency (Report ESA SP-1091), Paris.

Levesque, M. J., and Nerem, R. M. (1985). The elongation and orientation of cultured endothelial cells in response to shear stress. *ASME J. Biomech. Eng.* **176**, 341–347.

Levesque, M. J., and Nerem, R. M. (1989). The study of rheological effects on vascular endothelial cells in culture. *Biorheology* **26**, 345–357.

Lewis, M. L., Morrison, D. R., Barlow, G. H., Todd, P., Cogoli, A., and Tschopp, A. (1987). Cell electrophoresis and preliminary cell attachment investigations on STS-8. In *Space Life Sciences Symposium: Three Decades of Life Science Research in Space*, pp. 186–196. Office of Space Science and Applications, National Aeronautics and Space Administration, Washington, DC.

Littke, W., and John, C. (1984). Protein single crystal growth under microgravity. *Science* **225**, 203–204.

Lorenzi, G., Bechler, B., Cogoli, M. and Cogoli, A. (1988). Gravitational effects on mammalian cells. *Physiologist* **32** (Suppl. 1), S144–S147.

Mason, D. W. (1976). A diffusion-driven instability in systems that separate particles by velocity sedimentation. *Biophys. J.* **16**, 407–416.

Meehan, R., Taylor, G., Lionetti, F. J., Neale, L., and Curran, T. (1989). Human mononuclear cell function after 4°C storage during 1-G and microgravity conditions of spaceflight. *Aviat. Space Environ. Med.* **60**, 644–648.

Montgomery, P. O'B., Jr., Cook, J. E., Reynolds, R., Paul, J. S., Hayflick, L., Stock, D., Schulz, W. W., Kimzey, S., Thirolf, R. G., Rogers, T., Campbell, D., and Morrell, J. (1977). The response of single human cells to zero-gravity. In *Biomedical Results from Skylab*, pp. 221–234, Johnston, R. S., and Dietlein, L. F. (eds.). National Aeronautics and Space Administration, Washington.

Moos, P. J., Graf, K., Edwards, M., Stodieck, L. S., Einhorn, R., and Luttges, M. W. (1988). Gravity-induced changes in microtubule formation. In *Program and Abstracts, 4th Annual Meeting of American Society for Gravitational and Space Biology*, p. 55.

Morris, C. E., and Sigurdson, W. J. (1989). Stretch-inactivated ion channels coexist with stretch-activated ion channels. *Science* **243**, 807–809.

Morrison, D. R. (ed.) (1977). *Bioprocessing in Space*, National Aeronautics and Space Administration (Report NASA TM X-58191), Lyndon B. Johnson Space Center.

Morrison, D. R. (1988). Cellular effects of microgravity. In *Microgravity Science and Applications Flight Programs, January–March 1987. Selected Papers*, pp. 217–226. National Aeronautics and Space Administration (Report NASA TM-4069), Washington, DC.

O'Brien, R. W., and White, L. R. (1978). Electrophoretic mobility of a spherical colloidal particle. *J. Chem. Soc. Faraday Disc. II* **74**, 1607–1626.

Omenyi, S. N., Snyder, R. S., Absolom, D. T., Neumann, A. W., and van Oss, C. J. (1981). Effects of zero van der Waals and zero electrostatic forces on droplet sedimentation. *Colloid Interface Sci.* **81**, 402–409.

Plank, L. D., Hymer, W. C., Kunze, M. E., and Todd, P. (1983). Studies on preparative cell electrophoresis as a means of purifying growth-hormone producing cells of rat pituitary. *J. Biochem. Biophys. Methods* **8**, 273–289.

Pollard, E. C. (1965). Theoretical considerations on living systems in the absence of mechanical stress. *J. Theor. Biol.* **8**, 113–123.

Pollard, E. C. (1971). Physical determinants of receptor mechanisms. In *Gravity and the Organism*, pp. 25–34, Gordon, S. A., and Cohen, M. J. (eds.). University of Chicago Press, Chicago.

Prescott, D. M., Myerson, J., and Wallace, J. (1972). Enucleation of mammalian cells with cytochalasin B. *Exp. Cell Res.* **71**, 480–485.

Purcell, E. M. (1977). Life at low Reynolds number. *Am. J. Phys.* **45**, 3–11.

Raghava, Roa, K. S. M. S., Stewart, R. M., and Todd P. (1990). Electrokinetic demixing of two-phase aqueous polymer systems. I. Separation rates of polyethylene glycol-dextran mixtures. *Sep. Sci. Technol.* **25**, 985–996.

Richoilley, G., Tixador, R., Templier, J., Bes, J. C., and Planel, H. (1988). The Paramecium experiment. In *Biorack on Spacelab D1*, pp. 69–73, Longdon, N., and David, V. (eds.). European Space Agency (Report ESA SP-1091), Paris.

Sachs, F. (1988). Mechanical transduction in biological systems. *Crit. Rev. Biomed. Eng.* **16**, 141–169.

Sato, M., Levesque, M. J., and Nerem, R. M. (1985). *Proceedings of Biomechanics Symposium*, pp. 167–169, ASME summer meeting.

Schneider, W. C., Carr, G. P., Collins, M., Cowing, K., Hobisch, M. K., and Leveton, L. (1988). Flight programs. In *Exploring the Living Universe: A Strategy for Space Life Sciences*, pp. 154–170, National Aeronautics and Space Administration, Washington, DC.

Schnettler, R., Gessner, P., Zimmermann, U., Neil, G. A., Urnovitz, H. B., and Sammons, D. W. (1989). Increased efficiency of mammalian somatic cell hybrid production under microgravity conditions during ballistic rocket flight. *Appl. Microgravity Tech.* 2, 3–9.

Schopf, J. W., Galston, A. W. and Cowing, K. L. (1988). Gravitational biology. In *Exploring the Living Universe: A Strategy for Space Life Sciences*, pp. 101–111, National Aeronautics and Space Administration, Washington, DC.

Shank, B. B. (1974). Radiobiological experiments on satellites. In *Space Radiation Biology*, pp. 313–351, Tobias, C. A., and Todd, P. (eds.). Academic Press, New York.

Sievers, A., and Hensel, W. (1990). Gravity perception in plants. In *Fundamentals in Space Biology*, Asashima, M. and Malacinski, G. M. (eds.), Japan Scientific Societies Press, Tokyo.

Snyder, R. S., Rhodes, P. H., Herren, B. J., Miller, T. Y., Seaman, G. V. F., Todd, P., Kunze, M. E., and Sarnoff, B. E. (1985). Analysis of free zone electrophoresis of fixed erythrocytes performed in microgravity. *Electrophoresis* 6, 3–9.

Sprague, E. A., Steinbach, B. L., Nerem, R. M., and Schwartz, C. J. (1987). *Circulation* 76, 648–656.

Stathopoulos, N. A., and Hellums, J. D. (1985). Shear stress effects in human kidney cells *in vitro*. *Biotechnol. Bioeng.* 27, 1021–1026.

Surgenor, D. M. (1986). Blood formed elements in microgravity. *Blood* 68, 299.

Surgenor, D. M., Kevy, S. V., Chao, F. C., Lionetti, F. J., Kenney, D. M., Jacobson, M. S., Kim, B., Ausprund, D. H., Szymanski, I. O., Button, L. N., Curby, W. A., Luscinskas, F. W., Durran, T. G., Carter, J. H., Carr, E., Mozill, D. R., Blevins, D. K., and Laird, N. (1990). Human blood cells at microgravity: The NASA Initial Blood Storage Experiment. *Transfusion* 30, 605–616.

Talas, M., Batkai, L., Stoger, I., Nagy, K., Hiros, L., Konstantinova, I., Rykova, M. I., Mozgovaya, I., Giseva, O., and Kozarinov, V. (1983). Results of space experiments program Interferon I. *Acta Microbiol. Hungarica* 30, 53–61.

Taylor, G. R. (1977). Cell biology experiments conducted in space. *BioScience* 27, 102–108.

Tixador, R. G., Richoilley, G., Grechko, G., Nefedov, Y., and Planel, H. (1978). Multiplication de *Paramecium aurelia* au bord du vaisseau spatial Saliout-6. *Compt. Rend. Acad. Sci. Paris* 287, 829–832.

Tobias, C. A., Risius, J., and Yang, C.-H. (1973). Biophysical considerations concerning gravity receptors and effectors including experimental studies on *Phycomyces blakesleeanus*. *Life Sci. Space Res.* 11, 127–140.

Todd, P. (1977). Gravity and the cell: Intracellular structures and Stokes sedimentation. In *Bioprocessing in Space*, pp. 103–115, Morrison, D. (ed.). National Aeronautics and Space Administration (Report NASA TM X-58191), Johnson Space Center, Houston, TX.

Todd, P. (1985). Microgravity cell electrophoresis experiments on the space shuttle: A 1984 overview. In *Cell Electrophoresis*, pp. 3–19, Schütt, W., and Klinkmann, H. (eds.). deGruyter, Berlin.

Todd, P. (1989a). Essay: Gravity dependent phenomena at the scale of the single cell. *Am. Soc. Grav. Space Biol. Bull.* 2, 95–113.

Todd, P. (1989b). Physical phenomena and the microgravity response. In *Cells in Space*, pp. 103–116, Sibonga, J. D., Mains, R. C., Fast, T. N., Callahan, P. X., and Winget, C. M. (eds.). NASA Conference Publication *NASA CP-10034*. National Aeronautics and Space Administration, Moffett Field, CA.

Todd, P. (1990). Separation physics. In *Progress in Low-Gravity Fluid Dynamics and Transport Phenomena*, pp. 539–602, Koster, J. N., and Sani, R. L. (eds.). American Institute of Astronautics and Aeronautics, Washington, DC.

Todd, P. (1991). Gravity dependent processes and intracellular motion. In *Gravity and the Cell*, Am. Soc. Grav. and Space Biol. Bull. 4 (2), 31–39.

Todd, P. (1992). Physical effects at the cellular level under altered gravity conditions. *Adv. Space Res.* 12 (1), 43–49.

Todd, P., and Hjertén, S. (1985). Free zone electrophoresis of animal cells. I. Experiments on cell-cell interactions. In *Cell Electrophoresis*, pp. 23–31, Schütt, W., and Klinkmann, H. (eds.). deGruyter, Berlin.

Todd, P., Kunze, M. E., Williams, K., Morrison, D. R., Lewis, M. L., and Barlow, G. H. (1985). Morphology of human embryonic kidney cells in culture after space flight. *The Physiologist* 28 (Suppl. 6), 183–184.

Todd, P., Plank, L. D., Kunze, M. E., Lewis, M. L., Morrison, D. R., Barlow, G. H., Lanham, J. W., and Cleveland, C. (1986). Electrophoretic separation and analysis of living cells from solid tissues by several methods. Human embryonic kidney cell cultures as a model. *J. Chromatog.* 364, 11–24.

Tschopp, A., Cogoli, A., Lewis, M. L., and Morrison, D. R. (1984). Bioprocessing in space: Human cells attach to beads in microgravity. *J. Biotechnol.* 1, 281–293.

Van Alstine, J. M., Karr, L. J., Harris, J. M., Snyder, R. S., Bamberger, S., Matsos, H. C., Curreri, P. A., Boyce, J. and Brooks, D. E. (1987). Phase partitioning in space and on earth. In *Immunobiology of Proteins and Peptides IV*, pp. 305–326, Atassi, M. Z. (ed.). Plenum Press, New York.

INDEX

■ ■ ■ ■ ■